Woodwork Technology

THE MOTIVATE SERIES

Macmillan Texts for Industrial, Vocational and Technical Education

Woodwork Technology

J.K.N. Sackey

MACMILLAN

Macmillan Education
Between Towns Road, Oxford OX4 3PP
A division of Macmillan Publishers Limited
Companies and representatives throughout the world

www.macmillan-africa.com

ISBN 0 333 60053 3

First published 1999

Illustrated by 1-11 Line Art
Cover Illustration by Jerry Mason
Typset by EXPO Holdings, Malaysia

Printed and bound in Malaysia

2006 2005 2004
11 10 9 8 7

Contents

Acknowledgements

The author wishes to thank the following first year Technology Education students of the University College of Education of Winneba, Kumasi, Ghana for allowing the use of their design folios as illustrations: Anomale K. Danwono, David Kwaku Asare, Joseph Ameddi Tham. The author is also grateful to Miss Nada Frimpong and Miss Dora Fati, both secretaries of the University College of Education of Winneba, Kumasi for typing the manuscript. The family of the author deserves commendation for their encouragement and support. God bless all who offered help or advice.

Introduction

The rapid technological developments of our time have brought with them improvements and advances in the materials, tools and equipment of the woodwork trades. These changes have necessitated a constant re-examination of courses of study in schools and colleges dealing with this area of instruction. In this connection, when preparing the materials for this book, a number of syllabuses from various African countries and Britain were carefully studied in order to meet this need for up-to-date, comprehensive and thorough coverage of this area of study.

The book provides basic information about wood and wood products; non-wood materials; selection, safe use and care of hand and power tools. It also provides information and basic instructions in the areas of wood fabrication, fasteners, adhesives and hardware; wood turning, shaping, laminating, bending and surface decoration; basic upholstery; design principles and techniques, drawing and construction; occupational information and project works.

In order to help the reader to check his or her understanding of the text, a summary is provided at the end of every chapter of the salient issues covered in the chapter. This is followed by questions based on the chapter. Guides and answers to the questions are given at the end of the book. The meanings of technical and unfamiliar words used in the text are also provided in a list of keywords at the end of the book.

The subject matter has been treated in a simple, straightforward manner that will be readily understood by all readers. The level of coverage is such that it will meet the needs of students in apprenticeship programmes, junior and senior secondary schools, vocational and technical colleges, polytechnics and universities. It can also serve as a valuable reference for the do-it-yourself practitioner and all those engaged in the woodwork trades.

1 Safety in the workshop

Introduction

A **workshop** is a building or place where facilities such as machines, tools, and workbenches are provided to enable a worker or student to carry out his or her practical activities in a satisfactory manner. Most of the machines and tools used in the workshop, if not properly used or handled, may result in injury or damage to the machine or tool. Therefore, it is essential to observe safety precautions. Developing a safe working attitude and adopting safe methods are the surest ways of avoiding unnecessary accidents and dangers in the workshop.

General safety precautions

You must ensure that all passages to, from and in the workshop are completely clear. Do not leave any objects in passageways where people may fall over them.

Ensure that all pieces of timber, and both completed and uncompleted pieces of work are put away safely in a store room when you finish.

Do not spill oil or grease over the floor.

Clean the workbench, vice, tools and machines at the end of the lesson.

Return tools and items of equipment to their racks or boxes immediately you have finished using them.

Sweep the floor immediately after work and keep it tidy at all times.

Always clean timber of nails when dismantling any construction. Never leave protruding nails for someone to step on or catch themselves on.

Do not run in the workshop. You may knock into someone and cause a serious accident.

Do not try to carry long pieces of timber by yourself. The ends may easily knock someone and injure them. Get someone to hold one end while you hold the other.

Never wear loose clothing, such as loose long sleeves, which can easily be caught in a revolving machine part. Wear a well-fitting overall, overcoat or a close-fitting apron tied at the back.

Remove loose wrist-watches before using any machinery.

Wear footwear with thick soles and a strong upper to protect your feet from sharp objects on the floor or falling from above. Do not wear gym shoes, which can be easily pierced by sharp objects such as nails. Footwear with steel studs is likely to cause you to slip and fall.

Keep flammable liquids in sealed containers and keep them away from heat or flames.

Make sure you know where the first aid box is located.

Report any injury, however small it may be, to the teacher and to the person in charge of first aid.

If you feel unwell, report to the teacher or the person in charge of first aid.

In case of fire, raise the alarm and then walk calmly out of the building by the nearest exit door to the assembly point.

Safety in the bench shop

Never use defective tools, such as a hammer with a loose head that can fly off and hit someone.

Be careful when carrying tools. For example, hold a chisel close to the cutting edge when you are giving it to someone else, so that they receive it by the handle.

Never carry sharp-edged or pointed tools in the pocket of your overall or apron.

Always make sure that your work is firmly held in the vice or clamped to the bench.

When chiselling ensure that both hands are *behind* the cutting edge. Always chisel *away* from you.

Never test cutting edges by drawing your finger across them.

Return tools immediately after use to their proper places of storage.

Always make sure that you have the right tool for the job in hand.

Safety in the machine shop

Always follow the maker's intructions for the operation of a hand-operated power tool. They are meant to help you use it safely.

Always obtain permission before you use a machine. Ensure that you know how to stop it quickly.

Always check that the machine is properly adjusted and in good order before you use it. This will help to prevent accidents or injuries.

Make sure that the material is safe for use on the machine before you start the operation.

Always use the guards provided on a machine, so as to prevent accidents and injuries.

You must use **goggles** when there a possibility of pieces of material flying up, such as sawdust. This can be irritating to the eyes and may even cause blindness.

Do not use a machine with damaged or blunt cutters.

Never meddle with any machine. Report any machine that is not working properly to the teacher or whoever is in charge.

If you are working on a circular saw or a planer, always use a push-stick so that your fingers and hands are kept away from the running cutters.

Avoid talking to or distracting the attention of anyone operating a machine. If you are the operator yourself, do not talk to others while working.

If you are an observer, stand at a safe distance from the machine and never in line with its revolving parts or work.

Switch the machine off before you make any adjustment or check measurements on the work.

Never leave a machine running while you are engaged in doing something else. Make sure that you switch the machine off when you finish the operation and do not leave it until it has stopped running.

Report any conditions that could cause damage to the machine or injury to the operator.

When you start a machine, wait until it reaches operating speed before you apply the cutter. Avoid overloading the machine while cutting, particularly the lathe and the band saw.

Check the electrical connections and wiring of hand-operated power tools regularly as a badly connected wire can cause a short circuit or even electrocution.

First aid

When someone is injured, or suddenly falls ill, every second counts. In an emergency ensure that the injured person cannot come to any more harm and treat those injuries that threaten their life. Then call for medical help or send the person to the nearest clinic or hospital. You must not leave the injured person unattended until medical help arrives.

Most first aid is practical common sense that you can easily teach yourself. Note the following:

Make sure that you know where the first aid box is so that you can use it at once in an emergency.

When an accident occurs, inform the person responsible for first aid immediately.

If someone receives an electric shock, switch off the power immediately and call for medical help.

If someone has severe bleeding from a cut, take a *clean* piece of cloth and press it on the cut, or around the edges if a foreign body is in it. When the bleeding stops, take the pad away and remove any foreign bodies that come out easily. Gently clean the wound outwards with cotton wool or a swab soaked in warm soapy water. Use a fresh cotton wool for each cleaning. Gently dry around the wound with clean cotton wool and apply a plaster or bandage.

CHECK YOUR UNDERSTANDING

- Most machines and tools used in the workshop are dangerous if used improperly. Always handle them with care to avoid accidents. Observe all the necessary safety precautions.
- All passageways inside and leading to the workshop must be wide and kept unobstructed at all times. Never place anything in the passageways that people can fall over.
- The workshop, machines and tools must be clean and tidy at all times.
- Tools and equipment must be properly and safely stored in their racks or boxes immediately after use.
- The proper workshop outfit must be worn in the workshop.
- All injuries and accidents, no matter how small, must be reported and the appropriate treatment given and recorded.

REVISION EXERCISES AND QUESTIONS

1 Explain why safety in a workshop is so important.

2 What are the main safety rules you would recommend for a workshop?

3 State two safety precautions to be observed when handling sharp-edged tools.

4 State two safety precautions that must be observed when operating a machine.

5 Describe the steps that should be taken when someone receives a cut in the workshop.

6 Why should you not distract someone operating a machine in the workshop?

Woodworking hand tools 1

Introduction

Every school woodwork shop should be adequately equipped with a reasonable quantity and variety of tools and equipment for work to be done efficiently. It is necessary for you to be not only familiar with the names of the tools but able to identify and select the right tool for the job in hand, and use it safely and correctly. To make it easier to understand the correct applications of the various tools, they are grouped into classes as follows: holding and supporting tools; geometrical tools; percussion and impelling tools; boring cutting tools; shaving and paring cutting tools; and abrading and scraping cutting tools.

For convenience, shaving, paring, abrading and scraping cutting tools are discussed in chapter 3.

Holding and supporting tools

These tools and devices are those you will require to hold or fasten the workpiece securely so that you can work on it safely. They include: workbench, bench vice, bench stop, bench hook, bench holdfast, 'G' cramp, sash cramp, saw cramp or vice, mitre cramp, shooting boards, and mitre board.

The **workbench** (Figure 2.1) is an essential piece of apparatus in the workshop. You need it to support the workpiece from the time of setting out the job through to the end of all the necessary bench-work operations. It is in the form of a heavy table with a well (a little lower) at the centre of the top surface to accommodate some of the tools required. Some workbenches have built in cupboards for storing basic tools.

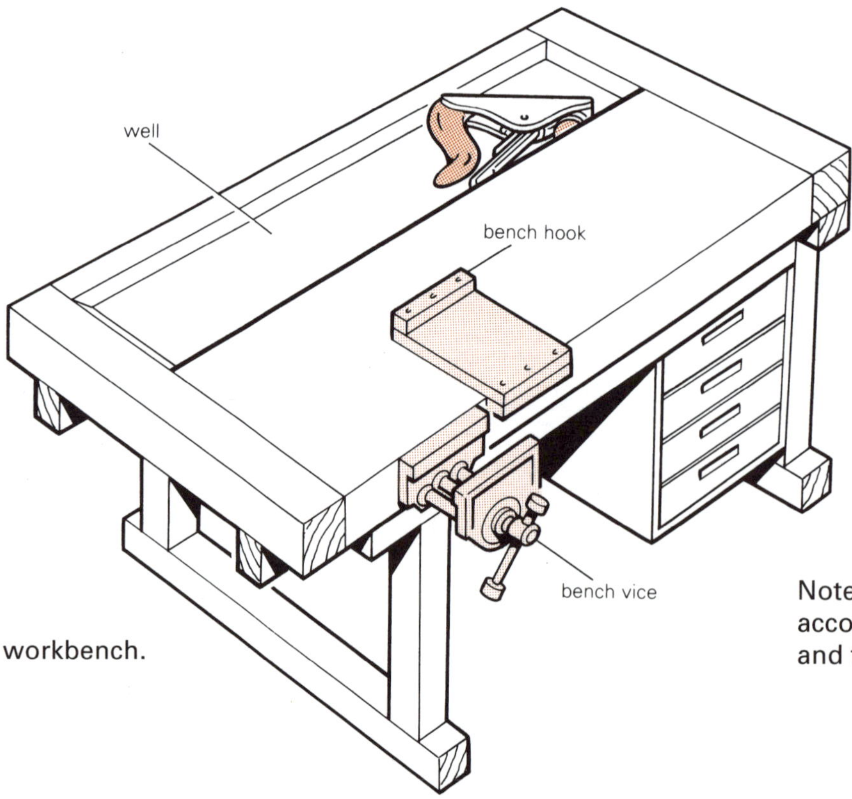

Note the well to accommodate tools, and the bench hook.

Figure 2.1 The workbench.

The **bench vice** (Figure 2.1) is made of cast iron, often with a quick release device that permits quick movement of the jaw and provides instantaneous grip. The vice is fitted to the side of the workbench so that the top of the jaw is level with the top of the workbench. A wooden cheek is fitted to the jaw to prevent the job from being damaged. You can use it to hold your job while you plane surfaces and edges, saw trenches and cut mortises or hold glued pieces of wood together while the glue is setting.

The **bench stop** (Figure 2.2a) is a device fitted through a hole, usually near the left hand end of the workbench, and is used to support a work-

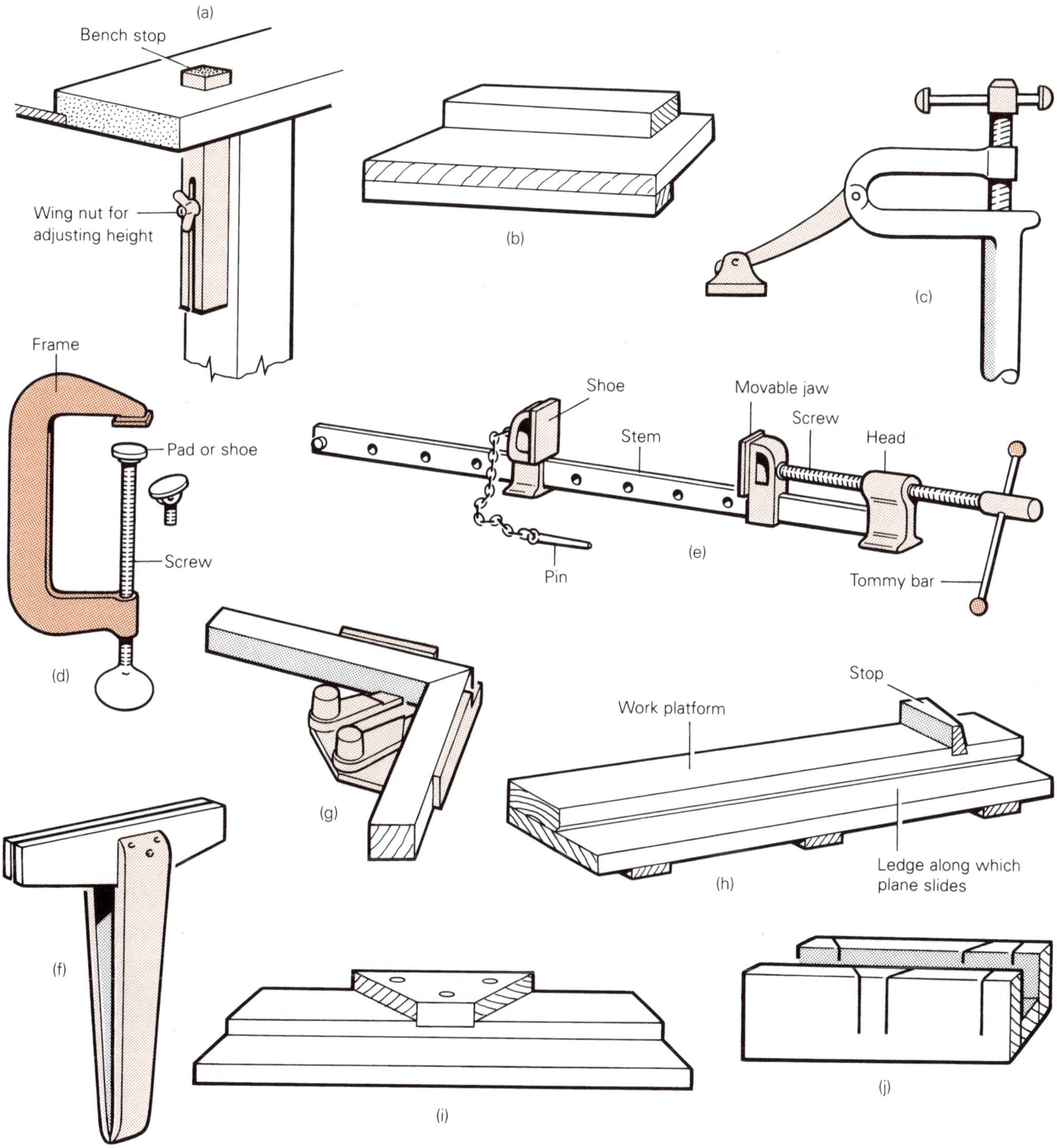

Figure 2.2 Holding and supporting devices: (a) bench stop, (b) bench hook, (c) holdfast, (d) 'G' cramp, (e) sash cramp. (f) saw cramp, (g) mitre cramp, (h) square shoot, (i) mitre shoot, (j) mitre box.

piece when planing its surfaces. A bench stop can be of wood or metal.

The **bench hook** (Figure 2.2b), also known as a **cutting board**, is a piece of board with a strip of wood screwed across the ends of the opposite faces. Its use is to support a piece of wood you wish to saw, cut or chisel. It prevents the top of the work-bench from being damaged.

The **bench holdfast** (Figure 2.2c) is a metal device fitted into a hole on the top surface of a workbench to hold down a job firmly. It enables workpieces of odd shapes to be held firmly on top of the workbench. It is also a useful holding device when you want to prepare mouldings or use rebate planes.

The **'G' cramp** (Figure 2.2d) is a metal cramp in a form of the letter 'G' used for holding a piece of wood (job) to the work-bench to enable such operations as sawing or chiselling to be performed. You can also use it to hold small parts of a job together when gluing a job. The size is determined according to the opening of the jaws (50–300 mm). The main parts are the frame ('G' shape), screw, pad or shoe and tommy bar.

The **sash cramp** (Figure 2.2e) is a long metal cramp with opening ranging from 45 cm to 195 cm. It has a movable jaw that is controlled by a screw, a shoe that slides freely on the stem and fixes to a position by a pin, head that contains the female thread, screw and tommy bar. It is often used while gluing, to hold and draw the parts being assembled together while the glue is setting. For example, you might use this cramp when gluing sashes, door frames, carcases etc.

The **saw cramp** or **vice** (Figure 2.2f) is a small metal vice that is fixed on a workbench to hold a saw when resharpening it.

The **mitre cramp** (Figure 2.2g) also known as corner clamp is a metal clamp used for holding and drawing together right-angled mitre joints while gluing and pinning.

Shooting boards (Figure 2.2h,i), are of two types: the square shoot and mitre shoot. Both may be made from wood or metal. The square shoot is used for holding a work-piece when shooting (planing) the ends square to a face or an edge. The mitre shoot is used for holding a work-piece so that the ends can be mitred (planed at 45° to the edge or face).

The **mitre box** (Figure 2.2j) is a device made of either metal or wood, used for holding a work-piece when cutting a mitre (at an angle of 45°). A tenon saw or a panel saw may be used for the cutting. The guides of the metal mitre box can be adjusted to the desired angle while those of the wooden mitre box are fixed at the required angle.

Geometrical tools

Geometrical tools are those that are used for measuring, marking out, setting out and testing of a job at various stages. Among the common tools are: the rule, straight edge, marking knife, wing compasses, callipers, try-square, mitre square, sliding bevel, combination square, marking gauge, mortise gauge, cutting gauge and panel gauge.

The **rule** (Figure 2.3a) is a measuring tool that is used for setting out and checking distances on a workpiece. It is also used for setting marking and cutting gauges. A rule can be made from flexible steel or wood, often boxwood. The common forms of rule are: 75 mm four-fold, 50 mm four-fold, 50 mm two-fold and 25 mm type. They are graduated in millimetres, inches or both. When you are setting out or checking a distance with the rule, place its edge against the workpiece, so that you can clearly see the graduation and mark off the exact distance with a pencil or marking knife on the work-piece.

The **straight edge** (Figure 2.3b) is made either from steel or wood. It has perfect straight and parallel edges. It is mostly used for testing the evenness of surfaces and edges of a work-piece or job.When testing or checking straightness of a job, you must bring the job to your eye level and see between the edge and surface of the job. If the surface is straight, you will see that the light between the edge and the surface being tested will be even and the straight edge will not rock. A dark spot indicates higher surface and must be planed down further to level up.

The **marking knife** (Figure 2.3c) is made of tool steel with one end angled and bevelled. The cutting edge is used for marking a cut line across the shoulders of joints such as tenons, and trenches to guide sawing or chiselling. The cut line is drawn in conjunction with a try square and must be square (at 90°) with the face edge.

A pair of **wing compasses** (Figure 2.3d) is a metal (steel) tool in a form of a pair of dividers. It has two legs that are ground to sharp points. The legs are opened to a required radius by means of a thumscrew. Its main application is for setting out arcs and circles on a workpiece or job.

Callipers (Figure 2.3e) are of two kinds: **inside callipers** and **outside callipers**. Both

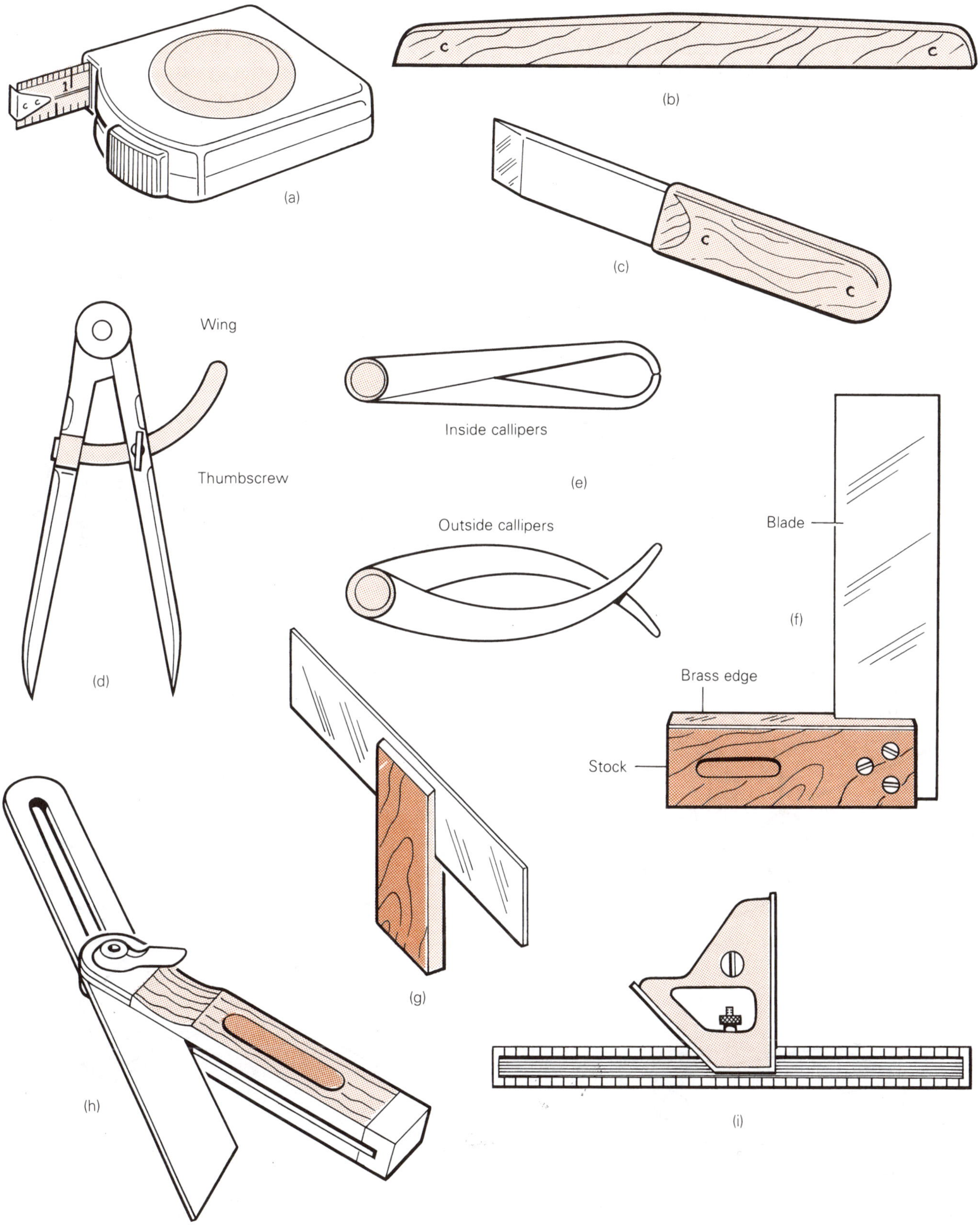

Figure 2.3 Geometrical tools: (a) rule, (b) straight edge, (c) marking knife, (d) wing compass, (e) callipers, (f) try-square, (g) mitre square, (h) sliding bevel, (i) combination square.

kinds are made of steel. The inside callipers are mainly used for checking diameters of holes or slots whilst the outside callipers are used for outside diameters of cylindrical objects. Callipers are classified as measuring tools and not marking out tools. They are indirect measuring tools since they have no graduation of measurement on them and so the actual diameter being checked must be read form a rule.

The **try-square** (Figure 2.3f), consists of two main parts, the blade made of tempered tool steel and the stock made of wood. The blade is riveted in the stock and is at right angles to the stock. The piece of wood from which the stock is made must be well seasoned so as to prevent warping and twisting. Rosewood or ebony is mostly used for making the stock. The inside edge of the stock is often covered with a brass plate in order to prevent wearing. A try-square is chiefly used for testing the squareness of surfaces and edges of

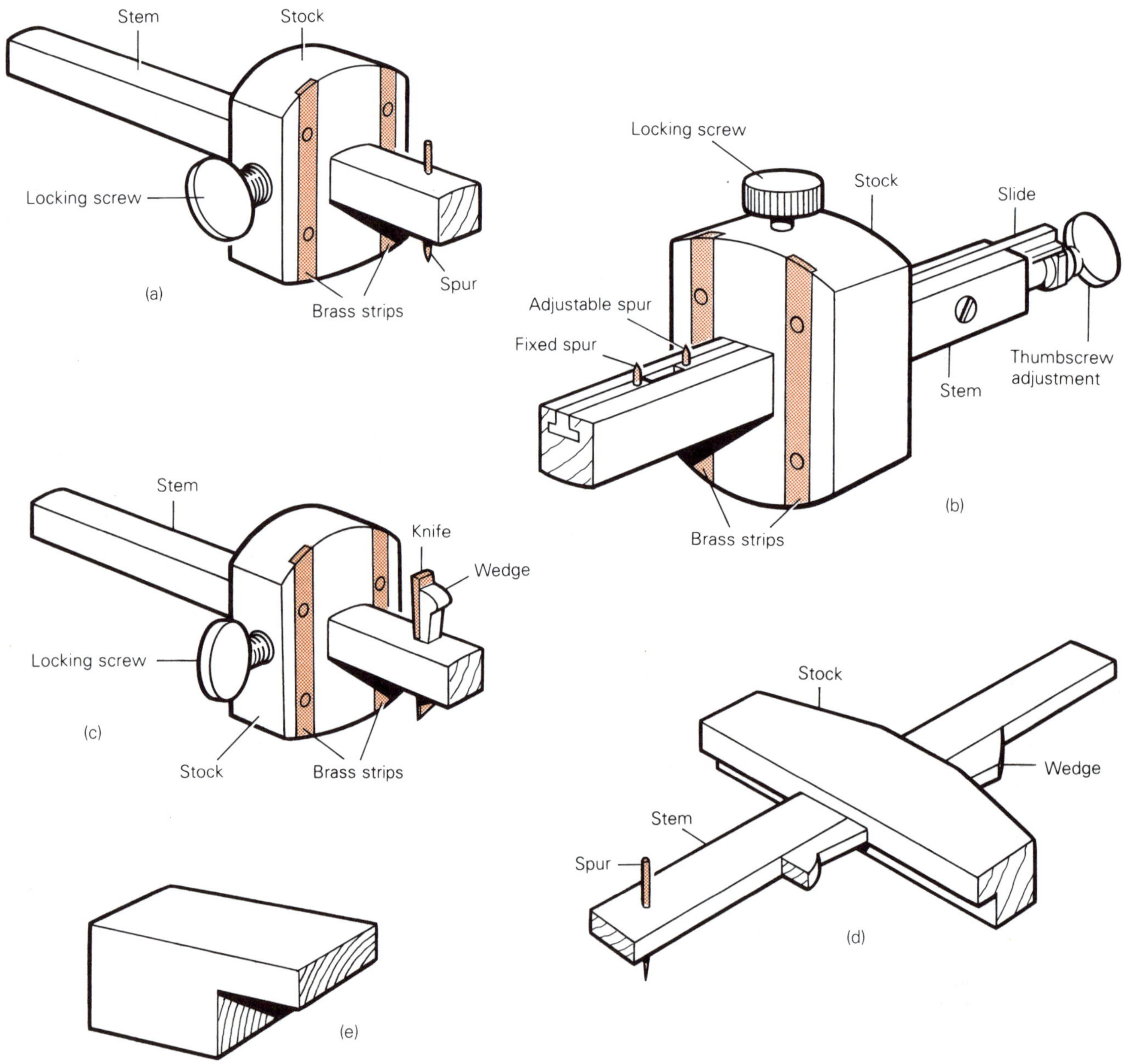

Figure 2.4 Gauges: (a) marking gauge, (b) mortise gauge, (c) cutting gauge, (d) panel gauge, (e) thumb gauge.

work-pieces, outside and inside corners of joints or carcases, and for marking lines at right-angles to a given surface or edge. (Never be tempted to use a try-square as a hammer, or the angle between the blade and stock will distort and it cannot be used to test squareness accurately.)

The **mitre square** (Figure 2.3g), sometimes called a **mitre set**, has a blade and stock as a try-square but the blade is fixed at 45° to the stock. It is used for testing and marking angles of 45°.

The **sliding bevel** (Figure 2.3h) comprises a blade and stock. The blade is slotted and is not permanently fixed in the stock but passes through a slot in the stock and is held at a required angle by a screw or lever. A sliding bevel is often used for testing and setting out bevels or angles other than 45° and 90°. (It can be used for marking or setting out angles of 45° and 90° but that is not its main application.)

The **combination square** (Figure 2.3i) consists of a steel graduated blade and a stock often referred to as the head. The head (stock) has two edges, 90° and 45°. The blade has a groove along its length that fits into a pin in the stock and slides to any length and held in position by a spring loaded screw. The 90° and 45° edges or faces of the stock enables the combination square to be used for testing and setting out angles of 90° and 45°. It has an additional head that, if fixed to the blade, can be used to mark a line through a centre of cylindrical object. When used in this way, it is referred to as a centre finder.

The **marking gauge** (Figure 2.4a) comprises a wooden (beech) stock that slides along the stem, a wooden (beech) stem that passes through the centre of the stock, a wooden (boxwood) thumbscrew that is used to hold the stock at a required distance, a steel spur (also called pin) that is fixed firmly into the end of the stem. It is used for marking lines parallel to a face or an edge of a workpiece. The required distance from the spur to stock is set with a rule. The appropriate methods for setting and gauging are illustrated in Figure 2.5.

The **mortise gauge** (Figure 2.4b) is similar to a marking gauge and has similar parts: stock, stem and thumbscrew, but two spurs instead of one. The spur near the end of the stem is fixed whilst the other is fitted to a sliding brass rod that moves when the thumbscrew at the other end of the stem is turned. It is often used for marking two lines parallel to a face or an edge of a workpiece, particularly when marking out mortises and tenons, or the pins and sockets of bridle joints. It is convenient to set the spurs to the chisel you will use – for example, the mortise chisel (see Figure 2.6). After the spurs are set, adjust the stock to bring the spurs to the correct positions on the workpiece. Push the gauge along gently, ensuring that the stock is always against the face or edge of the workpiece.

The **cutting gauge** (Figure 2.4c) also consists of a stock, stem, thumbscrew, but has a cutting blade instead of a pointed spur. The cutting blade is fixed parallel to the stock by a wedge. Its main

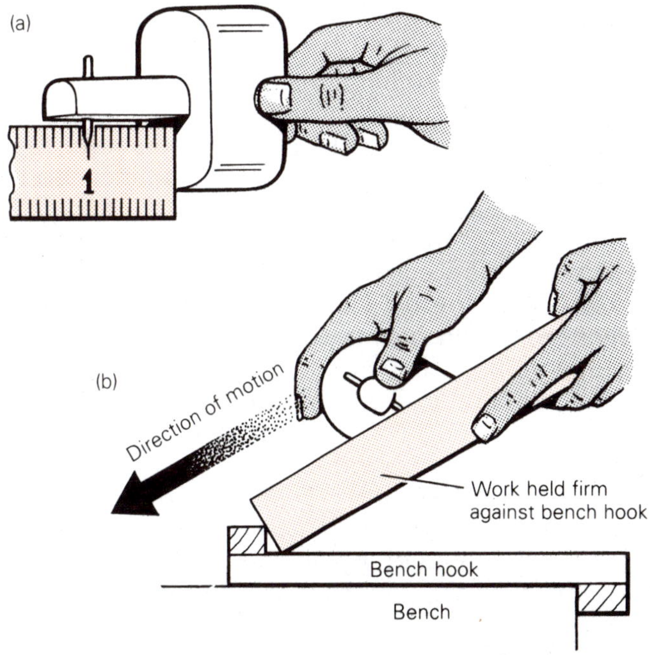

Figure 2.5 **Setting and using marking gauge: (a) setting, (b) using the gauge.**

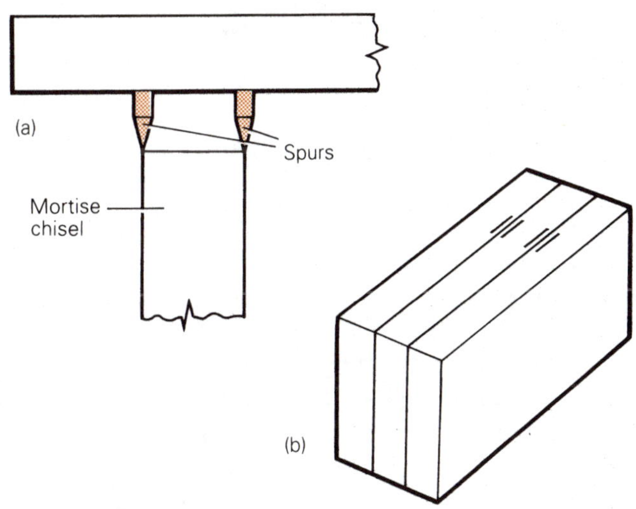

Figure 2.6 **Setting and using mortise gauge: (a) setting, (b) using the gauge.**

use is for cutting lines parallel to the end of a workpiece, i.e., cutting across the grain.

The **panel gauge** (Figure 2.4d) has the same parts as the marking gauge, except that the stem is longer. The long stem makes it suitable for marking the width of wide boards such as door panels, hence its name.

The **thumb gauge** is illustrated in Figure 2.4e.

Percussion tools

The word *percussion* means striking two object together and therefore percussion tools are those used for striking other tools. Those used in woodwork are hammers and mallets.

The common woodwork hammers are the Warrington hammer and the claw hammer. Both have two main parts: the head, made of forged tool steel, and the handle, made of wood (hickory) and wedged into the head.

The **Warrington hammer** (Figure 2.7a) is a light cross-pein hammer generally used for light hammering particularly in cabinet work. The cross-pein end is used for starting small nails and also for pressing down when gluing small inlays and veneers into position.

The **claw hammer** (Figure 2.7b) is heavier than the Warrington hammer and the head has a claw end instead of a pein end. It is mainly used for heavy hammering as in carpentry, for driving in large nails. The claw end is used for removing nails.

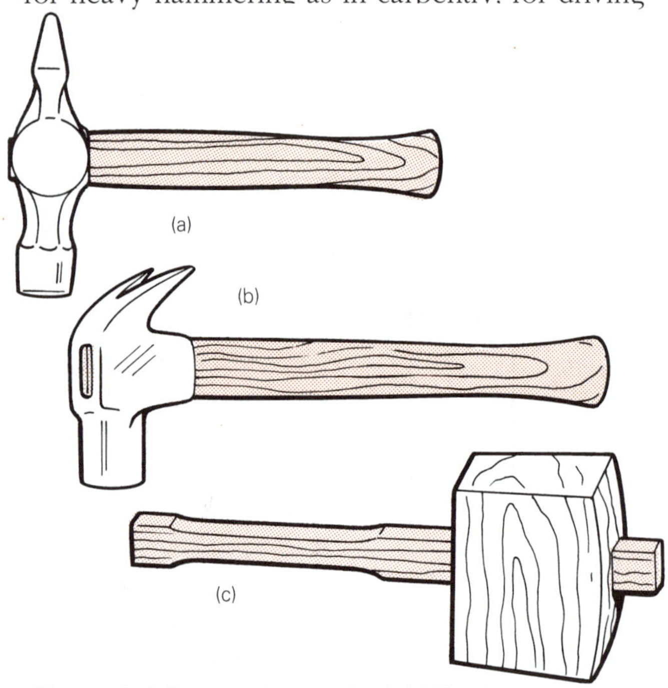

Figure 2.7 Percussion tools: (a) Warrington hammer, (b) claw hammer, (c) mallet.

The **mallet** (Figure 2.7c) may be said to be a wooden hammer. It has a wooden head, (100–175 mm) long and a wooden handle. The head is made of beech whilst the handle is made of either beech, ash, or hickory and fitted into a tapered mortise in the head. The general use of a mallet is for driving chisels. It may be used also when assembling work.

Impelling tools

To *impel* is to force, and impelling tools are those used for forcing other materials such as screws into a piece of work. The three common impelling tools are the screwdriver, nail punch and brace.

Screwdrivers are made of alloy steel in a form of flat, round or square. One end is ground into a tip to fit into the slot of the screw while the other end is shaped into a tang and fitted with a handle made from a tough hardwood such as beech, ash or hickory. The handle can also be made of plastic. The types of screwdriver commonly used in woodwork are the **London pattern** and the **cabinet pattern** (Figure 2.8a,b). The London pattern is flat in section and the cabinet is round in section. The ends of both types are ground flat. A third kind of screwdriver that fits screws with a cross-shaped slot is commonly referred to as a **Phillips pattern** screwdriver (Figure 2.8c).

When using a screwdriver, you must ensure that you select the correct size; the tip fitting properly to the bottom of the slot in the screw; start the screw in a pilot hole so that both your hands can be used. The wider the blade of the screwdriver the greater the gripping force and the longer the blade the greater the leverage for driving in large screws.

The **nail punch** (Figure 2.8d) is a piece of steel rod having one end turned to a taper and the striking end turned parallel and chamfered. The middle portion is knurled for firm gripping. It is used in conjunction with a hammer for driving the heads of nails below the surface of a job. There are different sizes to fit different sizes of nail.

The **carpenter's brace** (Figure 2.8e), though it can be said to be a holding and supporting tool, is better classified as an *impelling* tool since it is used to hold and force various bits into the wood. A brace consists of a head, crank, chuck and ratchet. The head is a block of hardwood (beech or teak) shaped so as to suit the palm of the hand

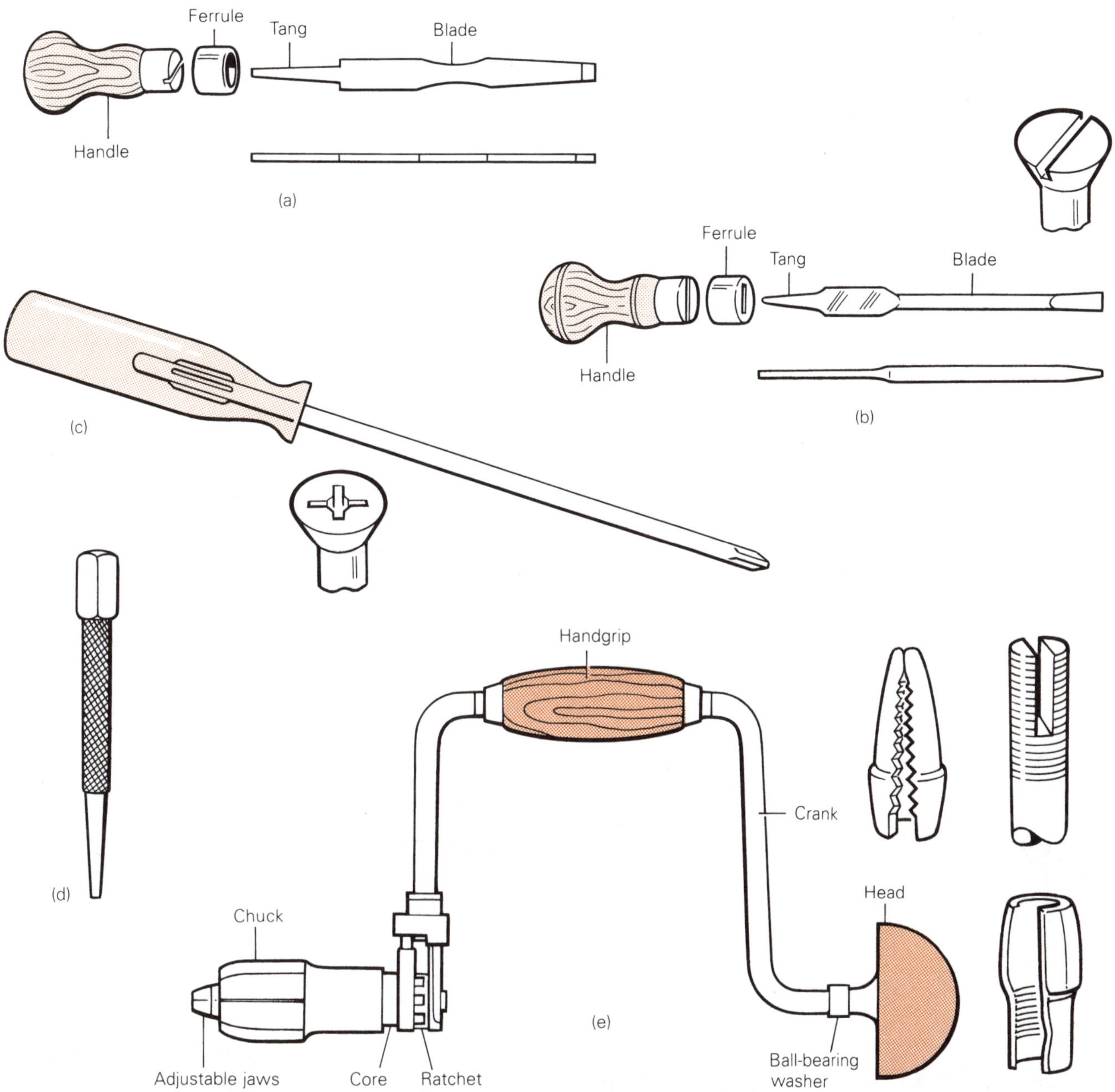

Figure 2.8 **Impelling tools: (a) London pattern screwdriver, (b) cabinet screwdriver, (c) Phillips screwdriver, (d) nail punch, (e) brace.**

and screwed to a steel sleeve also fitted over the end of the crank. The sleeve runs on hardened steel balls. The crank is made of a steel rod, bent to give a sweep of 200–350 mm. The chuck forms part of the jaw and is screwed and slotted to receive the socket and jaws. The jaws consist of two pieces of hardened steel loosely wired together and inserted into the socket. The ratchet is a device that enables the chuck to be turned in one direction only. The brace is used for holding bits and driving them into the work-piece to bore holes.

Boring tools

There have been advances in the methods of producing holes in pieces of wood. The portable hand-powered drill for example, often provides a quicker and more efficient way of drilling holes accurately. Nonetheless, the traditional methods are not entirely abandoned and are sometimes preferable. The tools you are likely to use are: bits, gimlets and bradawls.

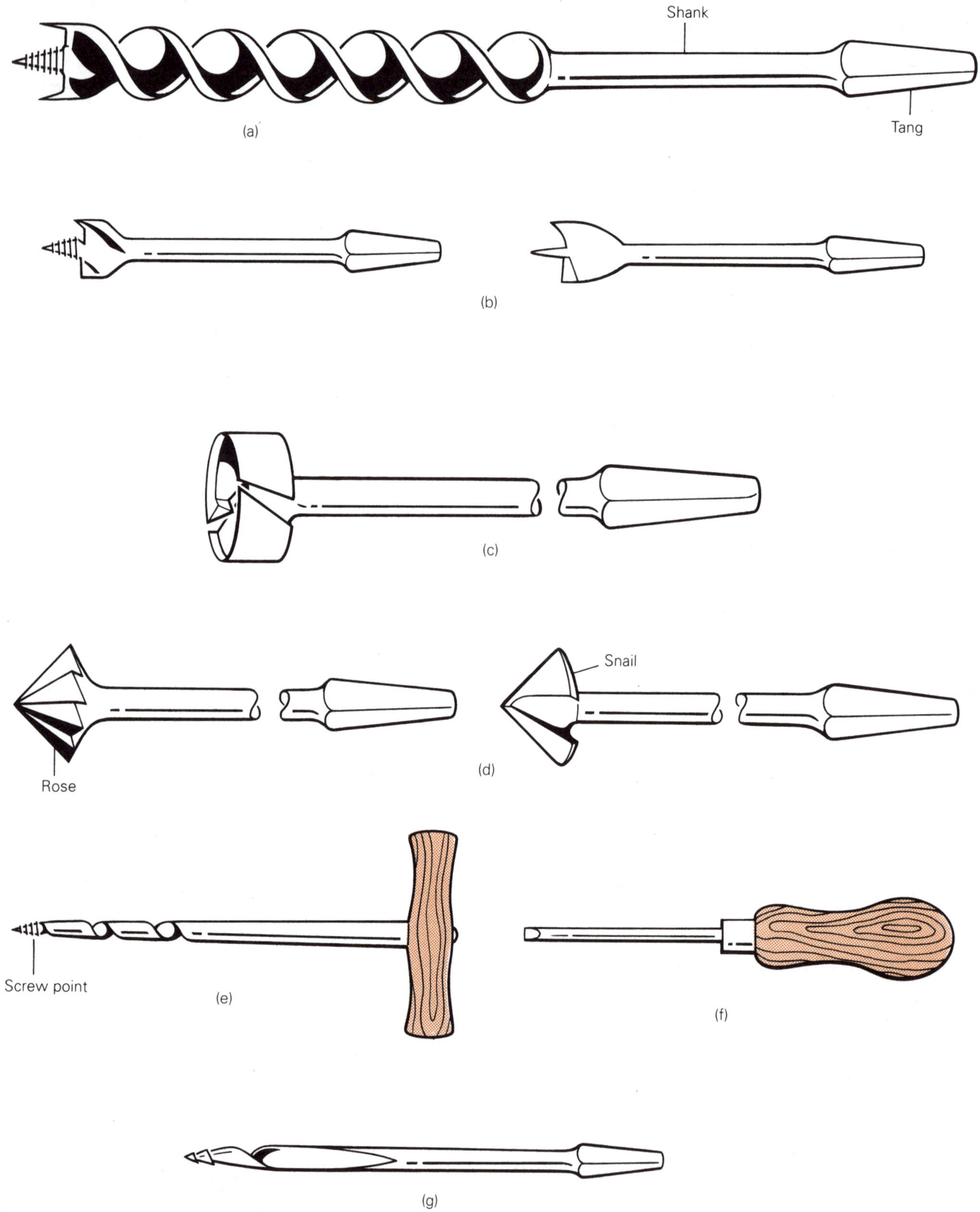

Figure 2.9 Boring tools: (a) auger bit, (b) centre bits, (c) Forstner bit, (d) countersink bits, (e) gimlet, (f) bradawl (g) cobra bit.

The traditional method of boring holes in a piece of wood depends largely on the use of many varieties of bits, designed for specific applications. All bits have two main parts, the body forming the cutting edges and the shank having a square, tapered tang that fits into the jaws of a hand brace.

Auger bits (Figure 2.9a) are forms of twist bits. The common type has a twisted or spiral body and is known as a **Jennings pattern**. The twisted body guides the bit to bore deep, straight holes and provides a passage for the chips to flow out. It has a threaded centre point that draws the bit into the wood, preventing the bit wobbling off the centre of the hole. Some bits have spurs on the outside of the cutting end that cuts through the fibres ahead of the cutter. The common sizes are from 6 mm to 30 mm and are used for boring straight deep holes in hard and softwood. In order to avoid splitting the wood when boring all the way through it you must watch out for the point showing at the other side, then reverse the work and bore from that side.

Centre bits (Figure 2.9b) are used for making shallow holes in thin pieces of wood.

Forstner bits (Figure 2.9c) are used for making stopped, or blind holes.

Countersink bits (Figure 2.9d) are of two forms: **rose** and **snail**. The rose has a series of flutes around the cone-shaped cutting point. The snail also has a conical point but there is only one flute forming the cutting edge. Both are used for enlarging the top of a hole to receive countersink head screws.

The **gimlet** (Figure 2.9e) is a small hand boring tool normally used for starting holes for nails and screws, particularly in a corner or any place where a brace cannot be used. The handle is fitted crossways to provide a firm grip to exert pressure.

The **bradawl** (Figure 2.9f) is a steel rod with one of its ends flattened to serve as a cutting edge. The other end is pointed and fitted into a round wooden handle. It is used to create small holes to take nails and small screws. When using it, place the cutting edge across the grain of the wood and force it in by twisting back and forth to break (cut) the fibres.

The **cobra bit** is illustrated in Figure 2.9g.

CHECK YOUR UNDERSTANDING

- Holding and supporting devices include: the workbench, used for supporting the job during all bench work operations such as sawing, planing, chiselling, rebating, grooving, etc.; 'G'-cramp, used for securing a job to the workbench; sash cramp, used for holding workpieces together when assembling and gluing; mitre cramp, used for holding and drawing together right-angled mitre joints; shooting board, used for holding workpieces when planing the end square to a face or an edge; mitre box, used for holding a workpiece when cutting a mitre; and saw vice, used for holding a saw when sharpening it.
- Geometrical tools may be categorised into measuring, marking out/setting out and testing tools.
- Hammers and mallets are referred to as percussion tools, because they are used for striking other tools.
- Impelling tools are those used for forcing materials such as screws and nails into a job. The common ones are the screwdriver, nail punch and brace.
- Among the tools referred to as boring tools are bits of various types, the gimlet and bradawl.

REVISION EXERCISES AND QUESTIONS

1. Name four holding devices in woodwork and state the specific use of each.
2. State the three groups of geometrical tools and give one example of each. State the application of each of the examples.
3. With the aid of sketches, show the difference between a try-square and a mitre square. When is each used?
4. What is the main difference between a marking gauge and a cutting gauge in terms of construction and application?
5. Make a labelled sketch of a claw hammer and state the function of each of the main parts. What advantage has the claw hammer over the Warrington hammer?
6. Which tools can be classified as impelling tools? Give three examples of impelling tools and state their main applications.

3 Woodworking hand tools 2

Introduction

This chapter continues the discussion of hand-operated tools commonly employed in the woodwork industry. The emphasis here is on the various categories of *cutting* tools: shaving tools, paring tools and abrading tools. The discussion centres on their type, description and uses.

Shaving tools

Shaving tools are those that are used to reduce a piece of wood to a required size or shape by the action of shaving off the waste. Generally, planes are used for this purpose. A plane comprises a wooden or metal body known as a **stock**, a steel blade known as a **cutting iron** and a **cap iron** that breaks the shavings to prevent the wood fibres from tearing up. The blade and cap iron are attached together with a screw and hold in position in the stock by either a wedge (wooden planes) or a metal cap (metal planes). The main function of a plane is to produce smooth surfaces or edges, plain or shaped, by taking off thin shavings. The size of a plane is determined by the length of the stock and the width of the cutting iron.

Planes may be divided into the following categories: bench planes, used for providing flat or plain surfaces; curve cutting planes, used for producing curved surfaces and edges; special purpose planes, used for producing rebates, tongues and grooves, mouldings, etc.; and multipurpose planes, that can be set up for a variety of special purposes.

Bench planes

Bench planes are made of either wood or metal and are of different kinds depending upon their specific applications or uses. They are the roughing plane, jack plane, trying plane, smoothing plane and block plane.

The **roughing plane** (Figure 3.1) is distinguished from the other bench planes by the position of the handle (also known as a **toat)** fitted in the front of the stock. It has a single cutting iron. This means that the cutting iron is not backed by

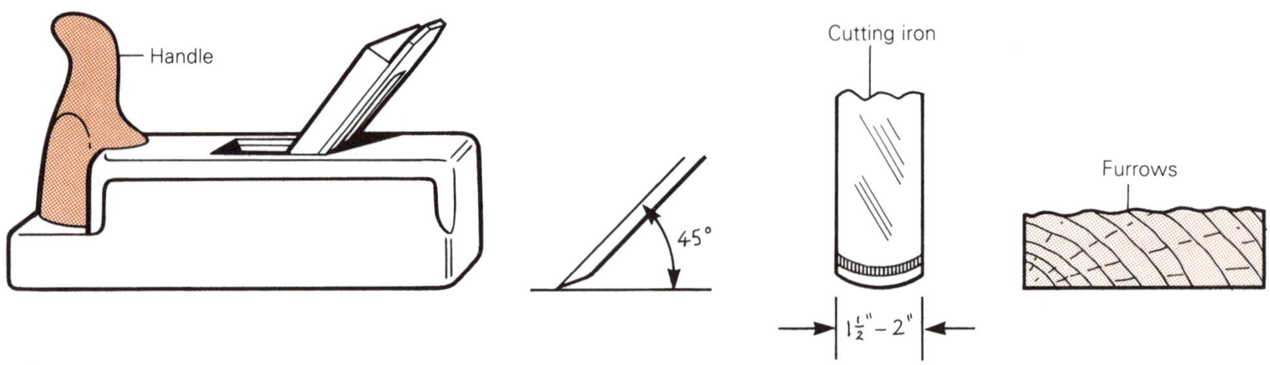

Figure 3.1 Roughing plane.

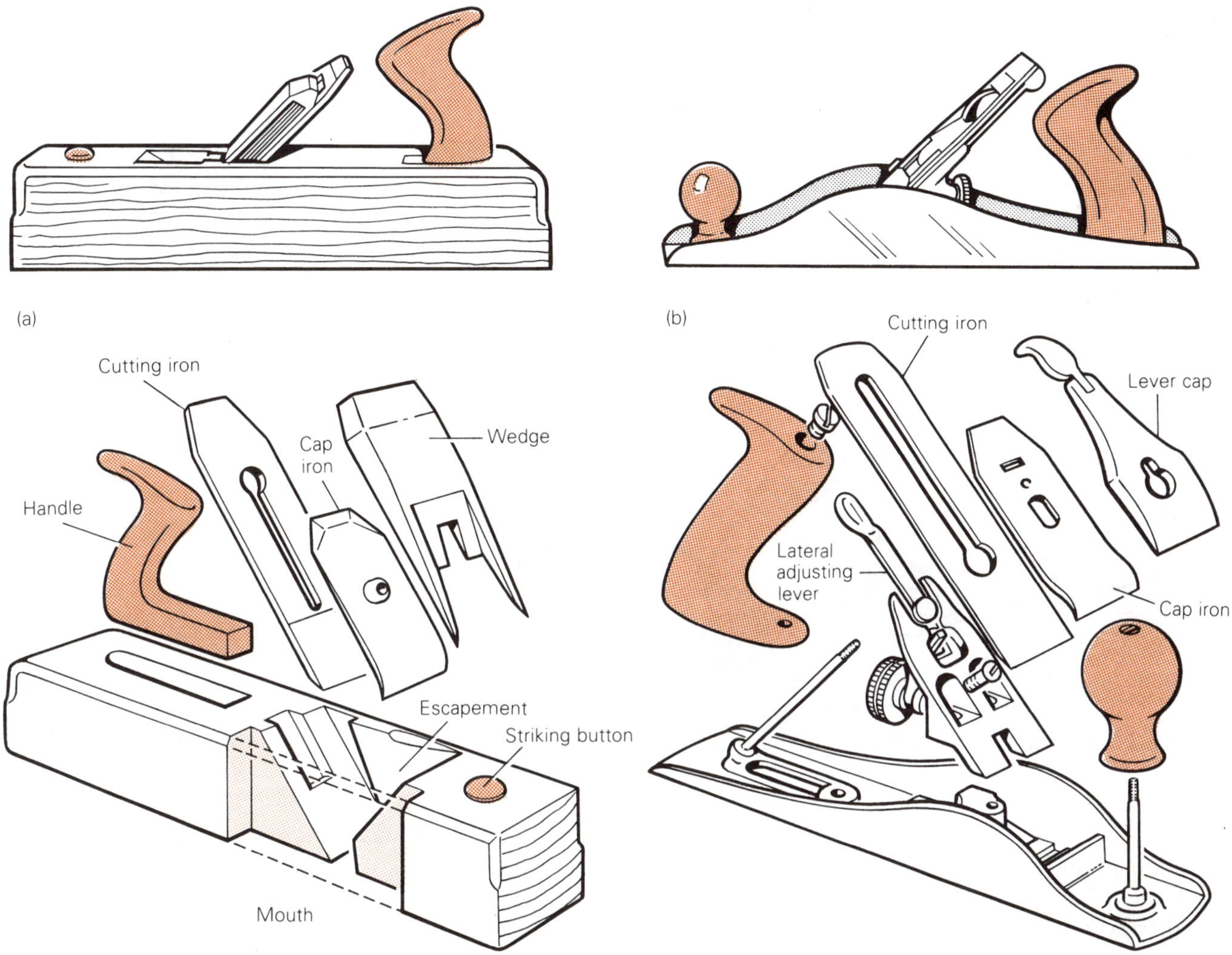

Figure 3.2 Jack plane: (a) wooden, (b) metal.

a cap iron. The average length of the stock is between 200 mm and 250 mm and the width of the cutting iron ranges from 37 mm to 50 mm. The **pitch**, or inclination of the cutting iron to the sole of the plane, is 45°. The edge of the cutting iron is fairly round and this enables thick shavings to be removed easily, though the surface produced is always furrowed or rough. Hence the name 'roughing plane'. It is used for reducing a workpiece quickly to the nearest required size, and is the first plane to use if you are to plane a sawn timber to size.

The **jack plane** (Figure 3.2) is similar to the roughing plane except that the handle is fixed into the stock *behind* the cutting iron and the stock itself is longer, about 350 mm. The cutting iron is fitted with a cap iron (back iron) which provides tension to the cutting edge to prevent vibration (chattering) and also breaks the shavings. The width of the cutting iron is between 50 mm and 56 mm. It is set at an angle of 45° to the sole of the stock. The cutting edge is slightly curved which allows fairly thick shavings to be taken off with less effort.

The other parts are: the **striking button**, a piece of rough wood let into the nose of most jack planes to serve as a striking point when removing the cutting iron or when reducing the set of the cutting iron; the **wedge**, a piece of rough wood (beech or danta) that fits into the grooves in the sides of the escapement to hold the cutting iron in position; the **escapement** or **throat** – the opening in the stock into which the cutting iron is fixed and through which the shavings escape.

The jack plane is mostly used for planing rough surfaces of sawn timber to almost the required size. See Figure 3.2a,b for the parts of the wooden and metal jack planes.

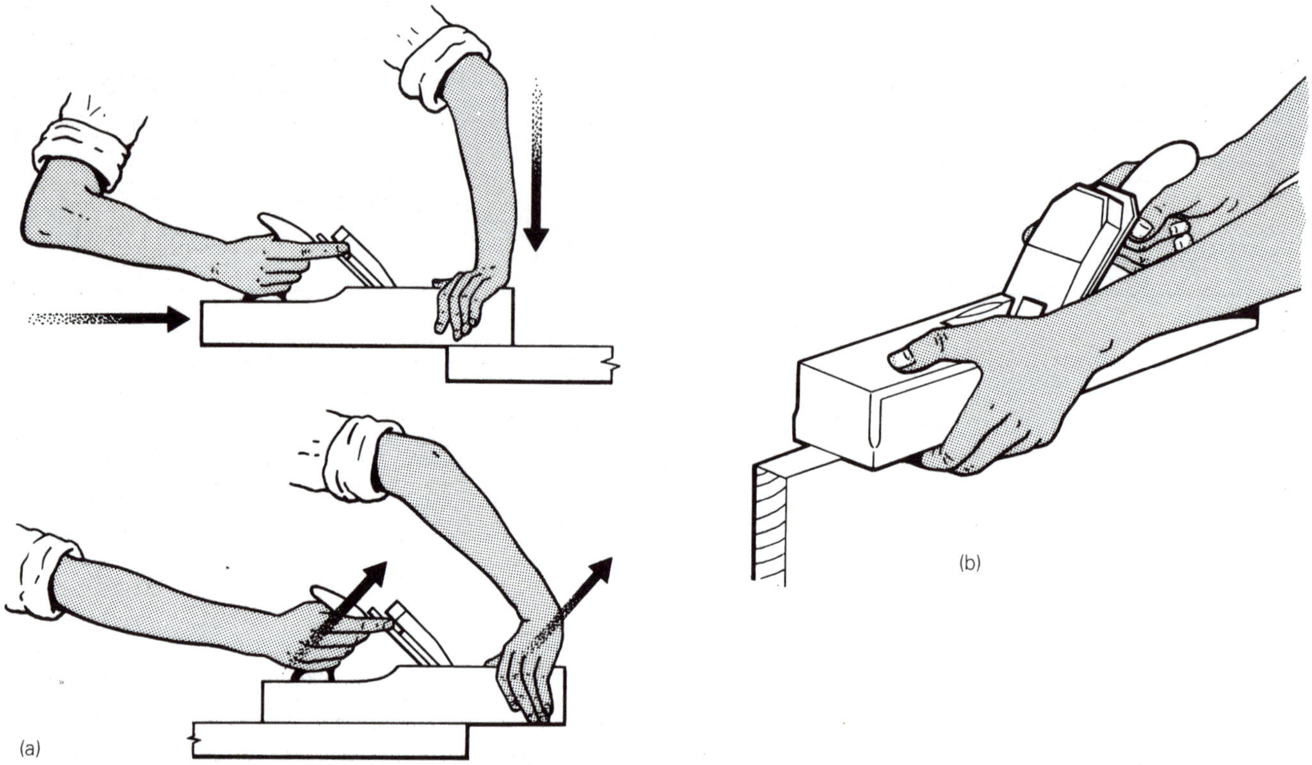

Figure 3.3 Using a plane: (a) planing a surface, (b) planing an edge. Note the position of the hands.

When planing a surface, hold the front part of the plane with your fingers on the left side and your thumb on the right side of the plane. When planing an edge, place your finger under the sole to run along the side of' the workpiece and your thumb on the top of the plane close to the throat (see Figure 3.3).

Procedure for setting the wooden jack plane

1. Set the cap iron about 1.5 mm away from the cutting edge of cutting iron. When tightening the cap iron (by the screw), hold the irons over the edge of the bench. Never hold them in your hand to avoid injury.
2. Place the stock in your left hand (if your are right-handed) and put your thumb in the throat and the fingers under the sole of the stock.
3. Insert the cutting iron in the escapement and hold it down with your thumb. Then insert the wedge.
4. To set the plane, turn over to the sole and check the projection of the cutting edge; make necessary adjustment and tap down the wedge slightly. If you want to increase the setting (projection of the cutting edge) gently tap the top of the cutting iron. To decrease the setting, tap the striking iron or button. Tapping the right or left side of the cutting iron will square up the cutting edge. When you are satisfied with the setting, tap down the wedge firmly.
5. If you want to remove the cutting iron, place the heel of the plane against your groin and strike the striking button sharply while supporting the wedge and the irons with your thumb to prevent them from jumping out. Never hold the handle to do this because it can break off owing to the short grain.

The **trying plane** (Figure 3.4a), also called a **jointer**, has the same parts as the jack plane except that the handle is closed and the stock is longer, being 500–750 mm in length. The stock is made of either wood or metal. The width of the cutting iron ranges from 56 mm to 60 mm and it has an almost straight cutting edge. The cap iron is set about 3–4 mm from the cutting edge.

This plane is often used to produce true and flat surfaces, or perfect straight edges on long boards. You should hold this plane in the same way as the jack plane. You must always keep the entire length of the sole in contact with the edge you are planing.

The **smoothing plane** can be of wood (Figure 3.4b) or metal (Figure 3.4c). The wooden type has no handle and is shaped so it can be used with one hand. The size of the cutting iron ranges from 31 mm to 56 mm with only the corners rounded slightly to avoid digging when planing. The body of the metal type is about 230 mm long and the width of the blade is 50 mm or more. Both types are used for cleaning up or dressing the surfaces and edges of workpieces ready for assembling, levelling up assembled joints and for cleaning-up a job before glass papering. The small size of both types makes it possible to concentrate on small cross-grained areas.

The **block plane** (Figure 3.4d) normally has a cast iron stock about 125 mm long with a cutting iron 35 mm wide. There is no cap iron. The cutting iron is set at an angle of 20° to the sole of the plane (i.e., the pitch is 20°). It is inserted in the stock with the bevel side up.

This plane is used for planing an end grain, as is necessary in finishing mitres, and also for planing chamfers. Owing to the low pitch of the cutting iron, vibration is minimised when planing the end grain. It can easily be held in one hand for planing. Considering its uses, a block plane can be classified as a special purpose tool.

Curve cutting planes

As the name implies, this class of planes comprises those used for producing smooth curved surfaces and edges. The two that are commonly used are the spokeshaves and the compass plane.

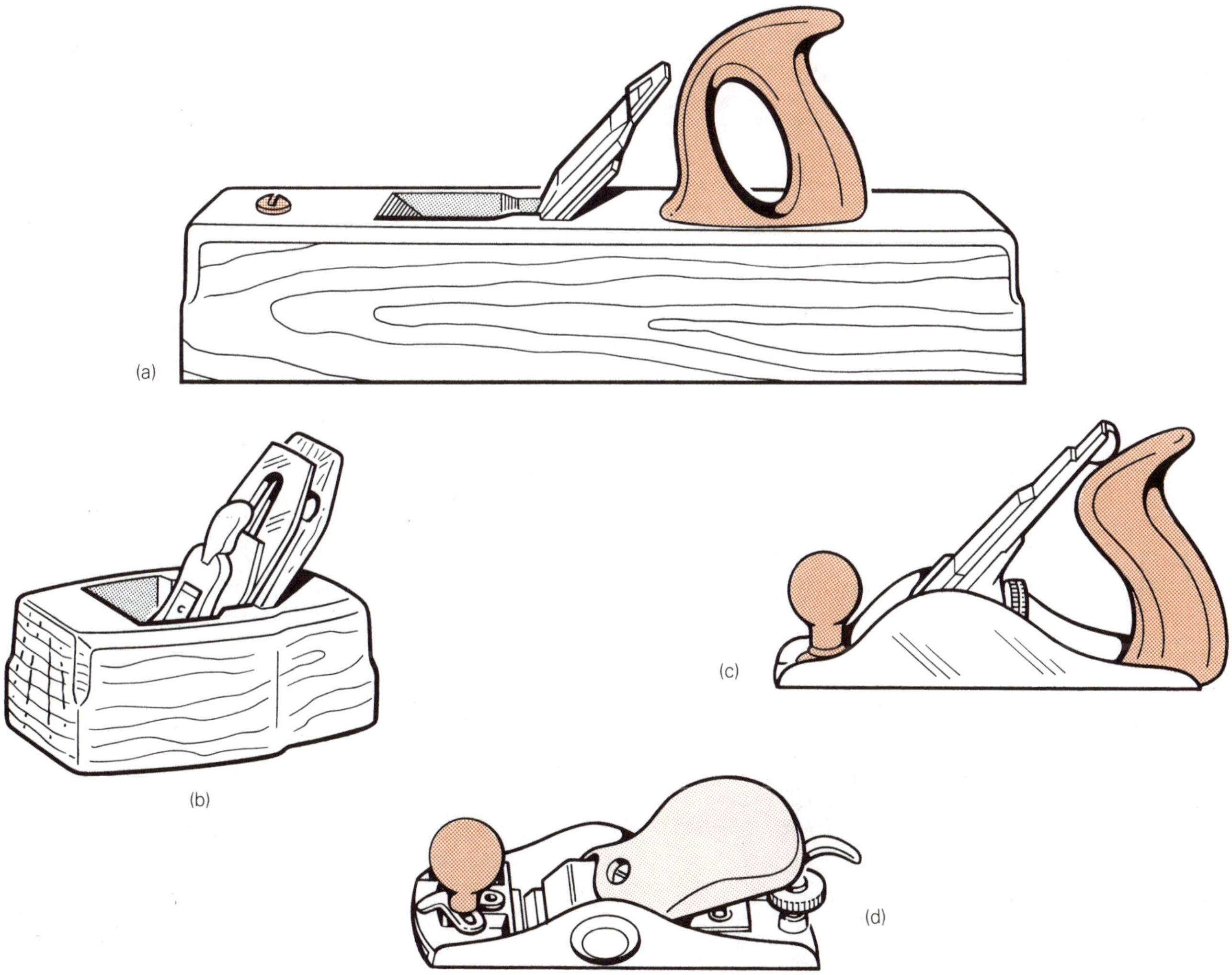

Figure 3.4 Other bench planes: (a) trying plane, (b) wooden smoothing plane, (c) metal smoothing plane, (d) block plane.

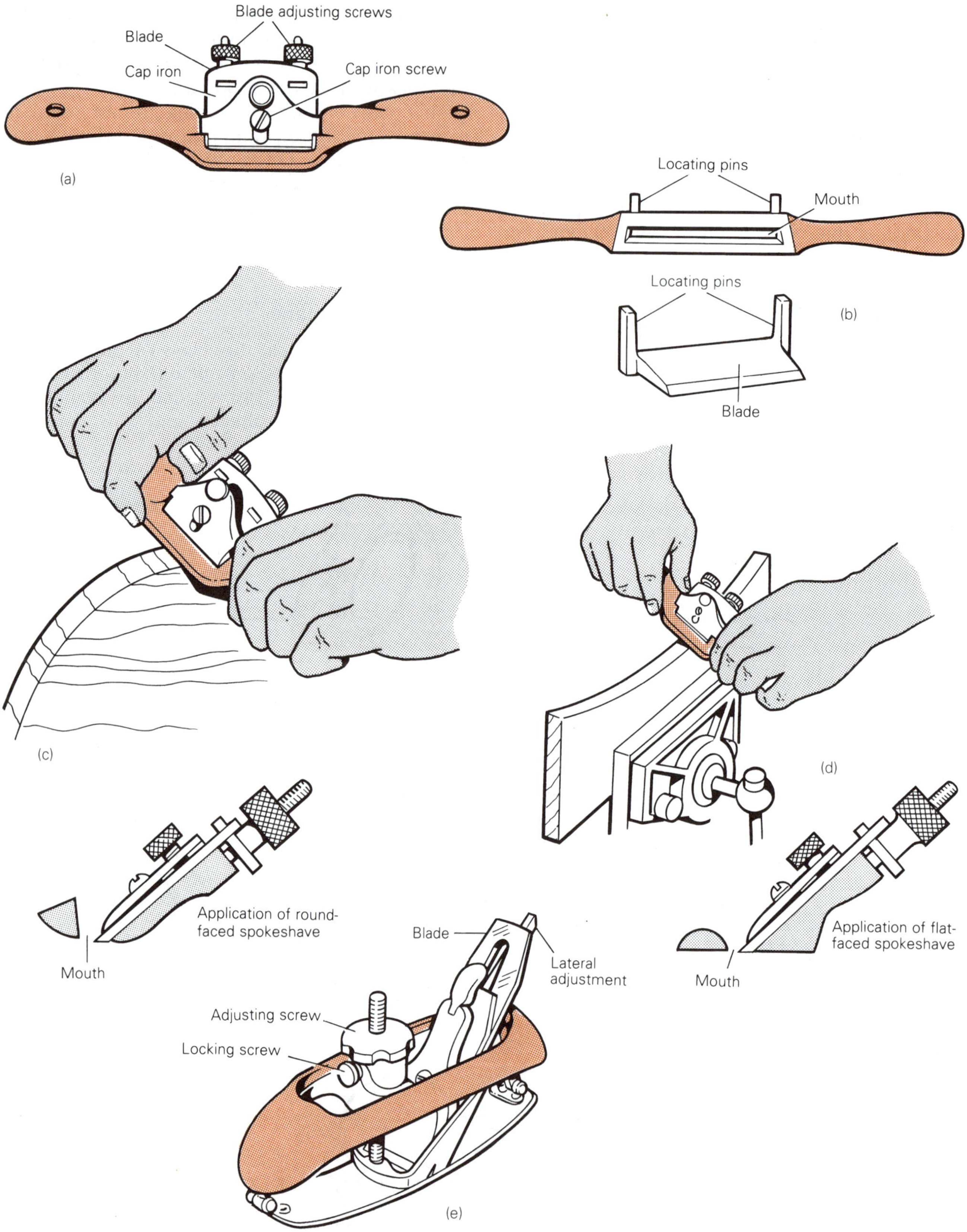

Figure 3.5 Curve-cutting planes: (a) metal spokeshave, (b) wooden spokeshave, (c) round-faced spokeshave, (d) flat-faced spokshave, (e) compass plane.

Spokeshaves (Figure 3.5a–d) are of two kinds: **flat-faced** and **round-faced** and they are made of either wood or metal. The stock of the wooden type is usually made of beech or boxwood and has a cutting iron of 75–100 mm in length. The cutting iron is held in the stock by two tangs. When inserted, the cutting iron lies flat on the surface and acts as part of the sole of the plane.

The metal type has a cast iron stock and a cutting iron of 34–53 mm. It is worth noting that the size of the metal spokeshave is governed by the width of the cutting iron while that of the wooden spokeshaves is determined by the length of the cutting iron.

The flat-faced type is used for smoothing flat or convex edges and narrow surfaces whilst the round-faced type is for smoothing concave edges They can also be used for planing stopped chamfers.

The **compass plane** (Figure 3.5e), also known as a **circular plane**, consists of a stock and a cutting iron. The stock can be of wood or metal. The wooden compass plane has a fixed curved sole, either concave or convex of a fixed radius. The length of the stock is normally about 250 mm. The width of the cutting iron is about 44 mm.

The metal compass plane has a *flexible* sole of about 250 mm in length. This is adjusted to the required shape or arc, be it concave or convex. The appropriate way to adjust the sole is to place the plane on the sawn curve and turn the adjusting screw until the sole fits in the curvature, and then tighten the locking screw. The main use of the compass plane is for smoothing curved surfaces and edges that are true arcs of a circle.

Special purpose planes

These planes, as the name implies, are used for specific operations such as rebating, tongue and grooving, trenching, moulding, etc. Those discussed are the rebate plane, side rebate plane, bullnose plane, shoulder plane, fillister planes, plough plane, matching planes, router plane and moulding planes.

The **rebate plane** (Figure 3.6) can have a wooden or metal stock. The length of the wooden type is about 220 mm and that of the metal type is about 150 mm. The mouth of the wooden stock (at the sole) can be square or skew (or inclined). The skewed mouth has the advantage of curling the shavings out of the throat. The width of the cutting iron is always the same as the thickness of the stock and ranges from 19 mm to 31 mm.

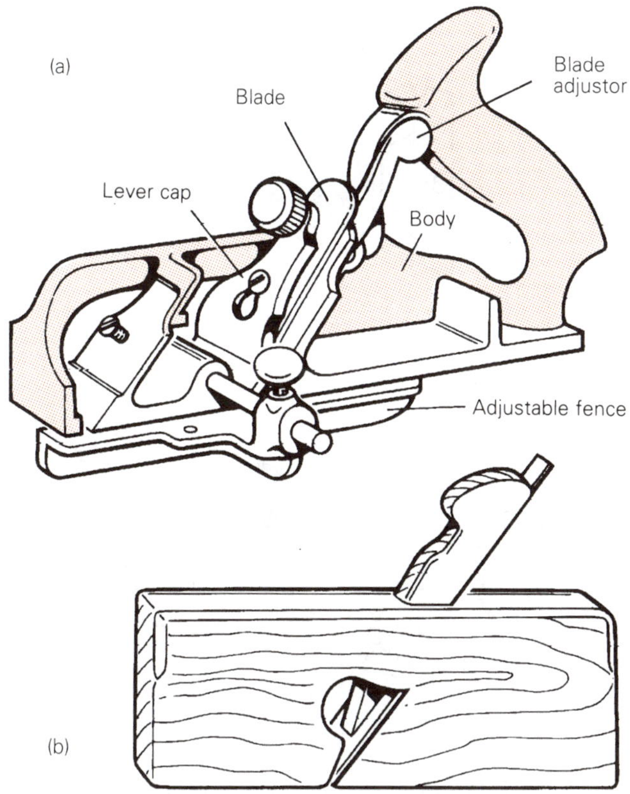

Figure 3.6 Rebate plane: (a) metal type, (b) wooden type.

There is no cap iron and the pitch is 50–55°. The bevel of the blade faces downwards.

The metal rebate plane is fitted with an adjustable depth stop to determine the depth of the rebate and an adjustable fence that determines its width. It also has a spur ahead of the cutter and when in operation this cuts the wood fibres before the cutter.

Both types are mostly used for planing or finishing rectangular recesses called rebates that are usually cut along the edge or across the end of a workpiece. When you want to cut a rebate across the grain, using the wooden type, you must first mark the shoulder with a cutting gauge and then saw it to the required depth before you plane it with the rebate plane. Where possible, you should nail a strip of wood to serve as a guide since the rebate plane has no fence or depth gauge of its own.

The **side rebate plane** (Figure 3.7a) is a small plane used for trimming the shoulders of rebates, trenches and grooves. It is normally fitted with two cutting irons, one for trimming the right-hand shoulder, the other is used for the left-hand shoulder. It has a depth gauge to maintain uniform depth.

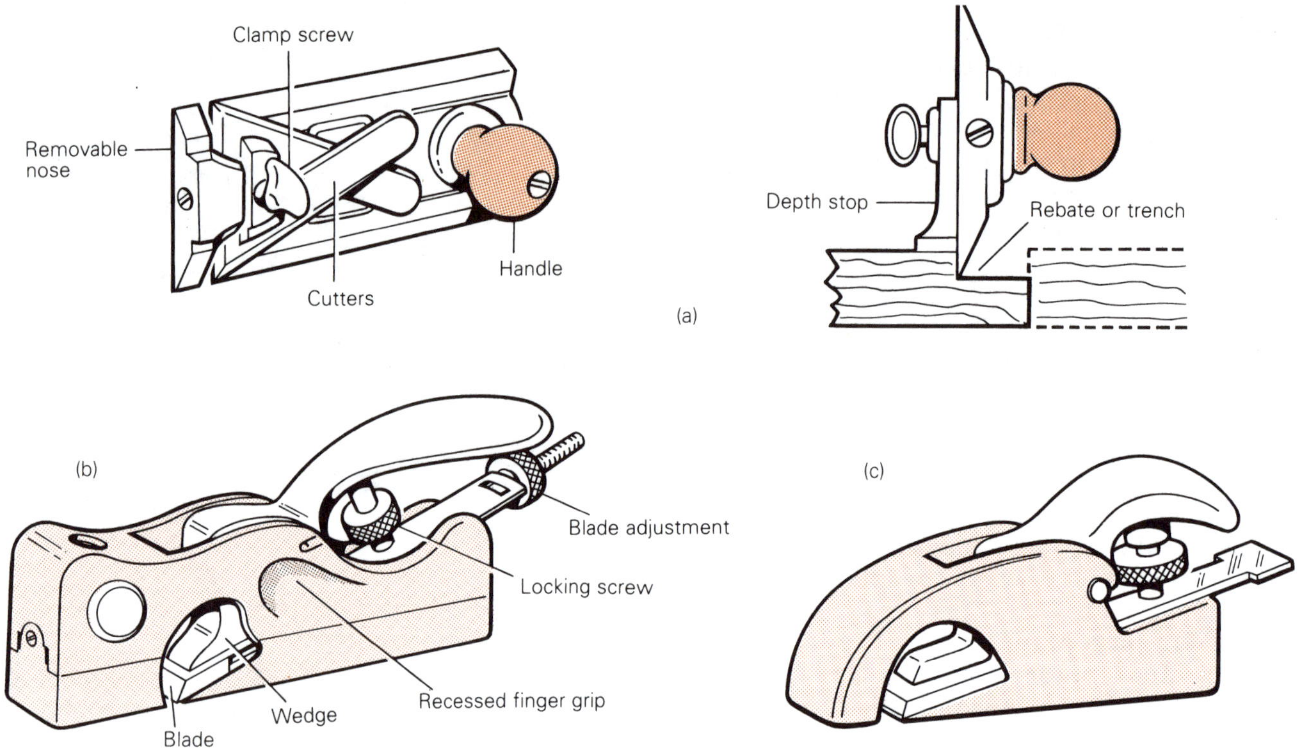

Figure 3.7 Other special purpose planes: (a) side rebate plane, (b) shoulder plane, (c) bullnose plane.

The **shoulder plane** (Figure 3.7b) is a metal plane with a stock and cutter of the same width. The cutter is set at a very low angle with the bevelled side facing uppermost. This plane is used mostly for trimming shoulders of tenon joints and finishing off through rebates. It can cut the grain because of its low angle.

The **bullnose plane** (Figure 3.7c) is similar to the shoulder plane but smaller, usually with a metal (cast iron) stock about 100 mm long and 25 mm wide. The cutting iron has no cap iron and is of the same width as the stock. It is fitted at 15° to the sole of the stock with the bevel side facing up. The bullnose plane is commonly used for cleaning-up through rebates, stopped rebates and stopped chamfers. With some bullnose planes, you can remove the nose to enable you to plane right into the stopped end of the rebate or chamfer.

Fillister planes (Figure 3.8a, b) are of two kinds: the side fillister and the sash fillister. The **side fillister plane** is similar to the rebate plane and has either a wooden (beech) or metal (cast iron) stock. The mouth can be square or skew. It is fitted with a fence to set the width of the recess and a depth gauge to ensure a constant depth of cut. Another useful fitting is the spur fixed in front of the cutter to cut through the fibres when rebating across the grain. The cutting iron has no cap iron and forms an angle of 50°–55° to the sole of the stock. It is used for planing a rebate on the side nearest to the operator.

The **sash fillister** is similar to the side fillister except that it has a fence attached to extended beams passing through the body of the plane and held in position by wedges. It is also used for planing rebates, but in this case, the rebate must be on the face or edge away from the operator.

The **plough plane** (Figure 3.9) is made of either wood (beech) or cast iron and it has a narrow sole of steel 3 mm thick. The wooden plough looks similar in construction to the sash fillister. It is fitted with a set of cutters ranging from 3 mm to 15 mm wide. It has a fence and a depth gauge similar to those of the sash fillister. The main use of the plough plane is for producing a rectangular recess or channel called a groove. Normally, a groove runs *along* the grain of the workpiece and is some distance away from the edge. The grooves can be of different sizes corresponding to the cutters supplied with the plane. To plough a groove, follow these steps:

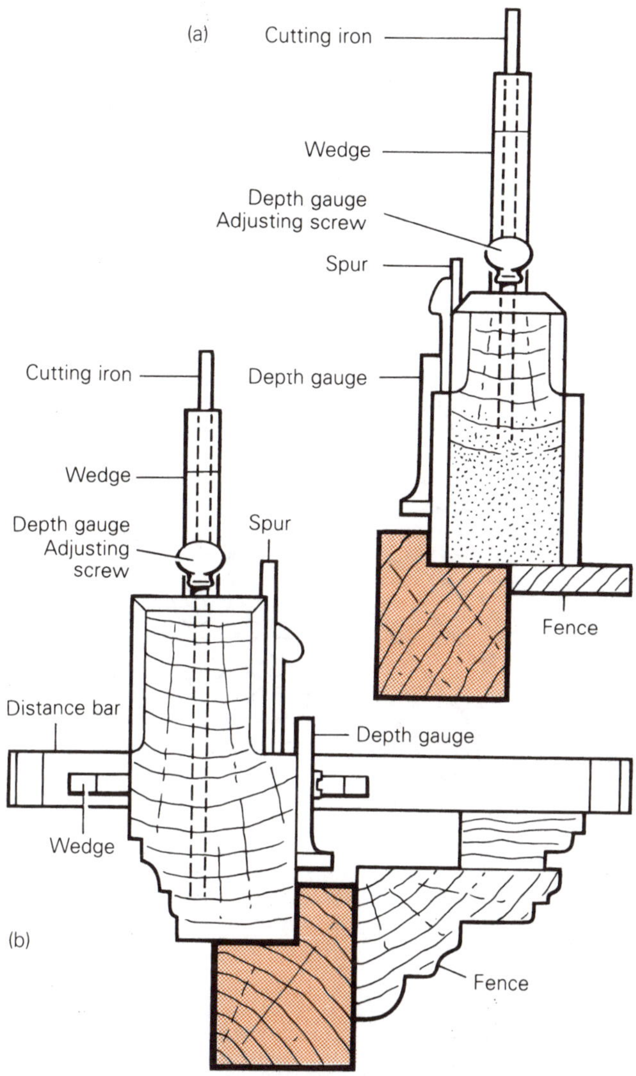

Figure 3.8 Fillister planes: (a) side fillister, (b) sash fillister.

1. Cramp the workpiece to the work-bench or else hold it firmly in the vice.
2. Position the cutter over the area to be grooved.
3. Set the fence against the face side or face edge.
4. Set the depth gauge to the exact depth of the cut required.
5. Plough (cut) the groove to the required depth. Start at the front and work backwards, towards you.

Match planes are two wooden planes that are used to perform complementary operations. One of the pair is used to plough a groove along the edge of a workpiece whilst the other is used to produce the matching tongue as in a tongue and groove joint (T&G). (A tongue is a projection left along the edge of workpiece when one or both corners of the edge are rebated.)

Moulding planes are wooden planes with variety of shapes of cutters purposely used to produce mouldings such as ovolo, bead, cavetto, rounded nosing, etc.

The **router plane** (Figure 3.10) is mainly used for levelling the bottom of trenches, the background of relief carving and recesses for inlays. A **trench** is a recess across the grain of a workpiece. The router plane is sometimes called the 'old woman's tooth' because the blade sticks out like the tooth of an old woman. The adjustable shoe is used to regulate the thickness of the shavings.

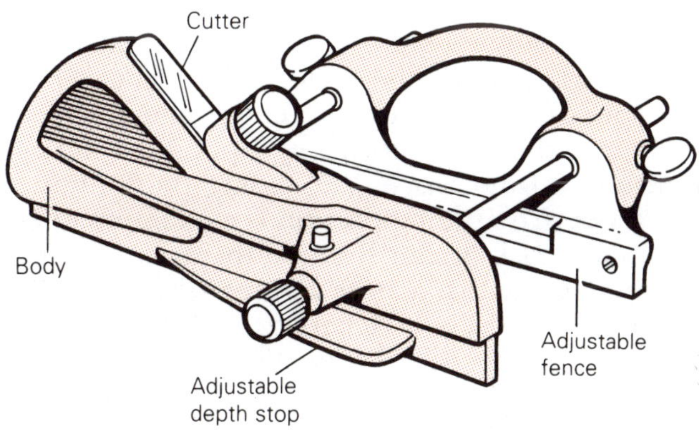

Figure 3.9 Plough plane.

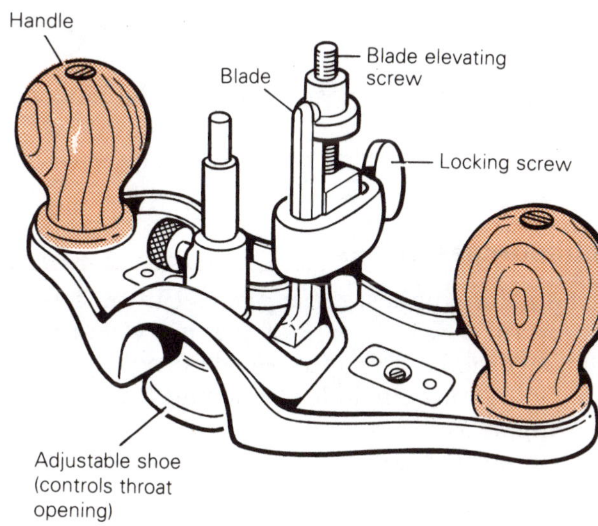

Figure 3.10 Router plane.

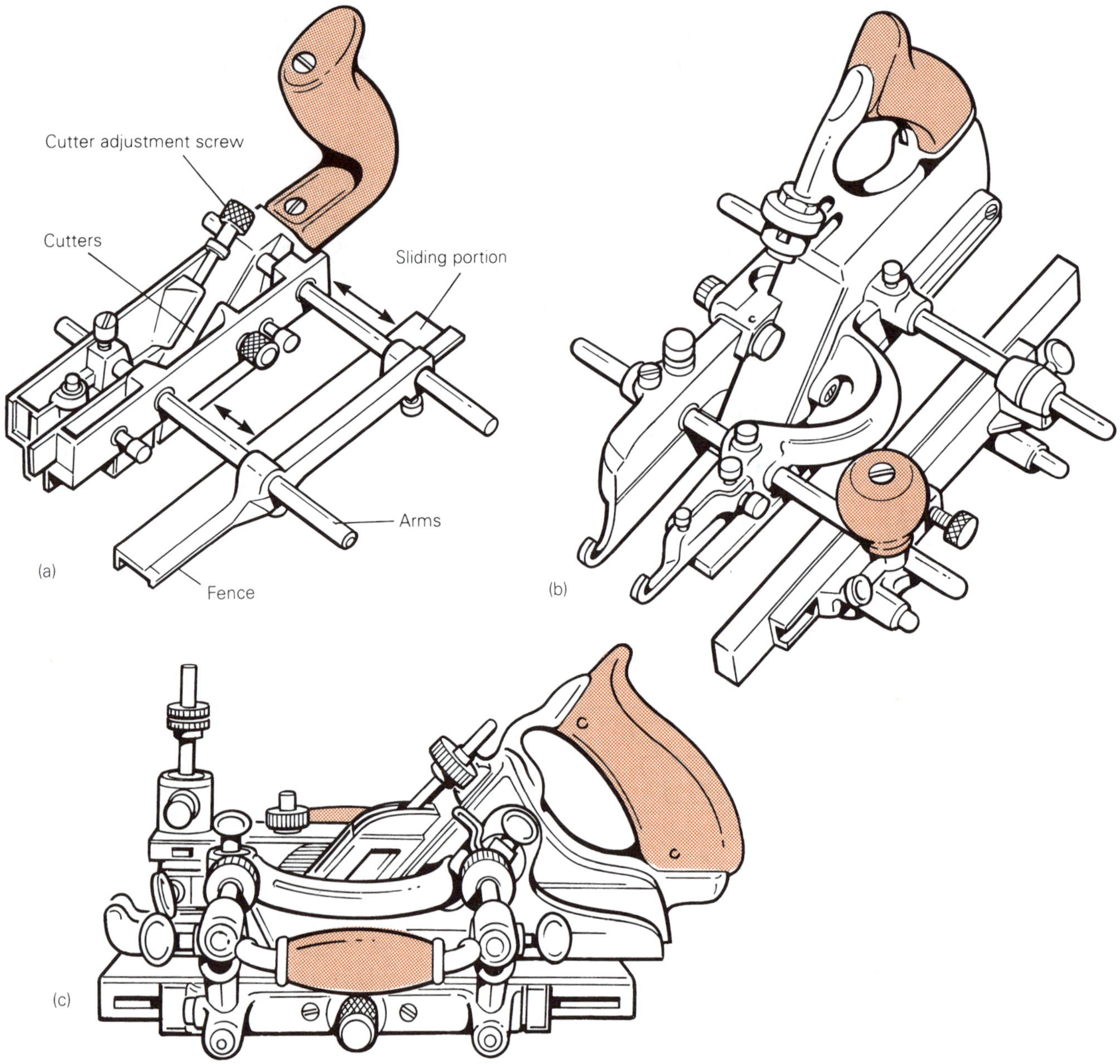

Figure 3.11 Multi-purpose planes: (a) combination plane, (b) multi-plane, (c) universal plane.

Multi-purpose planes

As the name implies, these planes can be set-up to perform a wide range of planing operations that are carried out by the special purpose planes. The operations include trenching, moulding, grooving, rebating etc.

The **combination plane** (Figure 3.11a) is one of the multi-purpose planes and it is considered to be the smallest and the lightest. It consists of the following main parts: the **main stock**, which holds and supports the cutter adjustment screw, the spur and the depth gauge; the **arms**, which are screwed into the stock and support the sliding portion and the fence; the sliding portion; and the **fence**. The sliding portion is a counterpart of the main stock. It can slide on the arms and the beading gauge that acts as a fence when beading tongued boards is attached to it. The fence is fitted onto the arms, slides along them, and is held in position by thumb screws. When the plane is being used the fence to fixed against the side of the workpiece.

The combination plane has a variety of shapes of interchangeable cutters that enable the plane to perform a variety of operations. It can be assembled as to perform the following operations: grooving, trenching, rebating, beading, tonguing and grooving.

The **multi-plane** (Figure 3.11b) looks like the combination plane and has similar parts but is longer in size. It is supplied with 23 standard cutters and additional 30 variety cutters can be purchased to enable it perform a variety of operations. It consists of the following main parts main stock, sliding portion, fence, depth gauge, arms and adjustable cam.

It can be set up or assembled to perform the operations of the following special purposes planes: plough plane, router plane, side fillister plane, sash plane, sash fillister plane, match plane, slitting plane, beading plane, round plane, hollow plane, nosing plane, fluting plane and reeding plane.

The **universal plane** (Figure 3.11c) is the largest and most versatile of the multi-purpose planes. It has similar parts to the multi-plane, with the following exceptions: the sole of the sliding portion can be raised or lowered to support the moulding cutters that may be higher or lower on one side or the other; an extra bottom (runner) may be attached to the sliding portion to give additional bearing on the workpiece when employing wide or irregular shaped cutters; an additional fence is provided at the other side of the plane; both fences can be lifted for angular planing. It is supplied with 55 standard cutters and 44 others may be bought as additional cutters. The 55 standard cutters consist of one fillister; one sash; eleven plough and trenching; eight beading; two match; four right-hand chambers; four hollow; one left-hand and right-hand chamber; four round; two quarter round for planning 6 mm hollows; two quarter hollow for planing; 6 mm rounds; two roman ogee, three reverse ogee; three grecian ogee; three sizes of bead with two beads each and two quarter round with bead.

> When you are using the plane hold it perfectly upright. To achieve this, you must ensure that the faces of the fences are at right angles to the cutter and parallel to the runners.

Paring tools

The term *paring* refers to a process of removing waste wood from a piece of timber using a chisel without a mallet. Chisels may be classified into firmer chisels and paring chisels. All of them consist of two main parts: the blade and handle. The blade is often made of tool steel (high carbon steel) and are hardened and tempered (heat-treated) to make them hard and tough. The tapered end (tang), fitted into the handle is often not heat-treated (left soft) so that it does not break off easily when in use. The handle is made of plastic or tough wood (ash or beech) and is shaped for easy handling.

Firmer chisels (Figure 3.12a) are of three kinds: the firmer chisel, the bevel-edged firmer and the mortise chisel. The **firmer chisel** has a rectangular blade with a tang or socket at one end to receive the handle. A shoulder at the base of the tang prevents it from being driven too far into handle. A brass **ferrule** is fitted on the end of the handle where the tang is fitted to prevent it

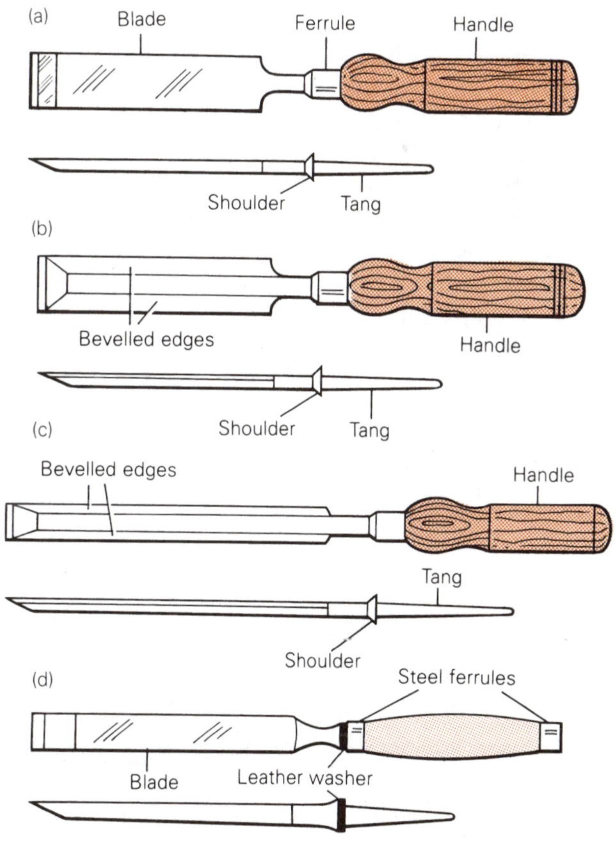

Figure 3.12 Types of chisel: (a) firmer, (b) bevelled-edge, (c) paring, (d) mortise.

splitting. The width of the cutting edge ranges from 3 mm to 50 mm. It is used for general chiselling.

The **bevel-edged firmer chisel** (Figure 3.12b), as the name indicates, has two edges of the blade bevelled along its length. The width of the blade ranges from 3 mm to 50 mm. It is generally used for paring and lighter chiselling and is also suitable for cleaning out corners of joints. A type with a shorter blade is called a **butt** chisel and is mostly used by carpenters and joiners for preparing recesses for butt hinges.

The **paring chisel** (Figure 3.12c) is similar to the bevel-edged but has a longer and thinner blade. The blade is between 3 mm and 50 mm wide. It is used for all forms of paring, particularly for cleaning out corners and finishing off joints. Never use a mallet on this chisel because it will break or bend due to its lightness.

> ▲ When paring, be careful to handle the chisel carefully, with the hands and fingers kept always behind the cutting edges.

The **mortise chisel** (Figure 3.12d) has a thicker blade to withstand the leverage when cutting mortises. A leather washer between the shoulder of the blade and the handle absorbs the blow from the mallet. The handle is fitted with iron ferrule at both ends to prevent splitting. It is used for heavy chiselling, deep work and cutting mortises.

Gouges (Figure 3.13) are similar in form to the chisel except that the blade is curved or fluted along its length. The two main types are the

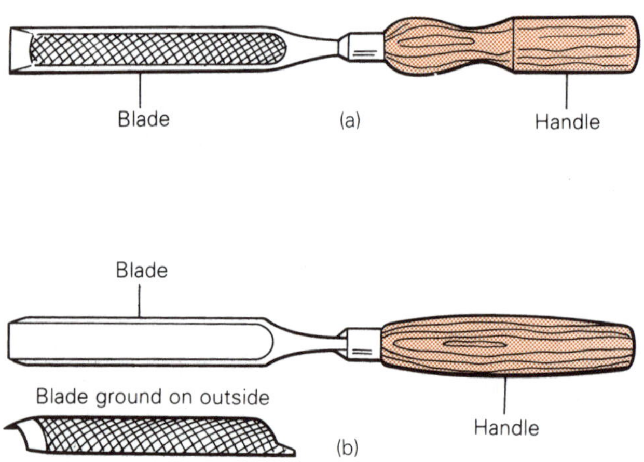

Figure 3.13 Gouges (a) scribing gouge, (b) firmer gouge.

scribing gouge, ground and sharpened on the *inside*, and the **firmer gouge** that is ground and sharpened on the *outside*. The scribing gouge, sometimes known as paring gouge, is used for producing concave cuts. The firmer gouge, also referred to as caving gouge, is mainly used for grooving, fluting and for general carving purposes.

Abrading tools

Abrading tools are those that cut by taking out many small pieces (particles) of the material at once. All types of saw cut by the action of the teeth removing small pieces in the form of saw dust, and so they are classified as abrading tools. Other abrading tools are files, rasps, grinding tools and honing tools.

Saws may be classified as hand saws, backed saws and curve cutting saws.

Hand saws are always used for straight cutting. The types of hand saw are cross-cut saws, rip saws and panel saws. All saws have two main parts: the blade (cast steel) and wooden handle (beech). The length of a saw is determined by the total length of the blade excluding the handle.

The **cross-cut saw** (Figure 3.14a) may be distinguished by its pointed teeth with knife-like edges. The front edge of each tooth forms an angle between 70° and 80° with the line joining the tips of the teeth. The length of the saw ranges from 550 mm to 700 mm and there are between 5 and 9 points per 25 mm. Its main use is for cutting thick boards across the grain. It can also be used for cutting a thin board along the grain (ripping). The cut made by a saw is called the **kerf** and is slightly wider than the blade due to the set of the teeth.

A **rip saw** (Figure 3.14b) looks similar to a cross-cut saw and it can be distinguished by the teeth having chisel edges with the front edge at 90° to the line joining the tips. The length ranges from 620 mm to 700 mm with a pitch of 3–6 points per 25 mm. It is mainly used for cutting thick boards along the gain. This type of sawing is known as **ripping**; from this the saw derived its name.

The **panel saw** (Figure 3.14c) is very similar to the cross-cut saw, having pointed knife-like teeth, but smaller is size. The average length is from 450 mm to 550 mm and it has fine teeth with a pitch of 10–12 points per 25 mm. It is generally used for sawing thin boards, usually across the grain. Other uses are for ripping tenons and

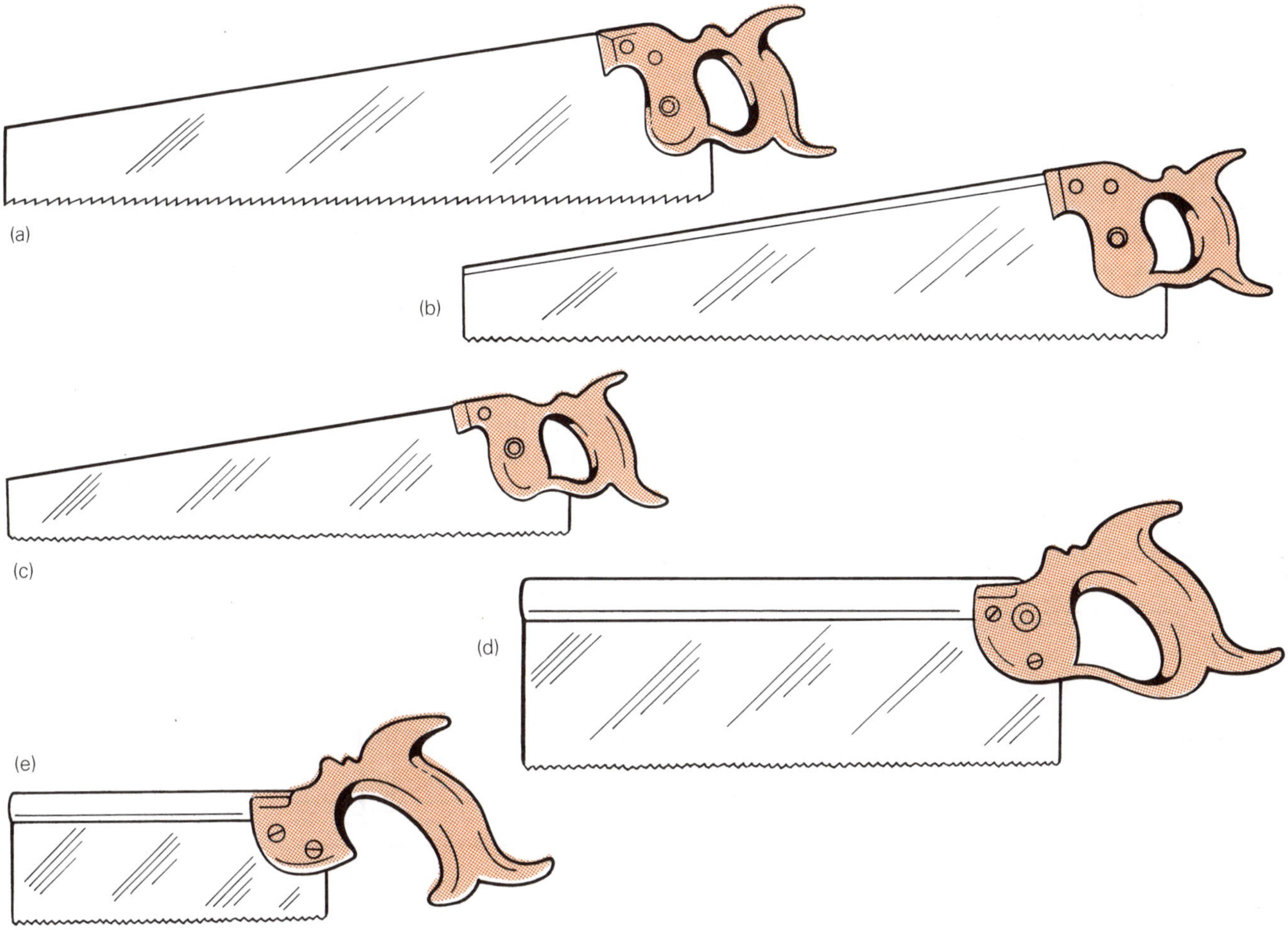

Figure 3.14 Hand and backed saws: (a) cross-cut, (b) rip, (c) panel, (d) tenon, (e) dovetail.

cutting plywood. When you are using it for cutting plywood, keep it almost flat on the surface (i.e., with the line of the teeth almost parallel to the surface) to avoid tearing the fibres.

Backed saws are of two kinds: tenon saws and dovetail saws. The **tenon saw** (Figure 3.14d) is between 200 mm and 450 mm in length, with a pitch of 12–14 points per 25 mm. The teeth are similar to those of the cross-cut saw. The back of the blade is stiffened with a brass or steel rib, folded over it to prevent it from **buckling** (crumpling or bending). It has a closed handle. It is generally used for joint construction, especially for cutting shoulders and cheeks of tenons.

The **dovetail saw** (Figure 3.14e) takes the same form as the tenon saw, except that it is smaller and has an open handle. The length ranges from 200 mm to 250 mm with a pitch of 18–22 points per 25 mm. It is commonly used for fine and accurate joint cutting, such as dovetail joints or for model making.

General hints on use of hand saws and backed saws:

1. Always use the correct saw for the job.
2. Hold the saw in one hand with your index finger against the side of the handle for a firm grip.
3. Begin sawing from the far edge and guide the saw with the thumb of your other hand.
4. Let a few teeth just touch the workpiece on the first few strokes.
5. Do not force the saw to cut. Exert very little effort.

Curve-cutting saws

There are other saws that are used for cutting curves and these are distinguishable by their thin, flexible blades that enable them to negotiate curves easily in the saw cut or kerf. The commonly used ones in woodwork are the coping saw, fret saw, bow saw, compass saw, key-hole saw and the nest of saws.

The **coping saw** (Figure 3.15a) is a small saw with a springy steel frame. It has a straight wooden handle like a file handle. The thin and narrow blade (about 3 mm wide and 170 mm long) is held by pegs between two studs that are loosely held in the frame to allow for rotation of the blade to any convenient position. The stud fitted into the handle is threaded so that when the handle is turned tension is given to the blade. This saw is mostly used for cutting sharp curves in a thin board near to the edge (Figure 3.16a). It is sometimes used for cutting off the waste from dovetail joints. In order to avoid buckling and breaking the blade, always fix the blade in the frame with the teeth pointing towards the handle so that it cuts on the *backward* stroke.

The **fret saw** (Figure 3.15b) looks similar to the coping saw, but has a longer rectangular springy frame. The length of the frame is between 300 mm and 500 mm and this determines the size of the saw. It has a very fine blade, between 125 mm and 150 mm long with very fine teeth. The blade is tensioned by a wing nut at the opposite side of the frame to the handle. The fret saw is suitable for cutting curves in thin boards. Owing to its long frame it can cut at a greater distance from the edge of the workpiece. To avoid breaking, the blade must be fixed with the teeth pointing towards the handle so as to provide a cut on the backward stroke (Figure 3.16b).

The **bow saw** (Figure 3.15c) has a wooden frame that resembles a bow, from which it derives its name. It has a thin blade but is wider (6 mm) than the blades of the coping and fret saws. The

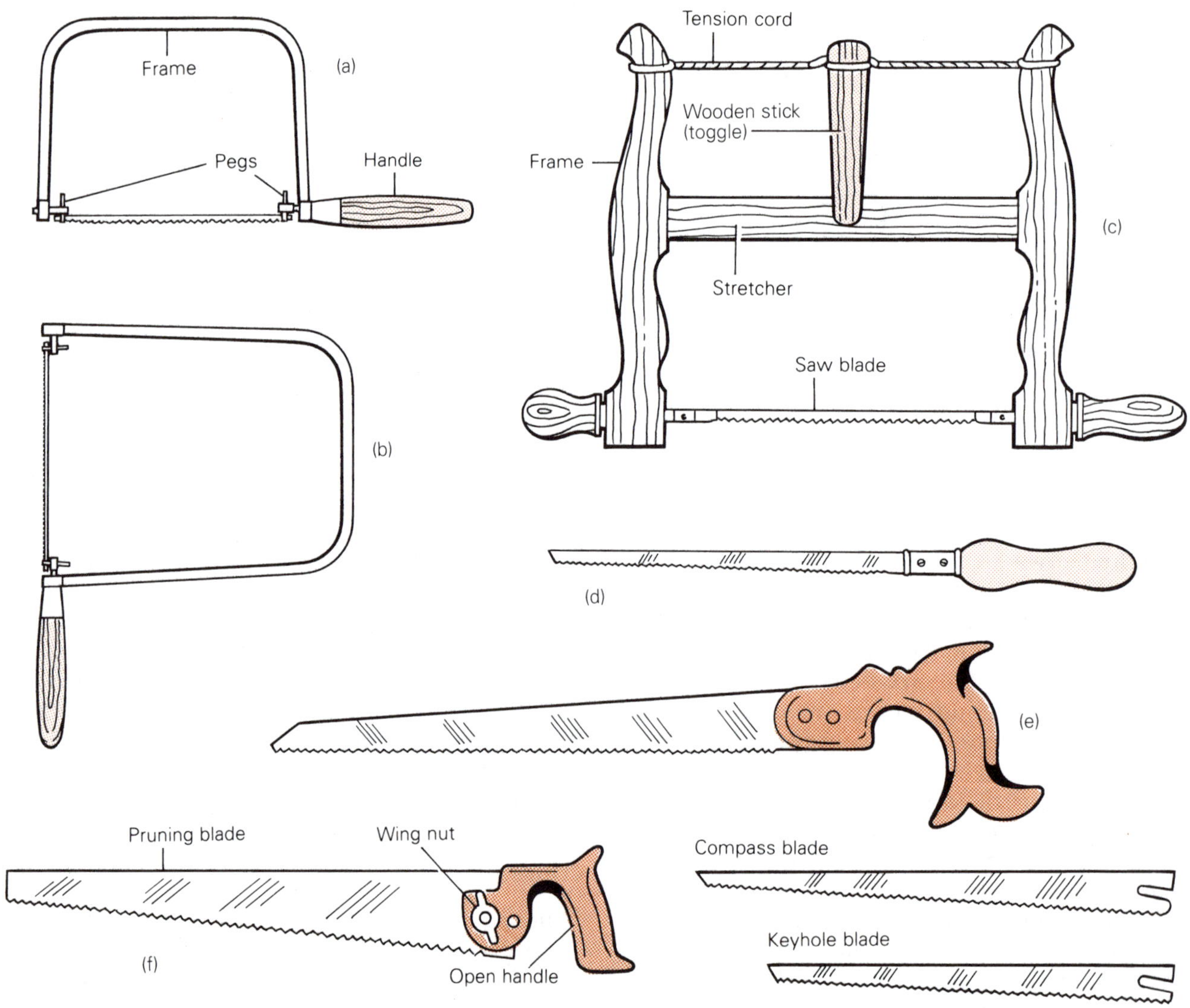

Figure 3.15 Curve-cutting saws: (a) coping, (b) fret, (c) bow, (d) compass, (e) keyhole (f) nest of saws.

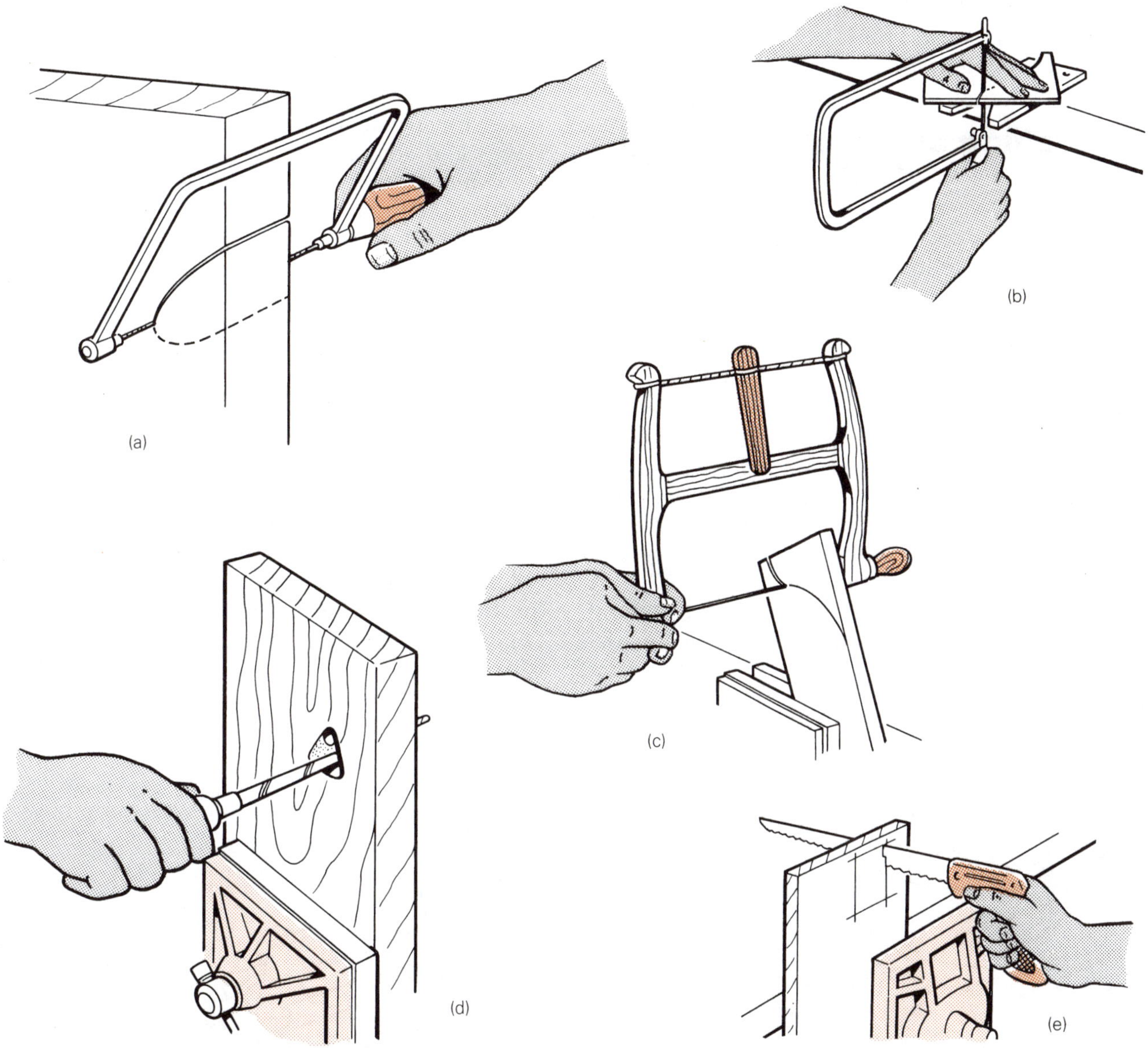

Figure 3.16 Applications of curve-cutting saws.

average length of the blade is about 300 mm and it is tensioned by means of a cord and winding lever (toggle) at the other end of the frame (top side). It is mainly used for cutting external and internal curves in fairly thick boards (Figure 3.16c).

The **compass saw** (Figure 3.15d) has a narrow, tapered blade of length ranging form 300 mm to 450 mm. The teeth are of the cross-cut type. The handle is open ended like that of the dovetail saw. It is used for cutting large internal and external curves When cutting an internal curve, you must first bore a hole in the workpiece to take the blade of the saw. It is also used for cutting straight internal cuts.

The **keyhole saw** or **pad saw** (Figure 3.15e) has a narrower blade than the compass saw. The length of the blade varies form 250 mm to 350 mm. The type with a shorter blade has a round handle and is known as a pad saw while that with the longer blade has a shaped handle, similar to the compass saw, and is referred to as a keyhole saw. Both are mainly used for cutting internal curves (Figure 3.16d).

The **nest of saws** (Figure 3.15f) comprises three separate blades, detached from an open handle: the **keyhole blade**, 250 mm long is used for small internal curves; the **compass blade**, 300–350 mm long is used for large internal and external curves; and the **pruning blade**, 450 mm long, is used for large flat curves and for straight cuts (Figure 3.16e).

Scraping and other abrading tools

Besides the saws, there are other hand tools whose cutting actions mean that they are also classified as scraping and abrading tools. These are the scraper, rasps, files, surform tools and abrasive papers and cloths.

The **hand scraper** (Figure 3.17a) is a flat, rectangular piece of high carbon steel (tool steel), 125 mm by 75 mm whose long edges are burred over to form the cutting edges. A scraper may be held in the hand for the scraping operation, fitted with a handle for easier manipulation, or even fitted into a spokeshave-like stock (this last type is known as a **cabinet scraper** - Figure 3.17b). A scraper is drawn over a surface of a job to scrape off or cut off a fine shaving, or to remove marks left by a plane. It may also be used to smooth a job before sanding.

Rasps (Figure 3.17c) are metal hand tools, usually half-round in section. They have raised teeth formed by punching triangular points on the surface. Rasps are graded according to the size of the teeth: i.e., a smooth rasp or a rough rasp. The place for a rasp in woodwork is for smoothing edges and shapes that cannot be worked easily by other cutting tools. A rasp cuts faster, but leaves a rough surface that must further be smoothened with a file or glasspaper. Care must be taken to prevent *splintering* (breaking off) when using a rasp.

Files (Figure 3.17d) also have a place in woodwork. Like a rasp, a file is used for smoothing edges and shapes where other cutting tools cannot easily be used. The teeth of files are produced by a series of cuts (single or double cuts) across the face, producing varying grades of teeth, (from fine to rough). There are different shapes and those commonly used in woodwork are: the **round file**, used for smoothing concave curves and enlarging holes; the **square file**, used for producing slots; the **half-round file**, used for smoothing concave edges and enlarging holes; and the **triangular file**, used to form sharp corners and grooves. The triangular file is also used for sharpening hand saws, hence it is also called the **saw file**.

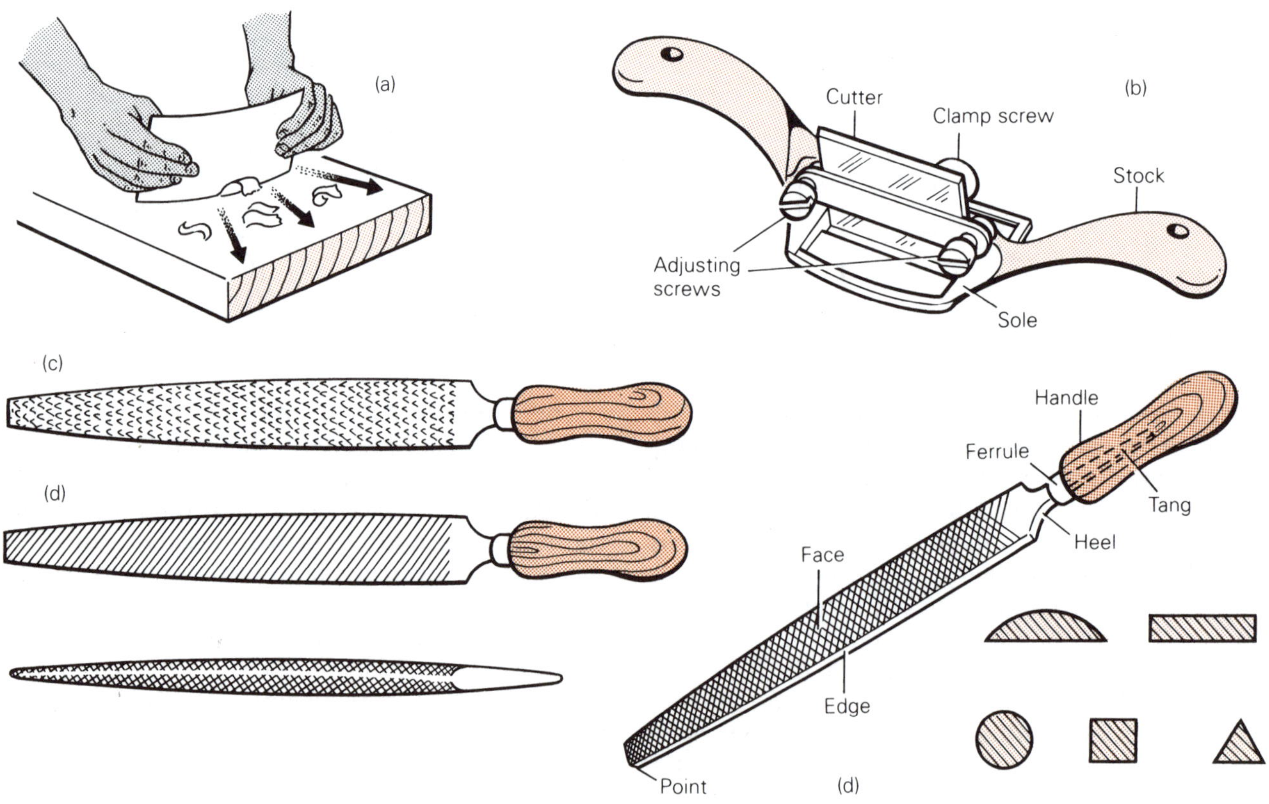

Figure 3.17 Abrading tools: (a) hand scraper, (b) cabinet scraper, (c) rasp, (d) files.

▲ To avoid injury, never use a file without a handle or with a loose handle.

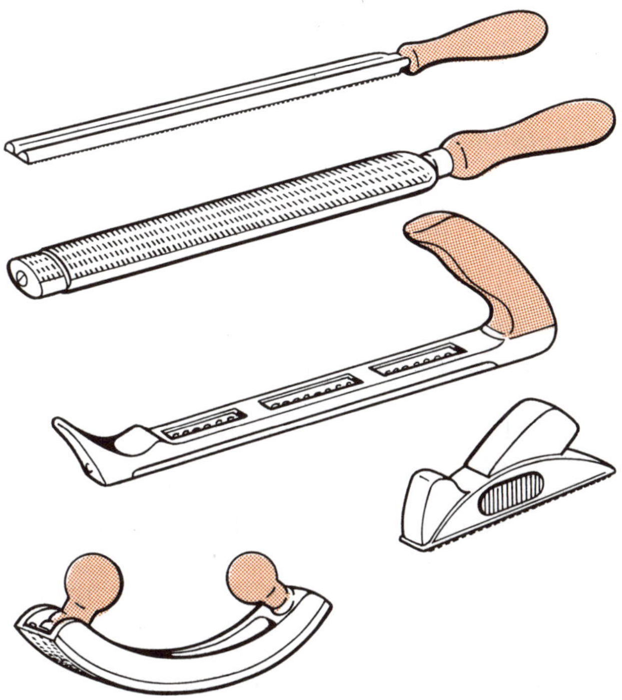

Figure 3.18 Abrading tools: surform tools or shapers.

Surform tools (Figure 3.18) have the features of a plane, a rasp and a file for the purpose of rapidly removing waste materials from flat or round surfaces. They consist of a stock, made from cast alloy, interchangeable alloy steel blades, and a handle, made of wood, moulded rubber or plastic. The interchangeable blades have rows of perforations or slots with non-clogging cutting edges, and are either half-round, flat, or convex. The surform tools, also known as shaper tools, are employed to remove waste materials quickly when shaping, trimming, forming and smoothing jobs made of wood, particle boards, plastics, hardboard, fibreglass etc.

Abrasive papers and **cloths** are papers and cloths coated with various abrasive materials such as glass, flint, garnet, aluminium oxide or silicon carbide for sanding or smoothing surfaces prior to the application of finishes. A paper coated with glass is called a glasspaper, a cloth coated with flint is a flint cloth, and so on. The coated abrasives are graded: **dead smooth**, **smooth** and **rough**. The grading can be done by numbers in order of coarseness (00, 0, 1, 1½, 2 and so on). The size of the abrasive grains is called **grit** and is specified according to meshes per 25 mm of the grading sieve.

The terms **open coat** and **close coat** refer to the spacing of grains on the paper or cloth. In the 'close', the grains are close together, making them suitable for hardwoods. In the 'open', there are spaces between the grains making the paper less liable to choke. It is therefore used more on softwoods, resinous woods and surfaces liable to cause choking.

Never apply abrasive paper on a surface before you plane or scrape the surface as the small abrasive materials imbedded into the wood and are likely to dull the cutting edges. Sanding must be done after **dressing** (smooth planing and scraping) the surface. Dust off all abrasive particles after sanding before you apply a surface finish (polish or lacquer).

Glasspaper is used for a hand smoothing of a piece of wood. It is made from an abrasive grain of crushed, sieved glass. It is supplied in sheets 300 mm by 250 mm. The grades are: 2/0 or flour, 0, 1, 1½, F2, M2, S2, 2½, 3. As a general guide, use M2 for preliminary smoothing followed by No. 1 or 1½. For delicate woods use No. 0 and when rotary movement is needed on a fine wood, use flour to finish. It is made in close coat only.

Garnet paper is made from abrasive grain of natural crushed garnet stone and supplied in sheets 275 mm by 225 mm, and in the form of rolls, belts and discs. Those supplied in sheets are graded as follows: 8/0, 7/0, 6/0, 5/0, 4/0, 3/0, 2/0, 0, ½, 1, 1½, 2, 2½, 3, 3½. The grades in rolls are: 5/0, 4/0, 3/0, 2/0, 0, ½, 1, 1½, 2, 2½, 3, 3½. In a comparison with a glasspaper, ½ is of the same coarseness as M2. They are made in both close and open coats and are used for both hand and machine sanding.

Flint paper has only limited use in woodwork. Flint is a natural product and is not so hard as garnet, aluminium oxide paper or glasspaper. Generally it is used more in other industries than in woodworking trades. Flint paper is made in sheets 300 mm by 250 mm and also in rolls. Grades in sheet and rolls are: 4/0, 3/0, 2/0, 1, 1½, 2, 2½, 3. In a comparison No. 1 flint paper corresponds to M2 glasspaper and No. 2/0 corresponds to No. 1 glasspaper. It is made in close coat only. **Flint cloth** is made in rolls and belts for machine work only, and the grades are 5/0, 4/0, 3/0, 2/0, 0, ½, 1, 1½, 2, 2½, 3.

CHECK YOUR UNDERSTANDING

- Planes are shaving cutting tools and are classified into: bench planes, curve cutting planes, special purpose planes and multi-planes.
- Bench planes are: the roughing plane, used for heavy and quick reduction of thickness of timber; the jack plane that is used for planning rough surfaces of sawn timber or wood; the trying plane, used after the jack plane to produce true and flat surfaces or perfect straight edges; the smoothing plane, used for dressing up surfaces and edges ready for glasspapering; and the block plane, used for planing end grain.
- Curve cutting planes consist of the spokeshaves and compass plane. Both are used for providing smooth curved surfaces and edges.
- The special purpose planes are those used for specific operations such as rebating, tongue-and-grooving, trenching and moulding.
- Multi-planes are those that can be set-up with other attachments to enable them perform a wide range of operations, including, trenching, grooving, moulding and rebating.
- Paring tools consist of chisels and gouges. The two main types of chisel are the firmer chisels and paring chisels. The firmer chisels are suitable for heavy chiselling with a mallet and the paring chisel is often used for light chiselling, by hand only.
- All types of saw are classified as abrading tools. The two classes of saw are the hand saws, used for straight sawing, and the curve-cutting saws.
- The hand saws are: cross-cut saw, rip saw and panel saw. The backed saws comprise the tenon and dovetail saws.
- The curve-cutting saws are: coping saw, fret saw, bow saw, compass saw, keyhole saw, pad saw and the nest of saws.
- Other abrading tools are scrapers, rasps, files, surform tools, and abrasive papers and cloths.

REVISION EXERCISES AND QUESTIONS

1. List the four classes of planes and state the main function of each class.
2. List the essential parts of the jack plane and explain the function of each part.
3. Explain the main difference in construction, between the wooden and metal compass plane.
4. In what ways is the fillister plane different from the rebate plane?
5. Make a neat sketch of a firmer chisel and indicate the main parts. State the function of each part.
6. Explain how a rip saw can be identified and state when it is used.
7. List the common shapes of files used in woodwork and state the main use of each.
8. Explain why it is not advisable to plane a surface of work that has been sanded with an abrasive paper.

4 Care and maintenance of hand tools

Introduction

A good craftsman must have good tools and be able to keep them in good condition at all times. Respect and care for your tools and learn how to select the right tool for the work to be done.

You must always make up your mind at the beginning of any job to do your best to produce a piece of quality work and realise that this cannot be done with blunt and badly maintained tools. Generally a fault in the finished work can be attributed to the tools being wrongly selected or wrongly used. It should therefore be your habit to sharpen and keep your tools in good condition at all times. Two distinct operations are required for maintaining sharp and keen cutting edges: grinding and sharpening.

Grindstones

Grindstones (Figure 4.1) are shaped blocks of sandstone or other abrasives mounted on a spindle. A supply of water should be available to cool the blade which becomes hot in the grinding process and to wash away the grindings of steel that could clog the grindstone and make it less effective. Stones run in a recess which collects the water: this must be drained off after use so as not to cause deterioration and softening of the stone.

Grindstones with carborundum stones mounted horizontally are now becoming common. The blade to be sharpened can be mounted in a tool rest, and held at the correct angle for grinding, while being washed with a special oil instead of water. Greater control is possible with this kind of machine (Figure 4.2).

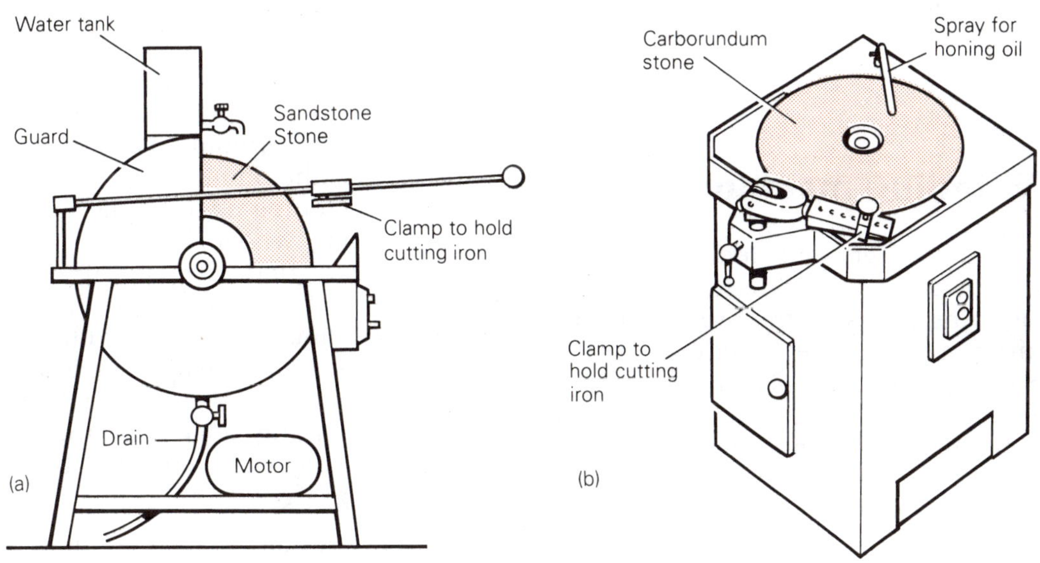

Figure 4.1 Grindstones: (a) sandstone type, (b) horizontal type.

Figure 4.2 Horizontal grindstone.

Grinding machine

A common means of grinding individual hand tools such as punches, chisels, scrapers, drills, plane cutting irons etc. is by the use of the **off-hand grinding** machine, often referred to as the **bench grinding machine** (Figure 4.3). The grinding operation is carried out on a set of **grinding wheels** (rough and smooth grades) attached to each end of the spindle. The wheel consists of particles of abrasive materials such as aluminium oxide or silicon carbide (artificial abrasives) held together by a bonding agent known as a **matrix**. Natural abrasives such as emery (60% pure) and corundum (90% pure) are also used for making the grinding wheel.

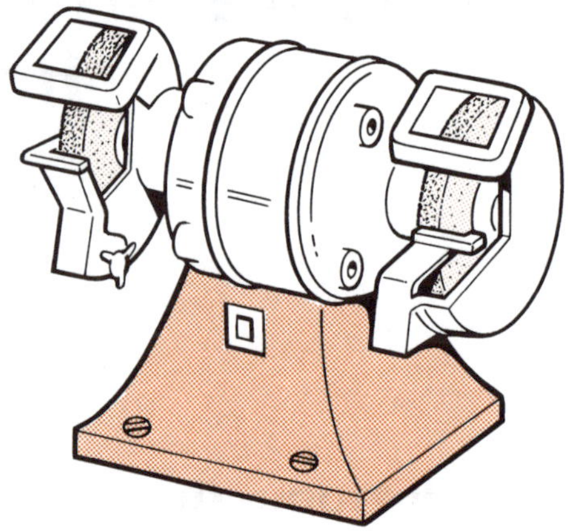

Figure 4.3 Bench grinding machine.

Sharpening processes

Grinding plane cutting irons and chisels

Grinding a cutting edge involves wearing away the worn edge to produce a new grinding bevel on a rotating abrasive wheel or grinder (Figure 4.4).

Before you use any grinding machine, ensure it is in good working condition and that there is no crack in the grinding wheels.

Use the smooth wheel to grind a cutting edge. Follow these procedures:

1. Set the tool rest close to the wheel to avoid the cutter or chisel from being trapped.
2. Obtain a coolant (water) to quench the cutting edge from time to time during the grinding operation.
3. Switch on the grinding machine and allow the wheel to gather momentum.
4. Place the cutting edge on the tool rest against the rotating wheel, holding it firmly at the appropriate grinding angle, 20°–25°.
5. Move the cutting edge across the wheel so that you are using the whole width of the wheel, to prevent the formation of grooves on the wheel.
6. Dip the cutting edge into the coolant frequently to prevent overheating that may soften the cutting edge.

Note that specially shaped grinding wheels may be required to grind hollow or shaped cutting

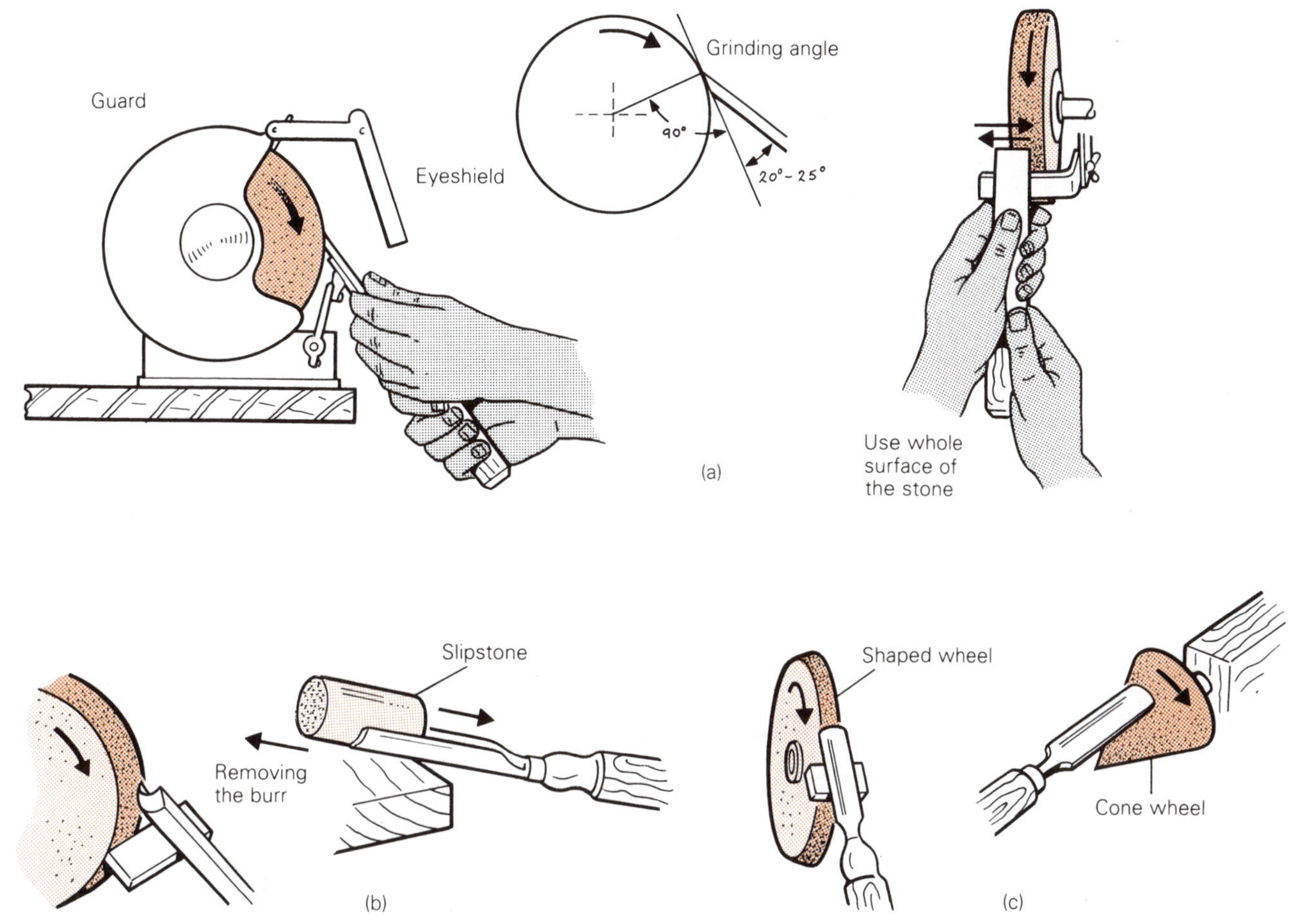

Figure 4.4 **Grinding of tools: (a) plane cutting irons and chisels, (b) carving gouge, (c) scribing gouge.**

edges such as the scribing gouges and moulding plane cutting irons. Those having the grinding surface on the outside (convex side) are ground on the normal flat grinding wheel.

> ▲ Ensure that the tool rest is set correctly, close to the grinding wheel; ensure that the guards are in place; use the eye shield or wear a pair of goggles; also wear a pair of gloves.

Hand sharpening plane cutting irons, chisels and gouges

The purpose of sharpening a cutting tool is to obtain a keen and sharp cutting edge for ease of cutting. The process involves rubbing the cutting edge on an oilstone, with oil being applied frequently to reduce friction and to float away the metal particles thus avoiding the pores of the stone becoming clogged. The process of sharpening is also known as **honing** and sometimes called **whetting**.

There are two main types of oilstones used for sharpening tool cutters: natural stones and artificial stones. The two common natural stones are the **Arkansas** and **washita** (Figure 4.5). The Arkansas are in two grades: hard and soft whilst the washita is of only the soft grade, even softer than the soft arkansas. The hard grade is suitable for sharpening tools such as carving tools and surgical instruments that need to be extremely sharp. The soft grades, having coarser and more porous grit, are faster sharpening, particularly for tools that do not require extreme sharpness.

The artificial stones are made from artificial abrasives, such as silicon carbide and aluminium oxide. The abrasives are crushed into particles and graded (fine, medium, coarse) and bonded together into stones of any desired shape. The

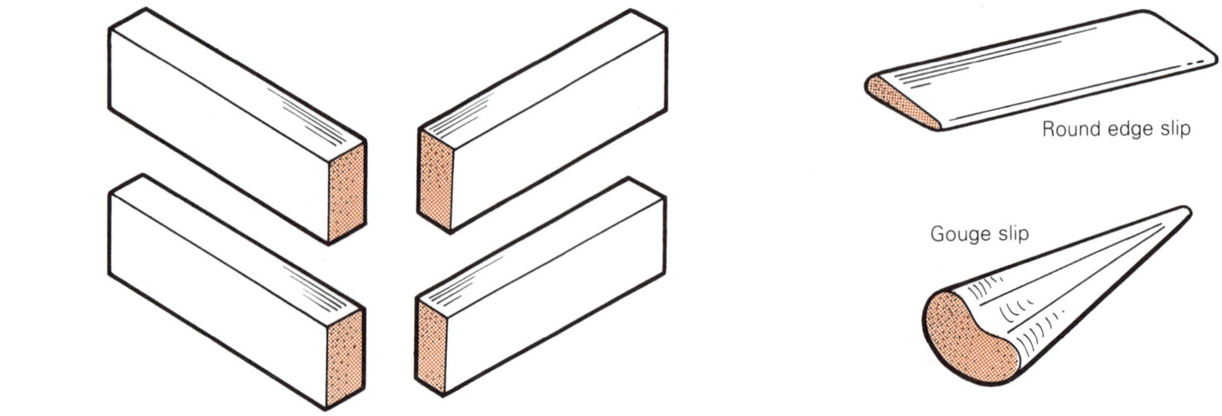

Figure 4.5 Oilstones and slips: (a) natural oilstones, (b) artificial stones, (c) slips.

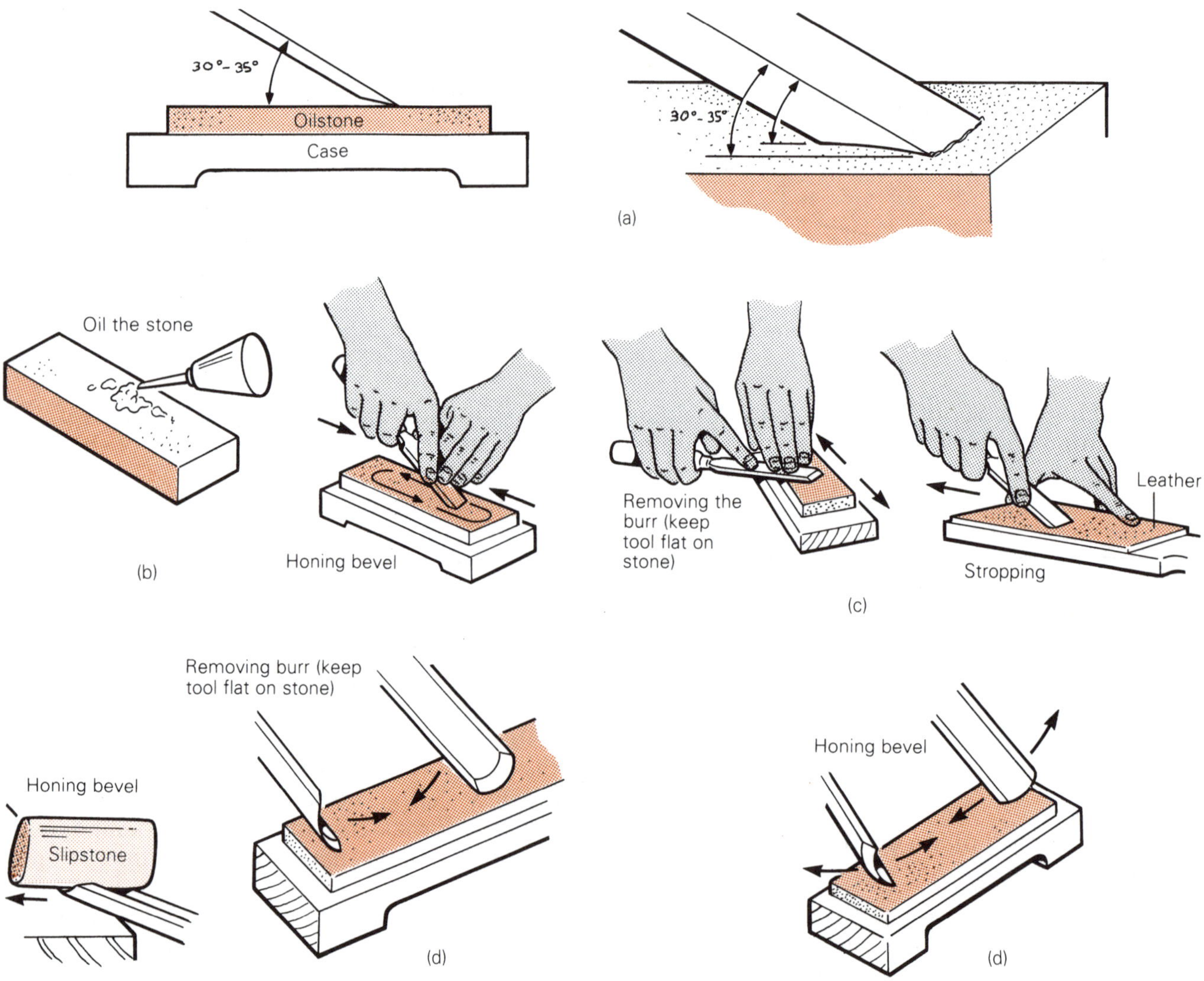

Figure 4.6 Sharpening chisels and gouges: (a) sharpening angle, (b) sharpening process, (c) sharpening a scribing gouge, (d) sharpening a carving gouge.

two commonly used artificial stones are the **carborundum** and **India** (Figure 4.5).

Oilstone slips or **slipstones** (Figure 4.5) are various shapes of oilstones purposely designed for sharpening curved or shaped cutting tools, such as scribing gouges or moulding plane cutting irons.

Procedure for using oilstone slips (Figure 4.6):

1. Hold the oilstone firmly in a vice or against a stopper.
2. Apply enough oil during the sharpening operation. For coarse grit/stone (i.e., soft stones) use water instead of oil.
3. Hold the cutting tool at a constant angle, 30°–35° to the oilstone and hone until a burr is created.
4. Remove the burrs that are formed at the cutting edge by turning over the cutting edge and rubbing it gently on the oilstone while keeping it perfectly flat. It is the removal of the burr that gives a keen edge.
5. Repeat this operation several times.
6. Use the whole surface of the oilstone to avoid making a series of grooves on the surface. This will also ensure even wearing of the oilstone.
7. Test the sharpness of the cutting edge against your finger nail: a sharp cutting edge should not show any thickness at the edge.

- Keep the surface of the oilstone clean and moist with clean oil always.
- Clean off the dirty oil immediately after sharpening a tool.
- Remove any irregularities on the surface of the oilstone by rubbing it on a wet sandstone, brick or on a hard surface with water or kerosene and emery powder.
- Store the oilstone in a covered box.

Procedure for sharpening scrapers:

A scraper, like all cutting tools requires its cutting edge to be sharp and keen in order to cut satisfactorily. The procedure is as follows (Figure 4.7):

1. File the cutting edge straight and square to the sides.
2. Smooth the cutting edge on an oilstone, by rubbing it straight forth and back, using the whole surface of the oilstone.
3. Remove all burrs formed, by placing the sides flat on the surface of the oilstone and rubbing it forth and back.
4. Flatten the edge by burnishing at 90° to the side.
5. Turn the edge over with the burnisher at 8° to 10° to the horizontal. This creates the burr that is the scraping edge.

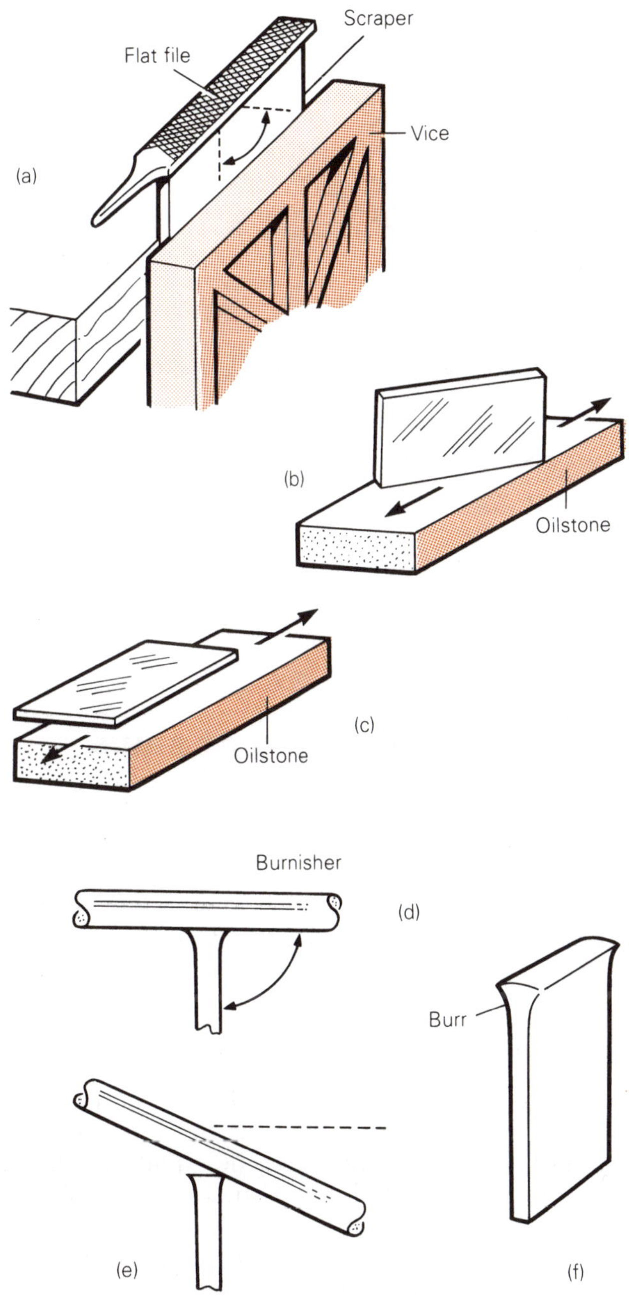

Figure 4.7 Sharpening a scraper: (a) filing the edge, (b) smoothing, (c) removing the burr, (d) flattening the edge, (e) turning the edge over, (f) final shape.

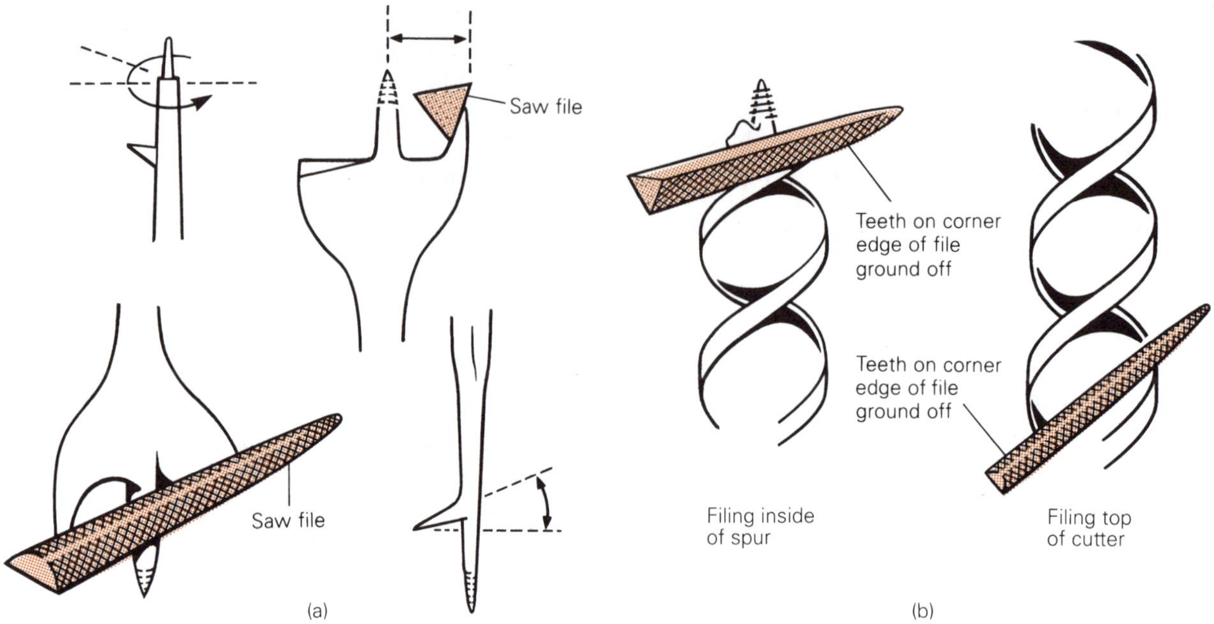

Figure 4.8 Sharpening of bits: (a) centre bit, (b) twist bit.

The procedure for sharpening bits (Figure 4.8):

1. Hold the bit vertically with the point on a flat surface.
2. File the top edge of the cutter at angle of 30°.
3. File the inside of the spur or scriber.
4. Finish with an oilstone slip.

Reconditioning of saws

For saws to cut satisfactorily, their cutting edges must always be kept sharp. It is essential to make reconditioning of saws a regular routine. The process of reconditioning a saw involves five stages: jointing, resharpening, setting, sharpening and side dressing (Figure 4.9).

Jointing

Jointing, also known as **topping** is the process of running a file along the tops of the projecting teeth of a saw until they become equal in height. Use a flat mill file holding it square to the blade. Before you perform this operation, fix the saw in a saw vice and allow it to project a little (not too much, to avoid vibration).

Reshaping

When the teeth of the saw become flattened as a result of the jointing process it is necessary to **reshape** them (Figure 4.9). Insert the saw in a saw vice and let the gullets (the indentations between the teeth) project about 3 mm; using a saw file, file straight across each gullet, at right angle to the blade. Ensure that the front edge of each tooth is inclined at 70°–80° to the line of the teeth. For a rip saw, file straight in every other gullet, inclining the file at right angles to the blade; turn the saw round in the vice and file the other gullets holding the file at right angles to the blade as before.

Sharpening

Sharpening (Figure 4.9) involves forming knife edges on the sides of the teeth and producing needle points at the tip of each tooth. Follow these steps: hold the saw in a saw vice, with about 3 mm of the gullet projecting above jaws of the vice; with a saw file (triangular), file each alternative tooth to the correct angle in a direction from the handle to the toe. The sharpening angle of a cross-cut saw ranges from 45° to 70° depending upon the purpose for which the saw may be used (i.e., lower angles for softwoods and greater angles for hardwoods). Rip saw teeth are sharpened at 90° to the edge of the blade. Turn the saw over in the vice and file the remaining teeth in the same manner as before (in a direction from toe to handle).

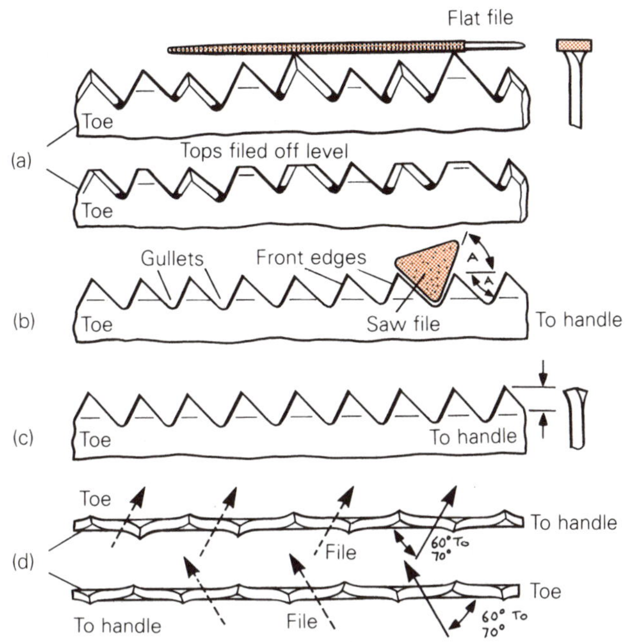

Figure 4.9 Sharpening hand saws: (a) topping, (b) reshaping, (c) setting, (d) sharpening or filing.

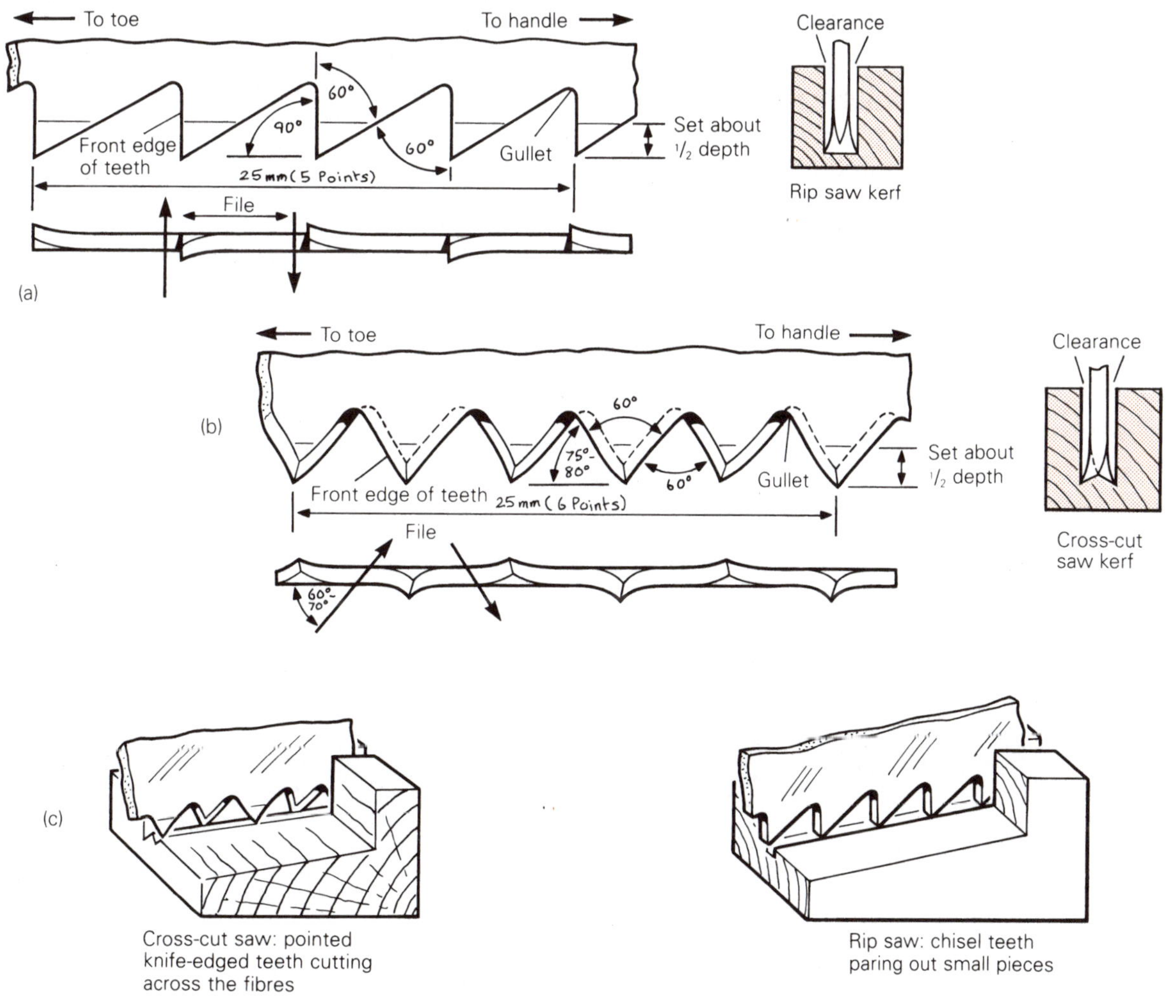

Figure 4.10 Shape of saw teeth: (a) rip saw, (b) cross-cut saw, (c) cutting actions.

Side dressing

Side dressing (Figure 4.9) involves removing any burrs or ragging that may be produced at the edges of the teeth due by setting and sharpening. This is done by rubbing an oilstone along the edges of the teeth to smooth them.

Setting

Setting is the process of bending adjacent teeth of a saw to opposite sides in order to widen the width of the cutting edge (see also Figure 4.10). This will make the kerf (saw cut) wider than the thickness of the blade to permit free movement of the saw and to prevent binding of the saw in the kerf.

The procedure is as follows: hold the saw in a saw vice, allowing more than half the depth of the gullet to project above the face of the vice; use either a pliers saw set or a special saw setting hammer to bend the top half of each alternative tooth to opposite sides of the blade.

CHECK YOUR UNDERSTANDING

- The two operations required for maintaining a cutting tool in good working condition are grinding and sharpening of the cutting edge.
- The grinding operation is carried out on a grinding wheel while the sharpening is performed on an oilstone.
- Both the grinding wheel and the oilstone are made from natural abrasives or artificial abrasives.
- The reason for sharpening a cutting tool is to provide a sharp and keen cutting edge, essential for producing work of quality.
- The process of reconditioning a saw includes jointing, reshaping, setting, sharpening and side dressing.

REVISION EXERCISES AND QUESTIONS

1. Briefly explain the essence of maintaining cutting tools.
2. Name two of the artificial abrasives that are commonly used for the manufacture of grinding wheels and oilstones and explain how the abrasive are bonded-together.
3. Briefly state the procedure for grinding a cutting tool.
4. What safety precautions are required when grinding a cutting tool?
5. State how you should care for an oilstone to prolong its usefulness.

Common processes and woodwork joints

Introduction

This chapter discusses a range of processes in woodwork using hand-operated tools. For each process, a definition and description of the steps are given. The chapter also describes the three types of joint used for cabinet construction: widening joints, angle joints and framing joints. Illustrations are provided to help you understand the procedures more easily.

Common processes

Marking out is the process of locating or marking the correct positions of joints on the work-pieces to be used for the construction. Sometimes the setting out is done on one length of the work-piece before it is cut into parts. The setting out may be done in the form of a working drawing, in pencil, on a piece of plywood. This is called a **working rod**. This method facilitates the marking out process, particularly where the same job will be produced in a large quantity. (See Figure 5.1).

Sawing is a process of cutting (off) a piece of board. When the cutting is along the grains of the board (along the length), the process is termed as **ripping** and when the cutting is across the grains (across the width), it is known as **cross-cutting**.

The procedure for sawing is as follows: (Figure 5.2)

1. Support the board firmly, ensuring that nothing is in the path of the saw.
2. Hold the saw with your forefinger extending

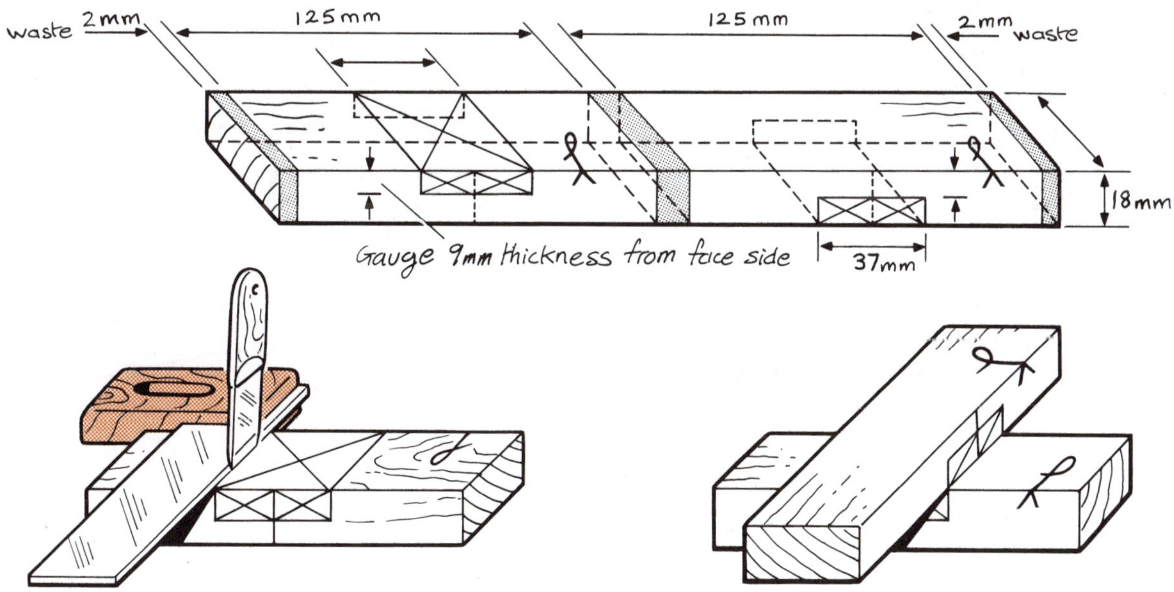

Figure 5.1 Marking out.

forward. This will provide firm control of the blade and prevent the hand grip twisting in your hand (Figure 5.2a).
3. Begin the cut with short backward strokes whilst guiding the blade with your thumb.
4. Use the full length of the blade after you have established the initial kerf. Move with slow and steady strokes (Figure 5.2b).
5. Twist the blade slightly to bring it back on course when the saw begins to wander off the marked line.
6. When the kerf begins to close on the saw keep it open with a small wedge.
7. Support the offcut with your free hand, particularly when approaching the end of the cut (Figure 5.2c).

Planing is the process of using a plane (hand or machine) to take off shavings from a piece of board, with the purpose of reducing its thickness, width or both, or smoothing the surface. A planed board is referred to as a **dressed board**.

The procedure is as follows:

1. Examine the two faces of the board and select the one that looks straighter and flatter as the **face side** (the first face to plane).
2. Hold the board against the bench stop securely, noting the direction of the grain so as to plane along the grain.
3. Hold the handle of the plane firmly and use the other hand to hold the front part of plane in such a way that your fingers are on the side of the plane away from you and your thumb on the side nearest you (Figures 5.3a and 5.3b). Hold the knob with all your fingers if you are using the metal plane (Figure 5.3c). Holding the plane in this manner enables you to exert adequate pressure for the cutting (planing) action to be effective.

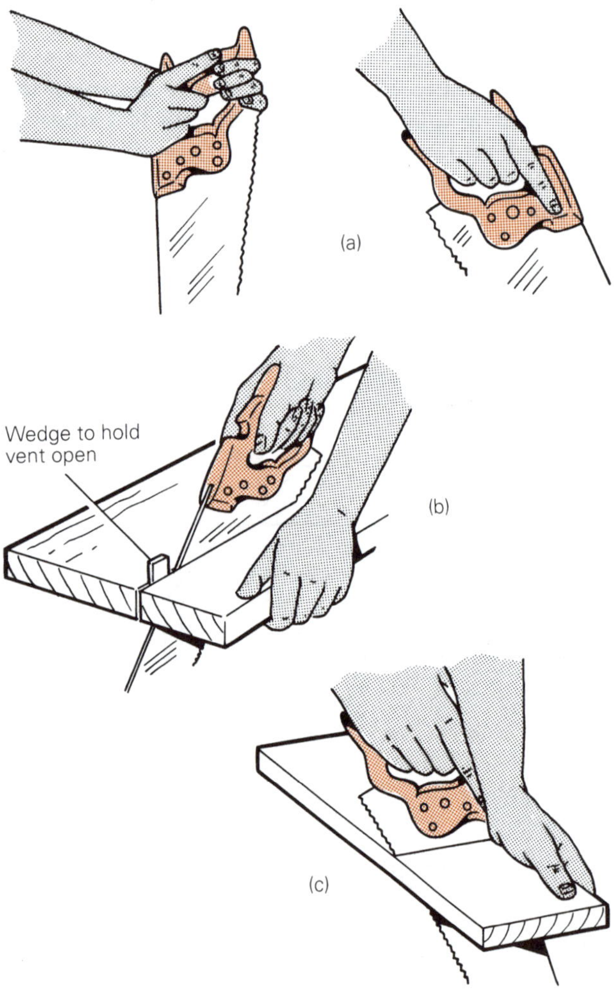

Figure 5.2 Sawing: (a) holding the saw, (b) ripping, (c) cross cutting.

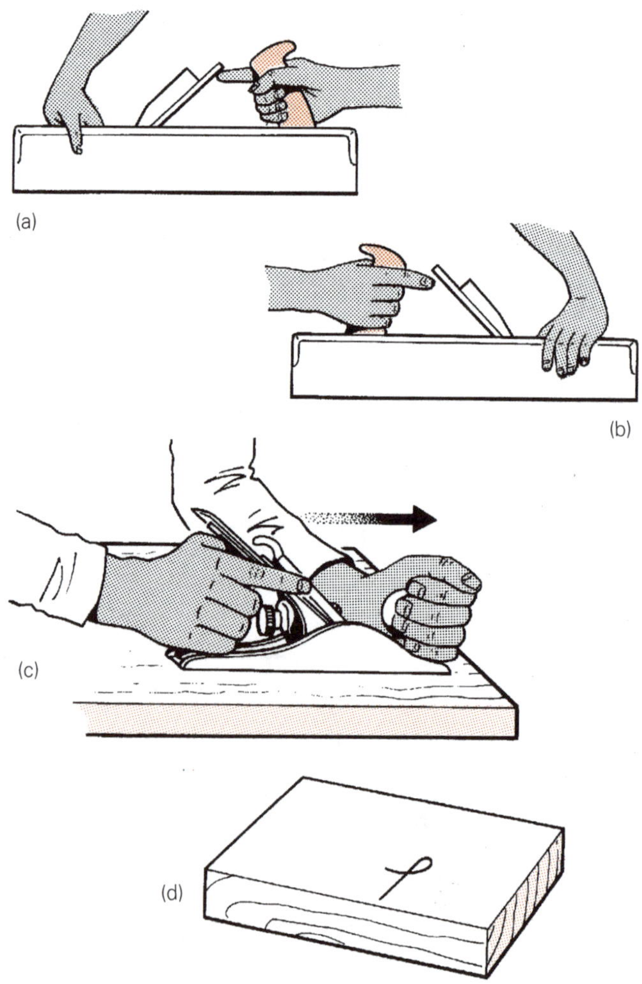

Figure 5.3 Holding and using a plane when planing a surface.

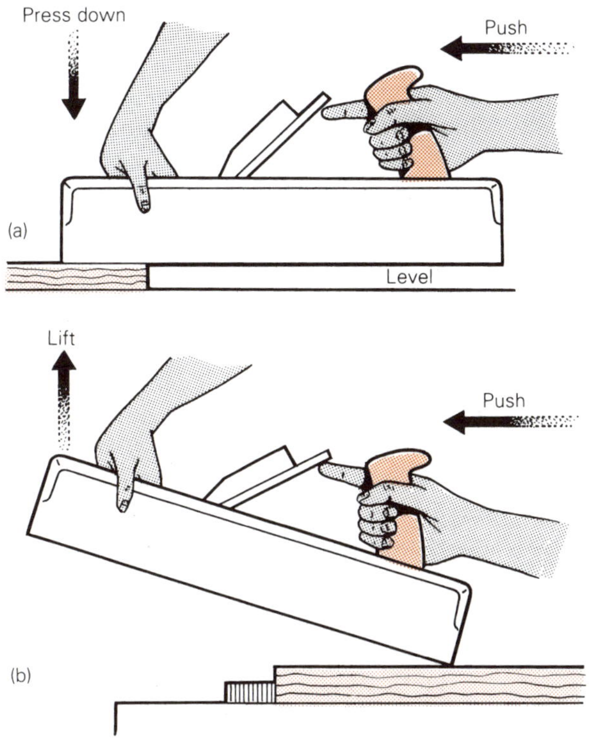

Figure 5.4 **Positions when planing: (a) starting, (b) finishing.**

4. Press down the plane and begin the stroke from the very end of the board to the other, maintaining the same pressure throughout (Figure 5.4). Plane the whole surface and check for straightness by laying the edge of a try-square or a straight edge along the surface. If the surface is straight, the edge of the try-square will touch all the surface. Check in several positions. When the face is judged to be straight, put the face side mark on it (Figure 5.3d).
5. Choose the adjacent edge of the board, hold the board firmly in the vice with this edge upwards and plane it straight and square to the face side. This time, the fingers of the hand holding the front of the plane must rest under the plane to run on the face of the board to serve as a guide. The fore-finger of the hand holding the handle must point forward in this case to direct the plane and keep it straight (Figure 5.5).
6. Test for straightness as before and use the try-square to check for squareness. To do this, hold the face of the stock firmly against the face side and the edge across the planed edge. Check at several positions. Place the face edge marks (Figure 5.5d) on the edge.
7. Set the marking gauge to the required width and gauge on both faces of the board. Gauge from the face edge, i.e., with the stock of the marking gauge running against the face edge.
8. Place the board in a vice and plane down to the gauge line. Check again for both straightness and squareness.
9. Set the marking gauge to the required thickness and gauge from the face side on the edges and ends.
10. Plane to the gauge line and check for straightness and squareness.

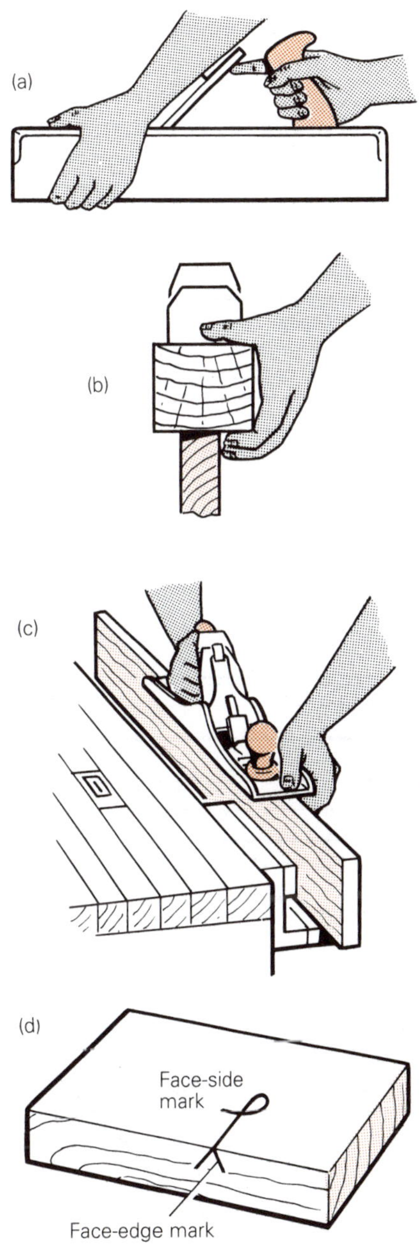

Figure 5.5 **Holding and using a plane when planing an edge.**

Mitring means producing a **mitre**, an angle joint where the ends of the two pieces forming the joint are cut at an angle of 45° and joined together so that a right angle is formed between the two meeting pieces, such as on the corner of a picture frame or a plinth. A mitre may be cut in a mitre box and planed in a mitre shooting board. (See Figure 2.2 on page 5)

Chamfering (Figure 5.6) is a process of planing off at 45° a definite amount of material from the edge of a piece of board. When the planing goes through the entire length of the piece, it is called a **through chamfer**. When the chamfer is stopped at either end it is called a **stopped chamfer**. A through chamfer can be planed with a jack plane, a smoothing plane or a block plane. For a stopped chamfer use a chisel, a spokeshave or a bull-nose plane, finishing the corners with a bevel-edged firmer chisel. Apart from making the edge of the piece safer to handle, chamfering is done to improve the appearance of the work and also to protect the edges themselves.

Note: when the side of a piece is planed at an angle other than 45° to the edge, such as on the square or straight edge, it is called a **splayed edge** (Figure 5.6). An edge is said to be **bevelled** when the whole of that edge is planed off to an angle other than a right angle.

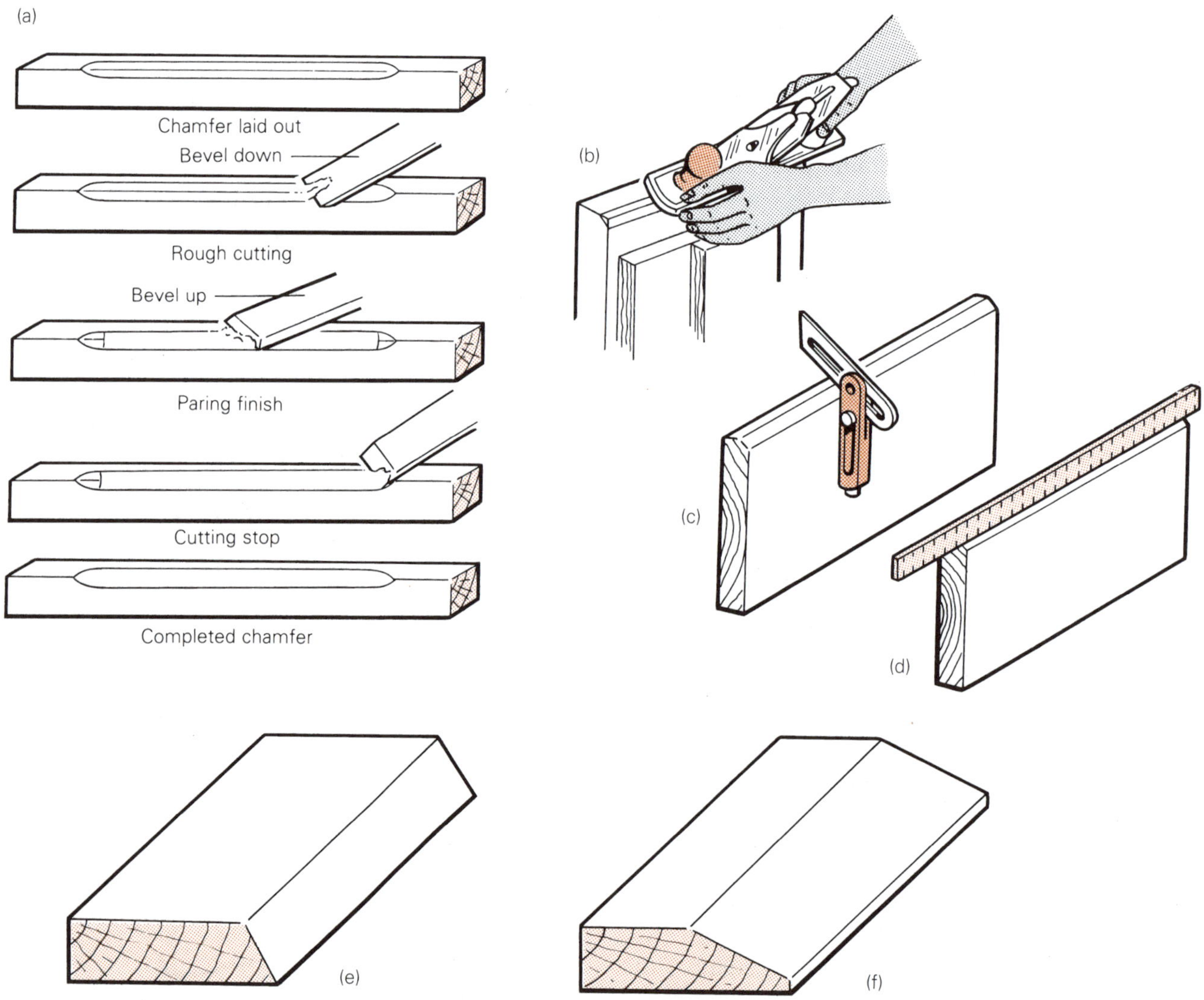

Figure 5.6 Chamfering and bevelling: (a) chamfering with a chisel, (b) chamfering with a plane, (c) testing flatness, (d) testing straightness, (e) bevelled edge, (f) splayed edge.

Trenching (Figure 5.7) is the process of cutting a recess across the grain of a workpiece, usually on the surface at a distance from the ends. The tools used are a saw, a chisel and a router plane.

The procedure is as follows:

1. Hold the workpiece securely in the vice or on the bench (using a bench hook).
2. Saw the sides (shoulders) with a tenon saw to the required depth (Figure 5.7b). A beginner should make sloping grooves along the marked lines on the surface prior to sawing the sides to avoid sawing over the marked lines.
3. Chisel out the waste to about 2 mm from the bottom. Chisel from each edge so as not to split the back edge (Figure 5.7c).
4. Place the router plane over the trench with the cutting iron in the recess (Figure 5.7d).
5. Set the cutting iron for a thin shaving and slightly less than the required depth.
6. Router (plane) to half the width to avoid splitting the wood at the end if it is a through trench.
7. Reverse the workpiece and router off the remaining half.
8. Now adjust the cutting iron to the required depth and finish off the routing.

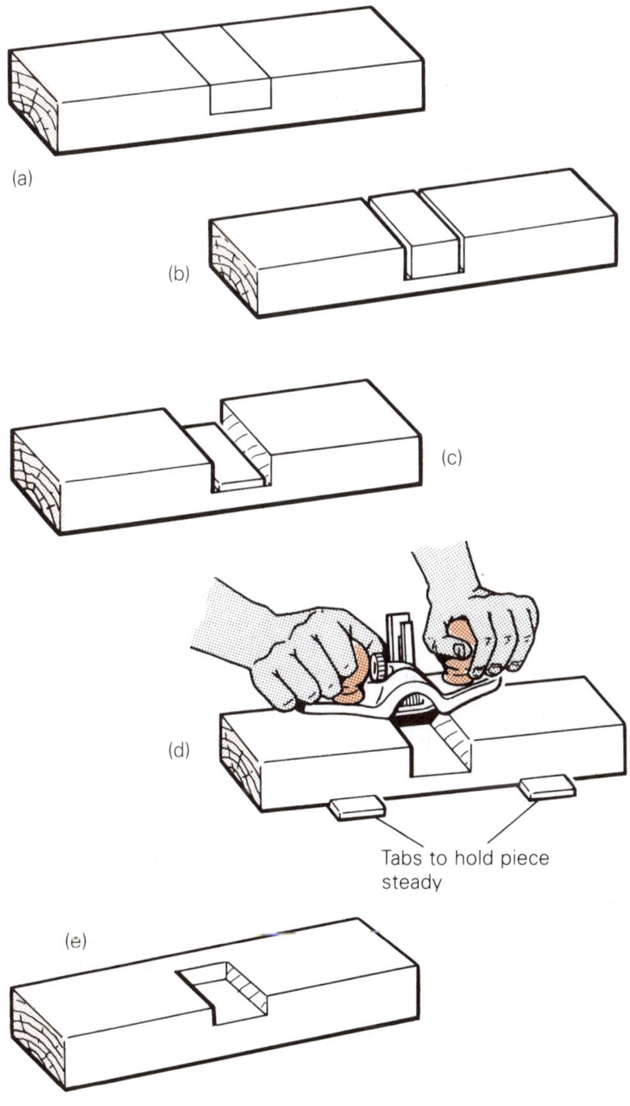

Figure 5.7 Trenching: (a) trench laid out, (b) saw the sides, (c) chisel the waste, (d) level off with router plane, (e) a stopped trench.

Note that the same procedure should be followed when preparing a stopped trench, except that there will be no need to reverse the workpiece.

Grooving is the process of cutting a recess on a surface or edge of a workpiece, usually along the grain (Figure 5.8a). The plough plane or the universal plane may be used. In each case, you should select a cutting iron with the same width as the width of the groove to be cut.

The procedure is as follows:

1. Gauge the position to be grooved, using a mortise gauge or a marking gauge. Gauge from the face side or the face edge of the workpiece.

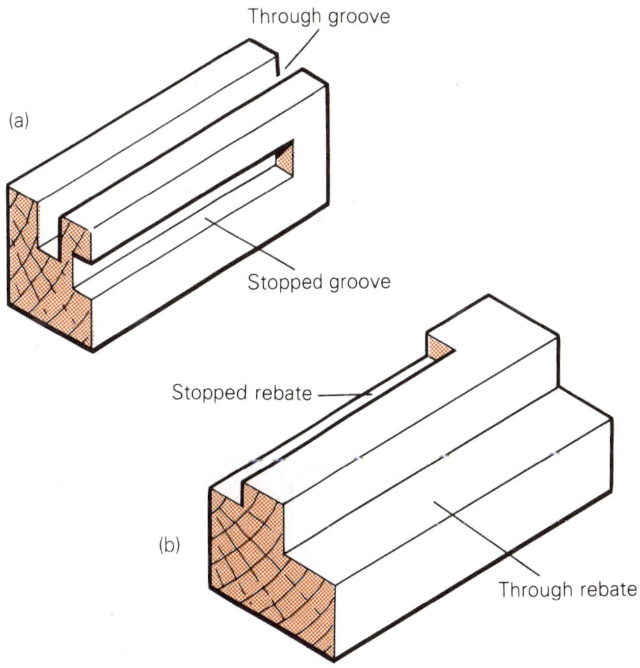

Figure 5.8 Grooving and rebating: (a) grooves, (b) rebates.

2. Clamp the workpiece onto the bench, using a vice, 'G' cramp or holdfast.
3. Fix the correct cutting iron, and adjust it to provide a convenient cut, clamping it securely.
4. Adjust the fence at the correct distance from the edge and secure it in position firmly (by the thumb screw).
5. Set and clamp the depth gauge to the required depth.
6. Starting at the front and working backwards, take successive cuts until the desired depth is attained.

Rebating is the process of cutting a rectangular recess (Figure 5.8b) along the edge or across the end of a board to form a shoulder. Any of these planes may be used: rebate plane, sash fillister plane, side fillister plane, combination plane or the multi-plane.

The procedure is as follows:

1. Use the marking gauge to gauge for the width and depth of the rebate. Draw a pencil through the marked lines to make them visible. You may use only the pencil to mark freely both the width and depth.
2. Hold the workpiece firmly in the vice, ensuring the edge is horizontal.

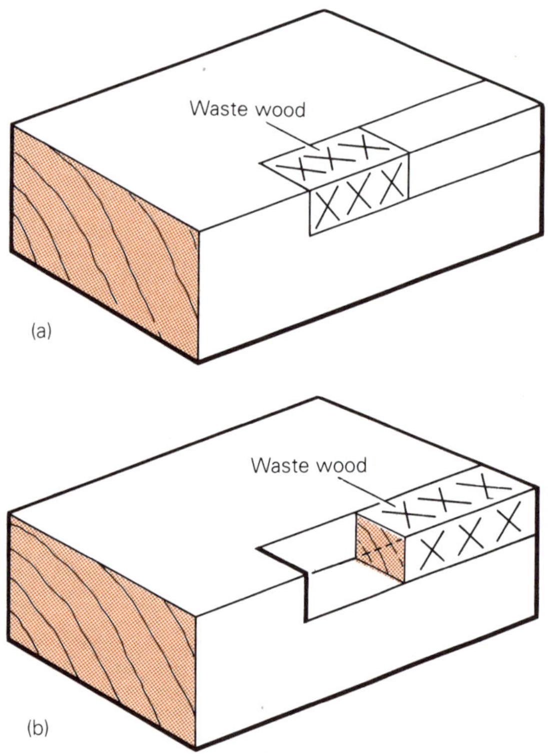

Figure 5.9 Cutting a stopped rebate: (a) cut out a piece, (b) remove the waste.

3. Adjust the fence of the plane to run along the face side at the correct width of cut (that is, the width of the rebate). Where the plane has no fence, for example the wooden rebate plane, fasten a strip of wood along the gauged line to serve as a guide.
4. Set the depth gauge to the desired depth of the rebate.
5. Adjust the blade for the appropriate cut. Check for lateral alignment.
6. Cut the rebate. Keep the fence steady against the face side of the workpiece at all times.
7. Keep the plane parallel to the edge of the workpiece. Remember to work from front to back as in grooving. Take successive cuts until the depth gauge touches and runs on the edge of the workpiece. Where the plane has no depth gauge, you must check the depth frequently, first by watching for the gauged line and finally by measuring with a rule at various positions to make sure that the depth is correct. (Note that, when rebating across the grain (on the end) of a workpiece, you must first chamfer the far end to avoid splitting. Preferably, use a plane with spur cutter that will cut across the grain so as to facilitate the planing. Finish off with a bullnose plane.)
8. To cut a stopped rebate, you must first cut out a small mortise. Sometimes it is easier to cut down to the required depth with a tenon saw and finish off with the rebate plane (Figure 5.9).

Housing is a term used to describe fitting the end of a piece of board into a trench in another piece, as commonly used for fixing shelves in cabinets (Figure 5.10). Various methods are used for preparing the trench and so the housing is named after the type of trench as follows: **through housing**, when the trench is through (across the width); **stopped housing**, when a stopped trench is used; **dovetailed housing**, when a dovetailed trench is used; and **shouldered housing**, when the trench is narrower than the thickness of the piece to be fitted in (shelf). A housing joint is prepared by first making the trench on one piece and then cutting the end of the other piece to fit the trench. It may be glued or nailed. A dovetailed housing may be left loose, (i.e., not glued) to enable the position of the shelf to be varied as the need arises. For the procedure to prepare a through housing joint see Figure 5.11.

Pinning is a method of joining the end of one piece of board to another in a similar manner to

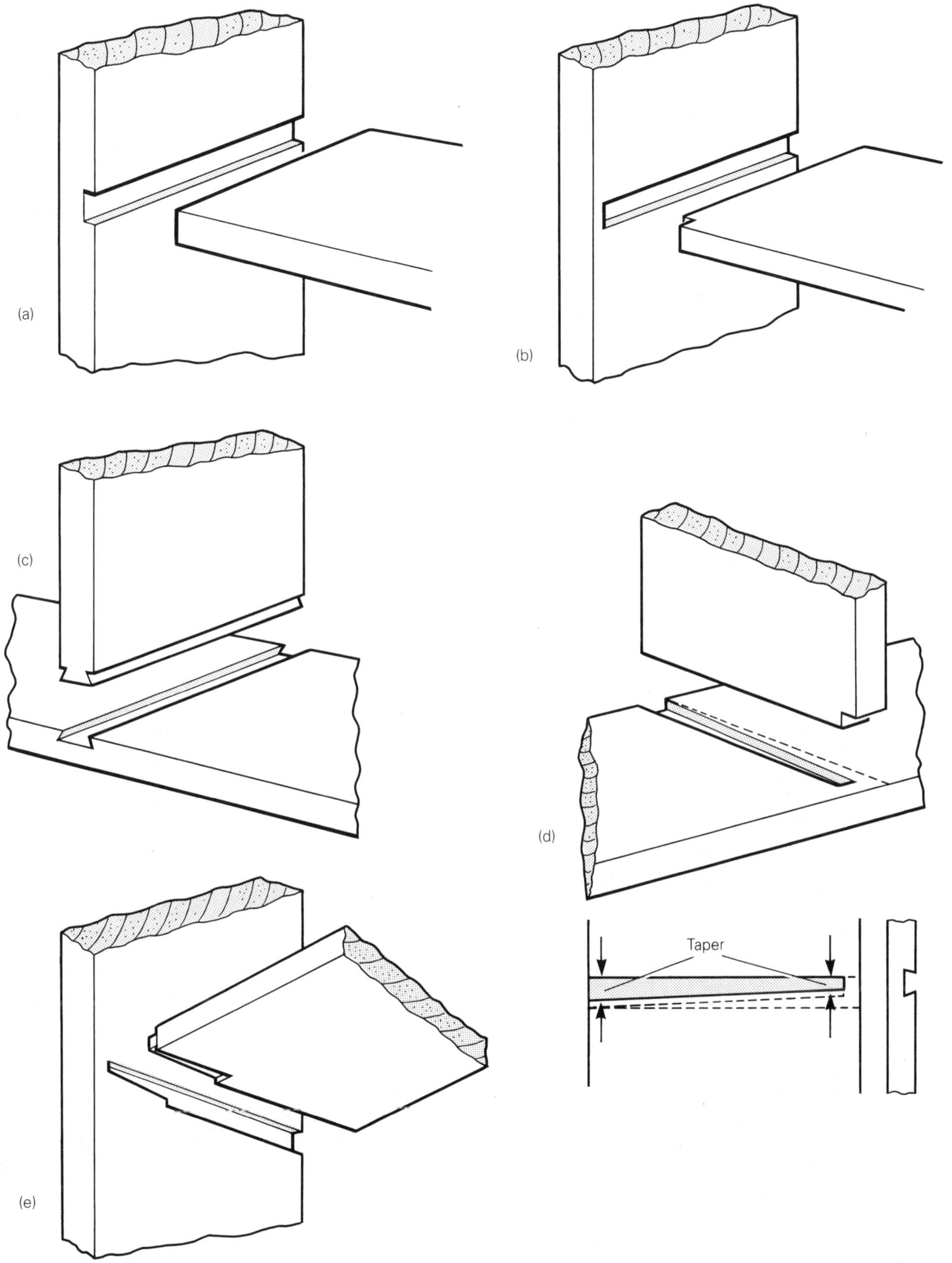

Figure 5.10 Housing joints: (a) through housing, (b) stopped housing, (c) dovetailed housing, (d) tapered bareface dovetail housing, (e) part dovetail housing.

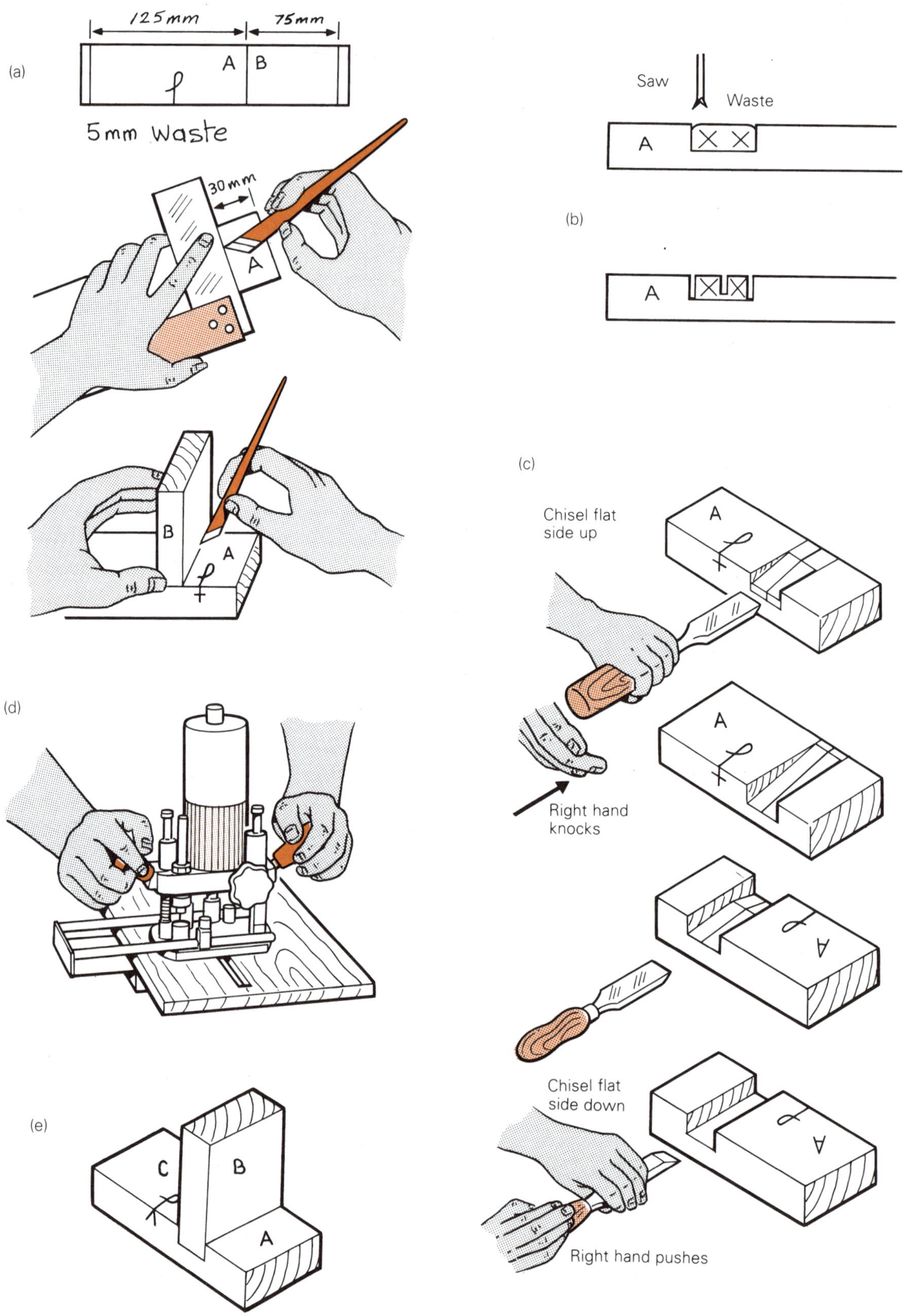

Figure 5.11 Preparing a through housing joint: (a) marking out, (b) sawing, (c) chiselling out waste, (d) using router plane, (e) the finished housing.

housing. The difference is that instead of fitting the end of the piece into a trench, as in housing, pins are made at the end of the piece and are fitted into a corresponding sockets on the other piece (Figure 5.12). Pinning is used in the same situations as housing, that is, for fixing vertical or horizontal members of a cabinet.

Follow this procedure when preparing a pinning joint:

Preparing the pins:

1. Measure and cut the piece to the required length, adding the dimension (length or projection) of the pins.
2. Square the ends and mark out the positions of the pins, square the lines on the two faces and the end from the face side. Use a mortise gauge and mark from the face edge. Mark the waste wood with pencil.
3. Hold the piece firmly in the vice.
4. Saw down the sides of the pins to the shoulder line. Saw on the *waste* side of the line.
5. Remove the waste using a coping saw and cut off the end shoulders with a tenon saw.
6. Remove the board from the vice and clamp it onto the bench, placing a piece of board underneath it to prevent damaging the surface of the bench. Clean up the shoulders with a bevel-edged firmer chisel.
7. Check for squareness at the shoulders.

Preparing the sockets:

1. Mark out the position of joint and draw parallel lines (equal to the thickness of the pins) across the face of the piece. The lines must be square to the face edge.
2. Place the pins (vertically) on the marked lines and trace with sharp pencil the locations of the pins. (If the pins are to go right through, you will have to *measure* and mark up both sides of the piece instead.)
3. Clamp the piece on to the bench and chop out the sockets to the required depth. If the pins are to go right through the piece then cut *half way*, before reversing the work and cutting through from the other side. Care is needed to ensure that the sides of the sockets are vertical.

Assembling:

1. Fit the pins in the corresponding sockets and press to fit.
2. Check for squareness and ensure the shoulders are properly fitted.

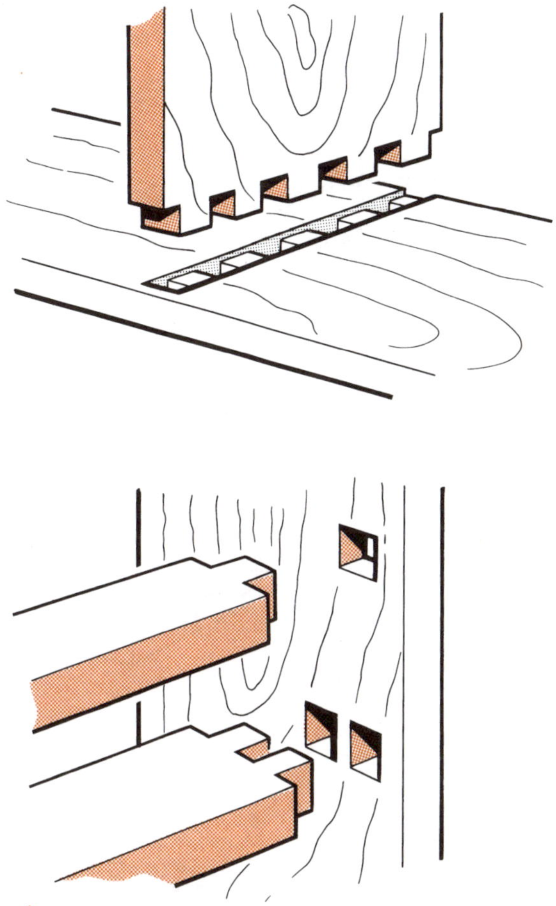

Figure 5.12 Pinning.

Dowelling is a process of joining pieces of boards with **dowels** or cylindrical pins (Figure 5.13). The dowels are usually glued into holes bored at the edge of each piece. Dowelling may be used to strengthen butt joints, mortise and tenon joints and bridle joints.

Always cut a groove along the side of the dowel. This allows glue and air to escape when fitting the joint.

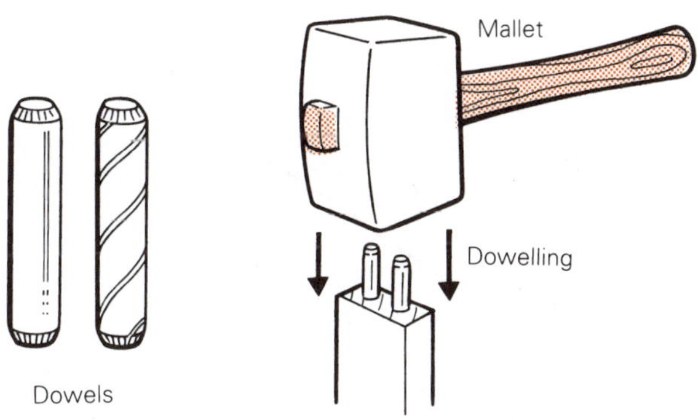

Figure 5.13 Dowelling.

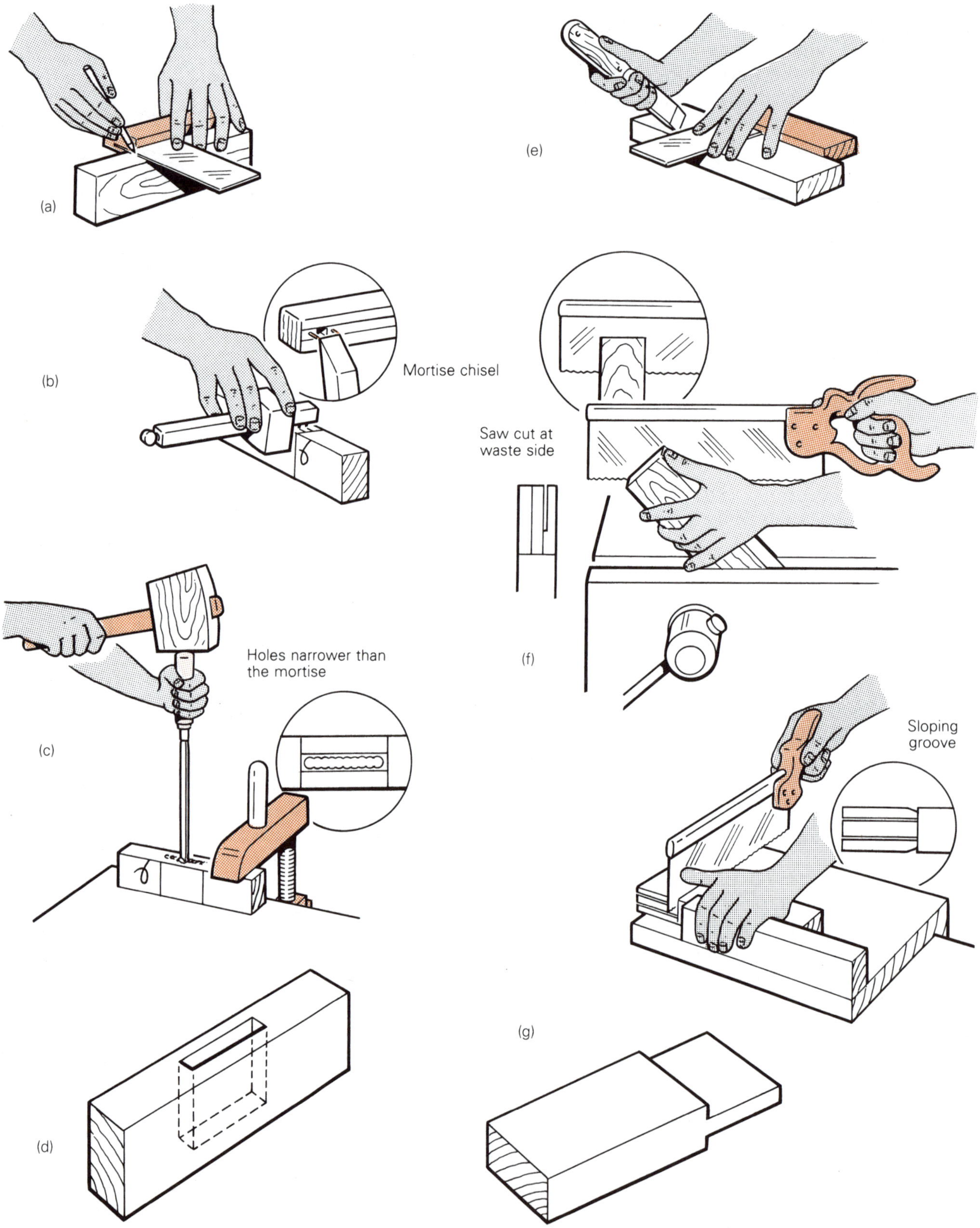

Figure 5.14 Preparing a common mortise and tenon: (a) mark out the mortise, (b) using the mortise gauge, (c) chisel out the waste, (d) the finished mortise, (e) mark out the tenon, (f) cut the tenon, (g) the finished tenon.

Mortising is the process of chopping out a rectangular hole across a workpiece, often, one-third of the thickness. A mortise can be **through** the whole width, **stopped** or **haunched**. A mortise is fitted with a male part known as a **tenon**.

The procedure for preparing a mortise is as follows (Figure 5.14):

1. Plane the workpiece to the required width, thickness.
2. Mark out for the mortise and square the lines across the face side and edges with pencil. Square the lines on only one edge if it is a stopped mortise (Figure 5.14a).
3. Set the mortise gauge to the thickness of the chisel to be used for chopping out the waste (about one-third of the thickness of the piece) (Figure 5.14b).
4. Gauge from the face side across the marked mortise at both edges. If it is a stopped mortise, gauge on one side only. Mark the waste portion 'X' 'X'.
5. Clamp the workpiece on the bench or hold it firmly in the vice. Put a waste piece under the cramp to avoid denting the workpiece. Similarly, a waste piece should be placed under the workpiece in the vice.
6. Chop out the waste with the same mortise chisel that you used to set the mortise gauge, working halfway through from both edges (if through mortise). You can bore or drill through first, before you chop off the remaining waste (Figure 5.14c). Leave about 6 mm at each corner of the mortise to avoid damaging the corners when levering out the waste. Remove this when the mortise is through and clean off the sides of the mortise with a bevel-edged chisel.

A **tenon** is a name given to a projection, one-third of the thickness of a piece of board left after the other one-third on each side is cut off to form shoulders. Where the thickness of the board is thin, it can be halved instead. A tenon of this type is called **bare-face tenon** (meaning that it only has one shoulder). Other types of tenon are **common tenon**, **haunched tenon**, **long and short shouldered tenon, stub tenon**. Examples are shown in Figure 5.15. Normally, a tenon is fitted into a mortise to form a joint called a **mortise and tenon joint**. A haunched tenon is fitted into a haunched mortise, while a long and short shouldered tenon must be fitted into a long and short shouldered mortise.

The procedure for cutting each type of tenon is similar to that of the common tenon (see Figure 5.14).

1. Plane the piece of wood to the required width and thickness.
2. Mark out the position of the tenon at one end of the wood and square the shoulder line across the two faces and two edges. Use the stock of the try-square against the face side and face edge when squaring the line (Figure 5.14e).
3. Set the mortise gauge to one-third the thickness of the board and gauge across the edges and the end to the shoulder line. Gauge from the face side.
4. Saw off the wastes and check the shoulders for squareness (Figures 5.14f).

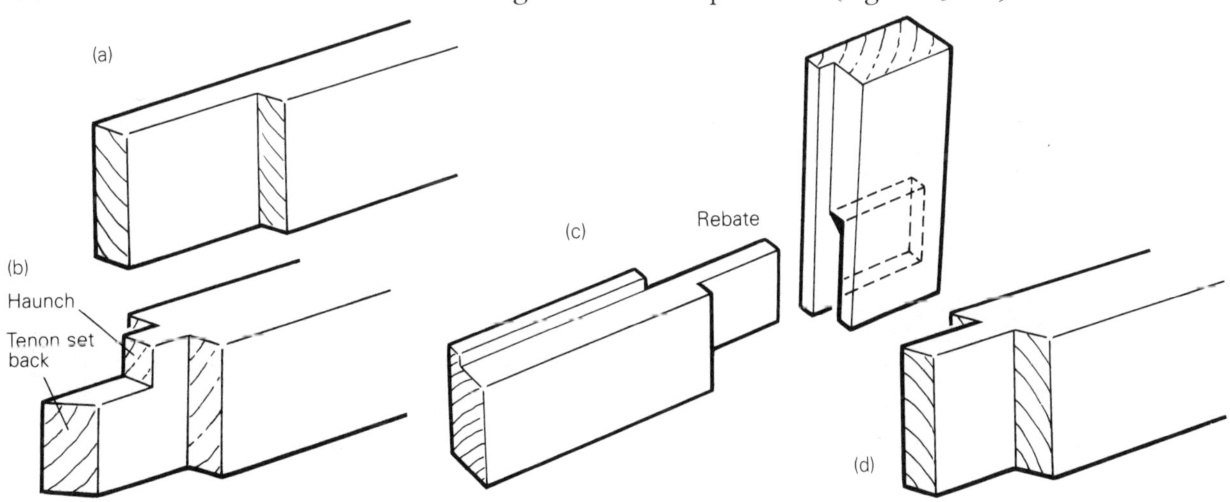

Figure 5.15 Types of tenon: (a) bare-face tenon, (b) haunched tenon, (c) long and short shouldered tenon, (d) stub tenon.

Common woodwork joints

Joints in woodwork are devices for holding parts of wooden artefacts or structures together. There are variety of them and those used for cabinet construction may be categorised into three: widening joints, angle or box joints and framing joints.

Widening joints

Joints that classified as widening joints (Figure 5.16) are those used for holding together a number of narrow boards edge to edge to provide a wide board. Among these are: butt joints, tongued joints, tongued and groove joints, rebated joints, dowelled joints and slot-screw joints.

A **butt joint** (Figure 5.16a) is the simplest form of widening joint, where the edges of the pieces to be joined together are planed straight and square to the sides, glued and cramped until the glue is set. To minimise *warping* (running out of flatness), arrange the pieces so that the direction of the grains of each piece run in the opposite direction. Butt joints are mostly used when preparing table tops and cores for veneers. Those used in cheap constructional work, such as making packing cases, are often held together by corrugated fasteners.

Tongued joints (Figure 5.16b) are formed by grooving along the edges of the pieces to be joined. Strips cut across the grain called **tongues** or **feathers** are glued into these grooves and the pieces are cramped together. The cross-grain tongues resist warping and *twisting* (running out of straightness), thereby providing extra strength

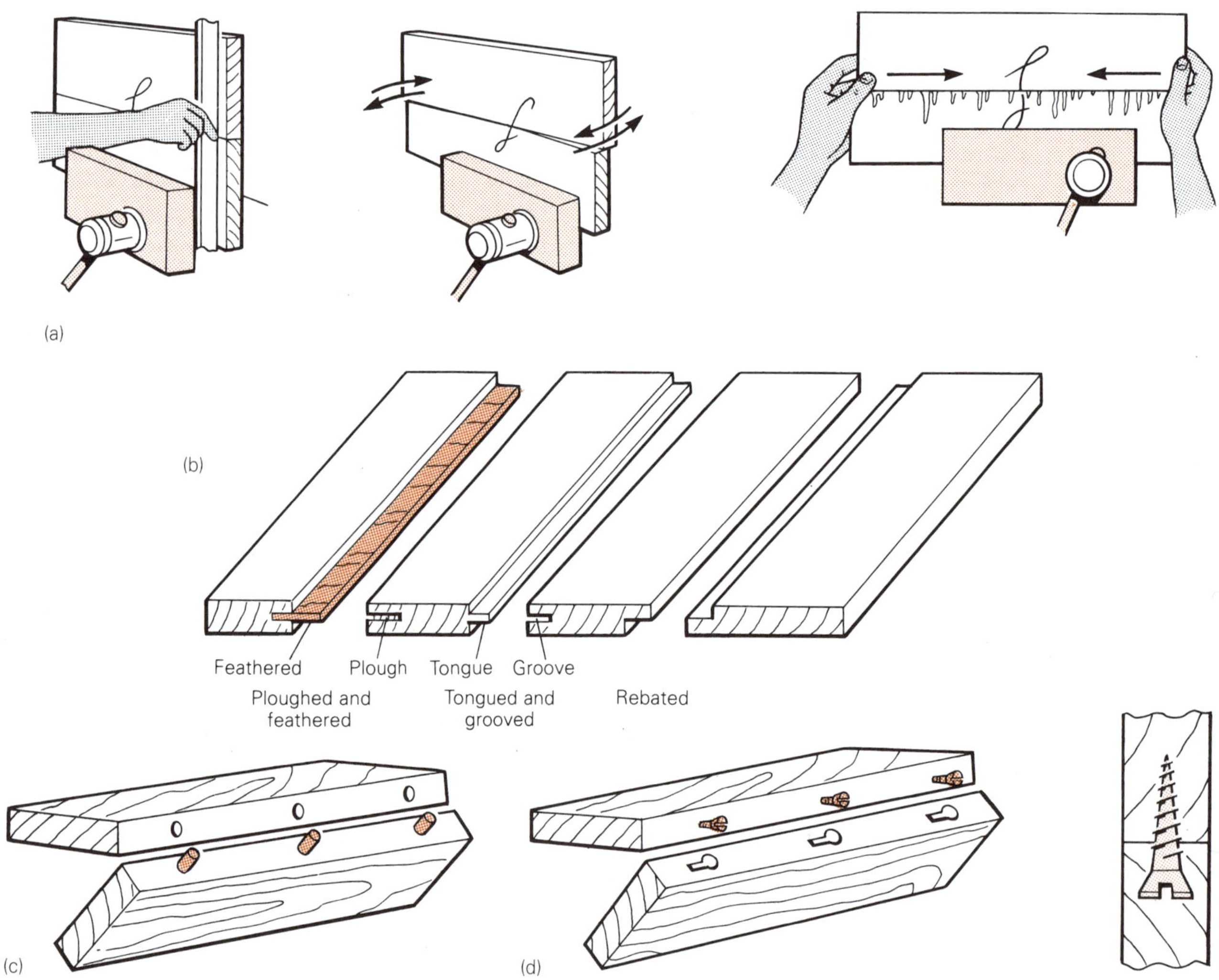

Figure 5.16 Widening joints: (a) butt joint, (b) tounged joints, (c) dowelled joint, (d) slot-screw joint.

to the board. This method is suitable for joining thick boards since the size of the groove and tongue must be about one-third the thickness of the board. Tongued joints are suitable for panels and sometimes for table tops.

The **tongued and grooved joint** (Figure 5.16b) is similar to the tongued joint except that the tongue in this case is formed along the edge of one piece and fits into a matching groove along the edge of the other. Matching planes are the ideal tools for preparing the tongue and groove. This joint is widely used for flooring boards, table or desk tops, packing cases.

The **rebated joint** (Figure 5.16b) consists of two or more boards, each with its edges rebated, glued and cramped together. This joint is also applicable for making floor boards, desk tops and packing cases.

The **dowelled joint** (Figure 5.16c. See also page 47) is made by fitting dowel pins into holes bored (drilled) along the edges of the pieces to be joined. These are then glued and cramped together. It is not ideal for thinner boards. It is used for table tops, etc.

The **slot-screw joint** (Figure 5.16d) is made by fixing screws along the edge of one piece and fitting them into prepared holes and slots made along the edge of the other piece . The holes are provided for the head of the screw and the slots take the shank of the screws. This joint is not usually glued and so it is suitable for timbers that do not glue readily.

Angle joints

Angle joints (Figure 5.17), sometimes referred to as **corner joints** are generally used for joining board when their faces meet at right angles with flush edges. They are mostly used in box-like constructions, such as drawers, boxes and solid cabinets. They are: the simple butt joint, plain mitred joint, rebated butt joint, rebated and mitre angle joint, housed and shouldered joint, comb joint and dovetailed joints.

The **simple butt joint** (Figure 5.17a) is a method of constructing boxes, such as packing cases by simply squaring or shooting the ends of the pieces, assembling them at right angles and gluing, screwing or nailing them together.

The **rebated butt joint** (Figure 5.17b) is similar to the simple butt joint except that one end is rebated and provides an additional surface (holding) for gluing, nailing or screwing. It is stronger than the simple butt joint. Its main application is in constructing of boxes, book shelves, carcases to be veneered, cheap drawers, etc.

The **plain mitre joint** (Figure 5.17c) is used when the end grain of each of the meeting pieces is to be concealed. It is essential that the meeting pieces are of same thickness and width. The procedure for mitring has been described on page 42. Plain mitre joints are not as strong as butt joints because the holding power of nails, screws or glue applied to the end grain is weak. They may be strengthened by a slip feather (i.e., a loose tongue) inserted in a groove cut along the joint, or by pieces of veneer, known as **keys**, glued in saw kerfs cut across the corner of the joint. They are used for plinths, and box construction.

The **rebated and mitred angle joint** (Figure 5.17d)is an improvement of the plain mitred joint. The meeting pieces are first rebated at the ends, one on the inside and the other on the outside and then mitred to conceal the end grains. The joint is fairly strong as there are two surfaces for nailing, screwing or gluing. It is used for boxes, plinths and cabinet construction.

The **housed and shouldered joint** (Figure 5.17e) is another version of the simple butt and rebated butt joints. The end of one of the meeting pieces is rebated to fit into a trench cut across the face of the other. This joint is sometimes referred to as a **tongued and trenched joint** or **dado joint**. It is used in drawer, box or cabinet construction.

The **comb joint** (Figure 5.17f) is made by cutting a series of alternate square pins and sockets of the same width at the ends of the meeting pieces and interlocking them. It is mostly used in box construction.

The **dovetail joints** (Figure 5.17g) are the strongest of all angle joints. Dovetailing is used to join two pieces of board so that one interlocks with the other and can be pulled out only in one direction. The joint is in two parts, the **tail** and **socket**. Both parts have sloping sides that look similar to the tail of the bird from which the joint takes its name. The common types are; single dovetail, common (through) dovetail, lapped dovetail and secret dovetail (Figure 5.18).

The **single dovetail joint** (Figure 5.18a) has one dovetail at the end of one of the meeting pieces which fits into a corresponding dovetail socket cut at the end of the other piece. This joint is often used for narrow pieces and it can be made through or lapped. It is chiefly used in carcase construction, such as brackets, top and bottom rails.

The **common dovetail joint,** (Figure 5.18b) also called the **through dovetail joint** has two or

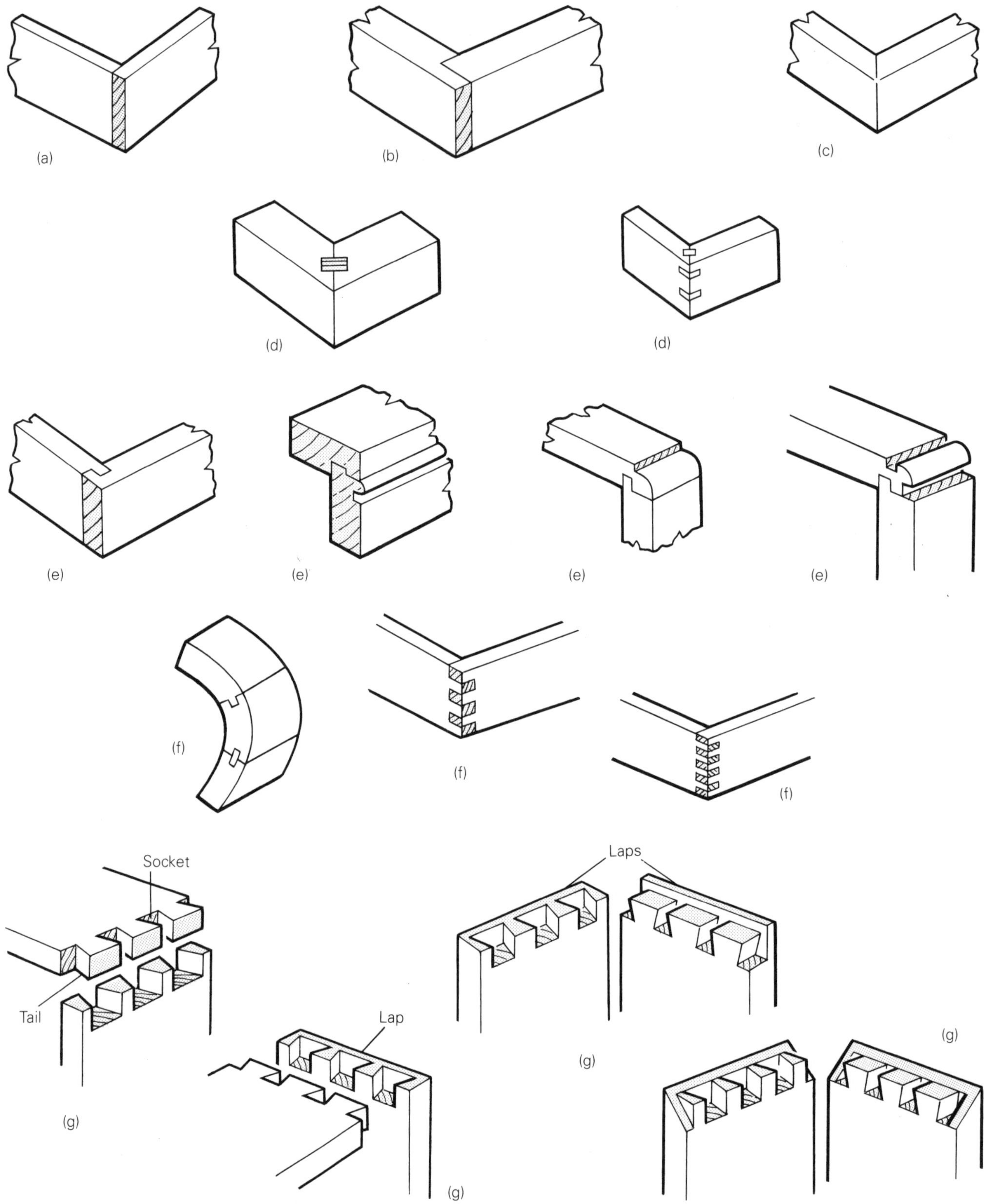

Figure 5.17 Angle joints: (a) butt joint, (b) rebated butt joint, (c) plain mitre joint, (d) rebated and mitred angle joints, (e) housed and shouldered joints, (f) comb joints, (g) dovetail joints.

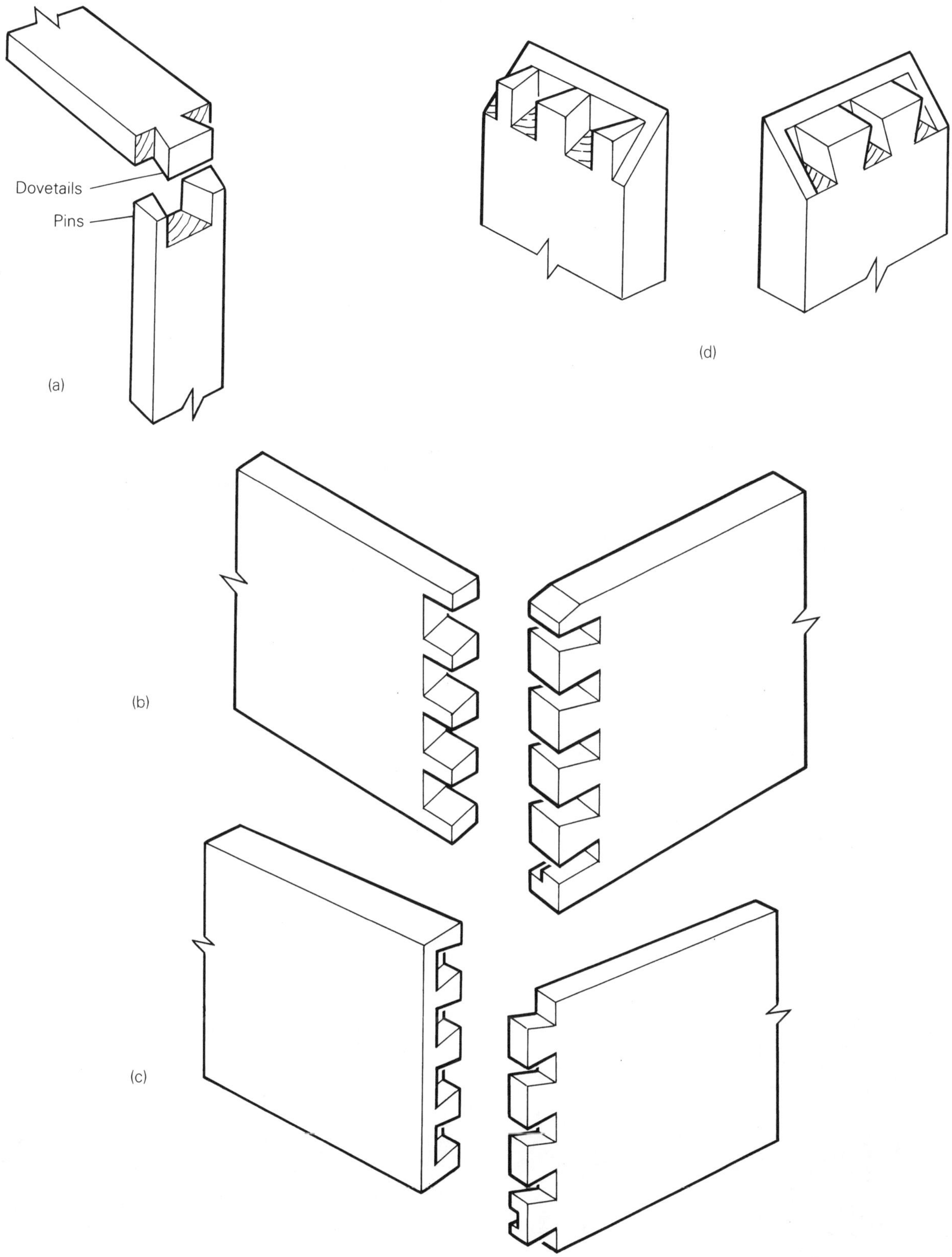

Figure 5.18 Types of dovetail joint: (a) single, (b) common, (c) lapped, (d) secret-mitred.

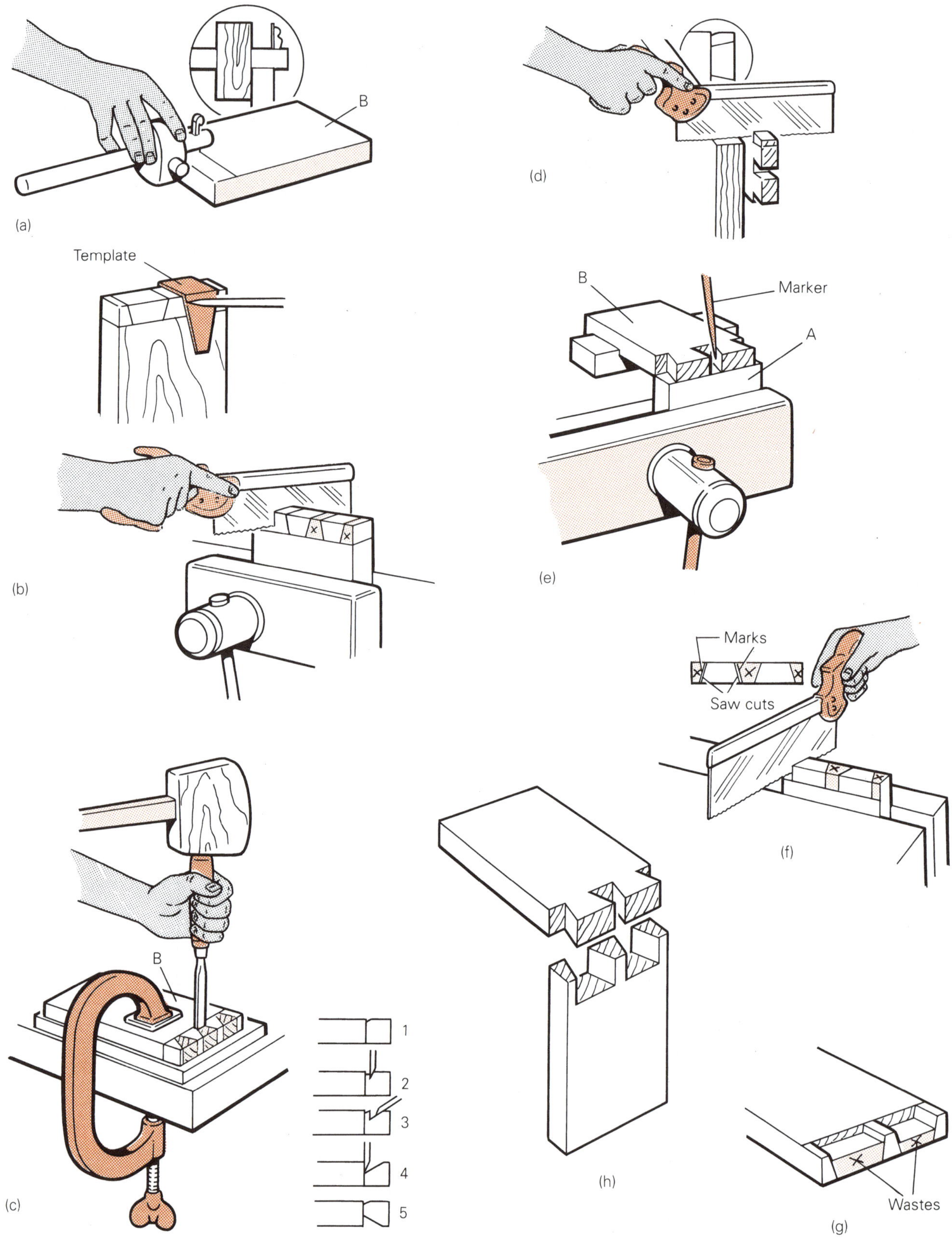

Figure 5.19 Preparing a common dovetail joint: (a) gauge the end of the piece to contain dovetails, (b) set out the dovetails, (c) chisel out waste, (d) cut off end pins, (e) Trace dovetails onto 'A', (f) saw the pins, (g) remove waste from pin sockets, (h) joint ready to assemble.

more dovetails at the end of one piece and a corresponding number of dovetail sockets on the other piece. When assembled, the end grains are shown on both sides of the joint. It is used for constructing boxes, carcases, and for joining the back of a drawer to the sides.

The **lapped dovetail joint** (Figure 5.18c) consists of two or more dovetails at the end of one of the meeting pieces, similar to those of the common dovetail, except that the length of the dovetails is reduced by an amount equal to one-third the thickness of piece. Similarly, the corresponding dovetail sockets are not cut through the thickness of the piece; a lap of one-third of the thickness is left out. The lap covers and conceals the end grains of the dovetails when the joint is assembled. It is chiefly used in joining the bottom piece of a carcase to the side pieces. In drawer construction, it is used in joining the sides to the front.

The **secret dovetail joint** (Figure 5.18d) has mitred laps at the ends of both meeting pieces (the dovetails and dovetail sockets) so that when assembled, no end grain is exposed and it looks like a plain mitred joint. It is mostly used in constructing high-class cabinets, jewel boxes, etc. where good appearance and strength are essential.

The procedure for the construction of the various dovetail joints is similar and is described here for three types.

A common dovetail joint is prepared as follows (see Figure 5.19):

1. Plane the pieces to the required width and thickness in one length and shoot the ends square with the face side.
2. Set out the position of the joints by gauging or squaring lines from the ends on both sides and edges (all round). The general rule is: the cutting gauge line on **piece 'A'** (pins) should be at a distance from the edge equal to the thickness of piece **'B'** (dovetail) and that on piece **'B'** should also be at a distance equal to the thickness of piece **'A'**. Allow 3 mm waste on each piece for cleaning up after assembling (Figure 5.19a).
3. Set out the position and shape of the dovetail on the face side of piece **'B'**(Figure 5.19b).
4. Saw the sides of the dovetail, *sawing on the waste side but to the line.*
5. Chisel out wastes of dovetail sockets on piece **'B'**, chiselling vertically and obliquely alternately and cut off end pins (Figure 5.19c,d).
6. Place piece 'B' over the end of piece **'A'** (held in a vice) with the face sides of both pieces on the same side (inside or outside) and, with the face edges flush, trace the shape of the dovetail with a marker onto piece **'A'** (Figure 5.19e).
7. Remove piece '**B**' and square the sides of the pins down on both sides.
8. Saw the sides of pins down to the gauge lines, *sawing on the waste side but to the line* (Figure 5.19f).
9. Pare sloping recesses on each shoulder of the sockets.
10. Chisel out wastes of pin sockets on piece **'A'**, chiselling from both sides (Figure 5.19g).
11. Assemble and saw off the waste ends (Figure 5.19).

The slope for the dovetail needs to be 1 in 6 for a softwood and 1 in 8 for a hardwood. You should be able to use dovetail templates instead of a bevel gauge to mark the slopes themselves (Figure 5.20).

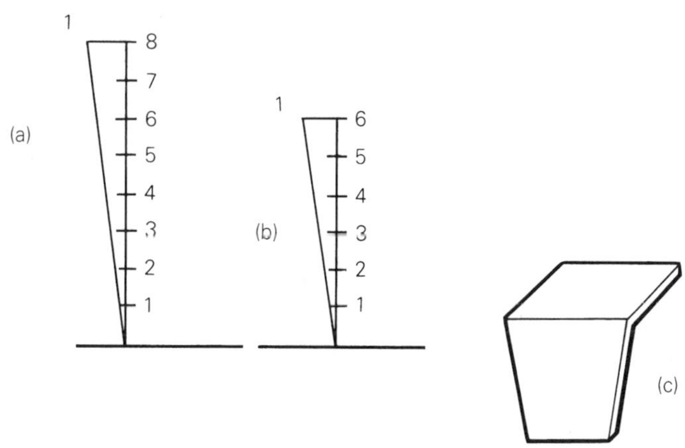

Figure 5.20 Correct slopes for dovetail joint: (a) hardwood, (b) softwood, (c) dovetail template.

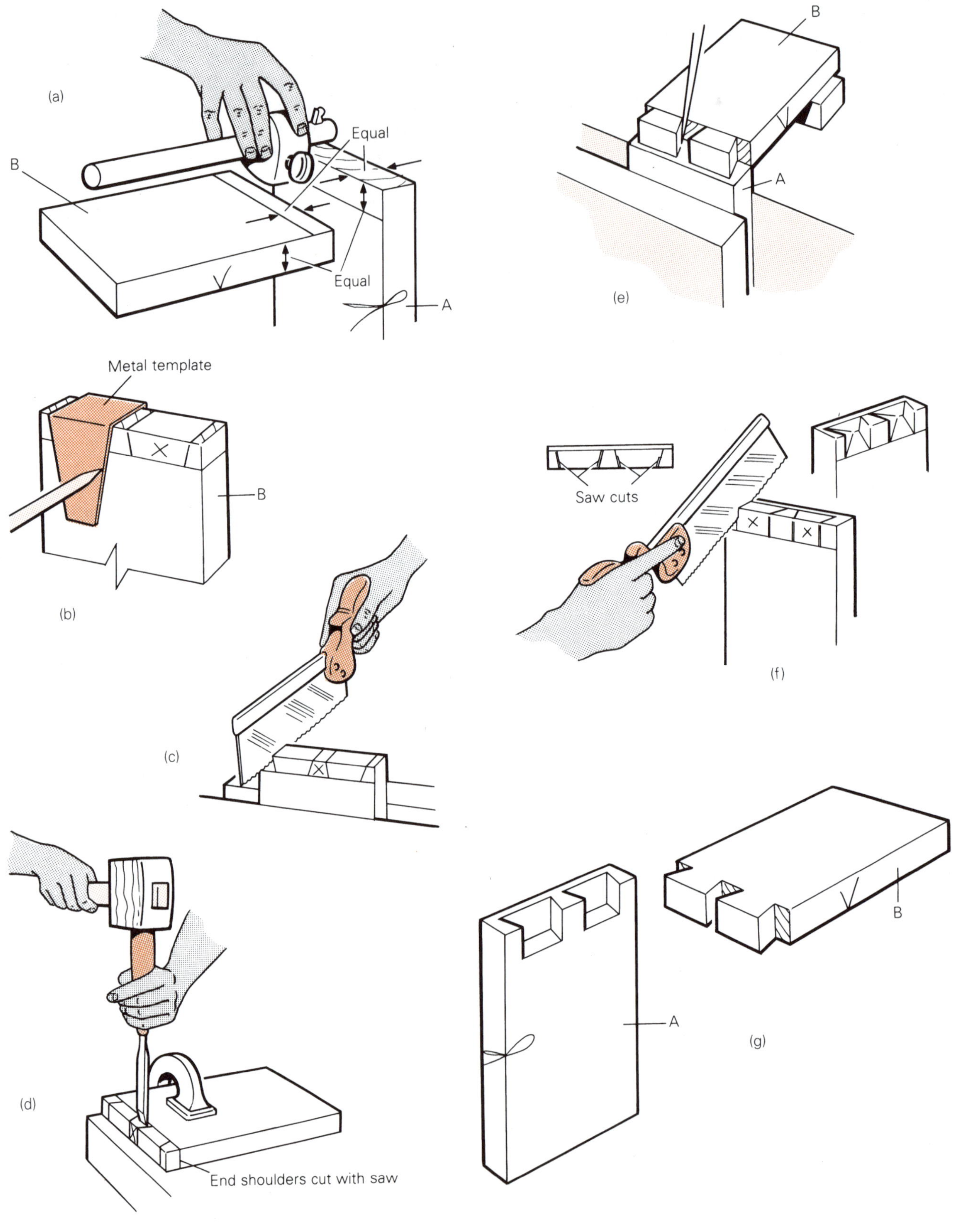

Figure 5.21 Preparing a lapped dovetail: (a) gauge piece 'A', (b) mark out dovetails on 'B', (c) cut dovetails, (d) chop out waste, (e) trace dovetails onto 'A', (f) saw pin sides, (g) joint ready to assemble.

To construct a lapped dovetail joint, follow these steps (Figure 5.21):

1. Prepare the pieces to the required width and thickness in one length and shoot the ends square to the face side.
2. Set the cutting gauge to three-quarters the thickness of the piece **'A'** and gauge from face side along the end of the piece, leaving a lap of one-quarter the thickness on the outside. Using the same setting, gauge from the edge across both sides and edges (Figure 5.21a).
3. Using a dovetail template or a sliding bevel set to the required gradient (see Figure 5.18), set out the dovetails on the face side of piece **'B'** (side piece). Mark 'X' on the wastes between the dovetails (Figure 5.21b).
4. Hold the piece in the vice and cut the sides of the dovetails, sawing on the waste side and to the lines (Figure 5.21c).
5. Remove the piece from the vice and cramp it on the bench. Chop out the waste of the dovetail sockets, chiselling vertically and horizontally alternately (Figure 5.21d).
6. Hold piece **'A'** (front piece) in the vice and place piece **'B'** over the end of piece **'A'**. Ensure that the face sides are on the inside and, with the face edges flush, trace the shape of the dovetail with a marker onto piece **'A'** (Figure 5.21e).
7. Saw the sides of the pins, sawing on the waste side to the line (Figure 5.21f).
8. Cramp piece **'A'** on the bench and chisel out the pin sockets, working from both sides.
9. Assemble the joint by a tap of the hand (Figure 5.21g).

Procedure for constructing a secret mitred dovetail joint (Figure 5.22):

1. Plane the pieces to the required width and thickness in one length.
2. Gauge for the position of joint (Figure 5.22a).
3. Cut the rebates at the end of both pieces (Figure 5.22b).
4. Set out the pins and saw the sides of the pins (Figure 5.22c).
5. Chisel out the wastes of the pin sockets (Figure 5.22d).
6. Set out or place the pins over the position for the dovetails and trace the shape of the pin sockets with a sharp pencil or knife (Figure 5.22e).
7. Chisel out the dovetail sockets (Figure 5.22f).
8. Cut and pare out the mitre on the laps (Figure 5.22g).
9. Assemble the joint.

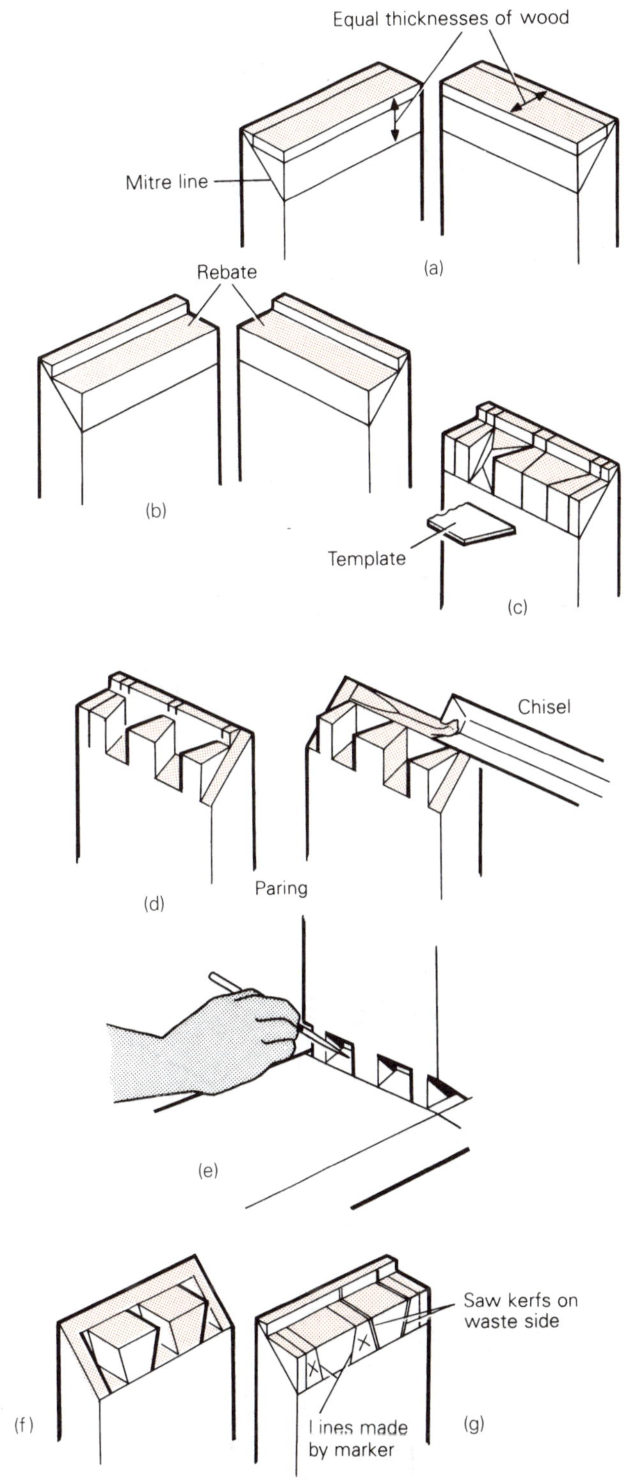

Figure 5.22 Preparing a secret mitred dovetail joint.

Framing joints

Framing joints (Figure 5.23) are those that are commonly used in forming frame-like

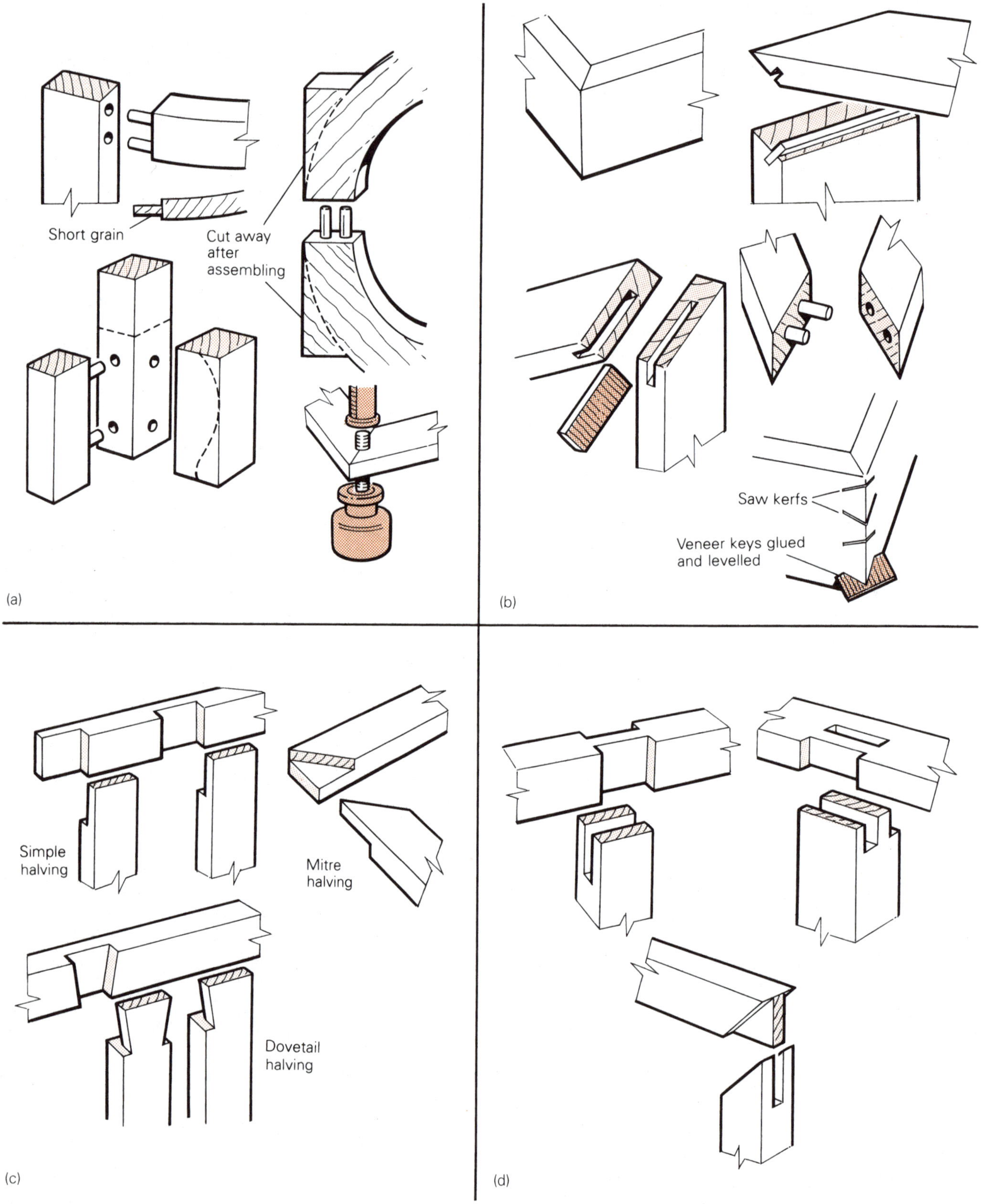

Figure 5.23 Framing joints: (a) dowelled joints, (b) mitred joints, (c) halving joints, (d) bridled joints.

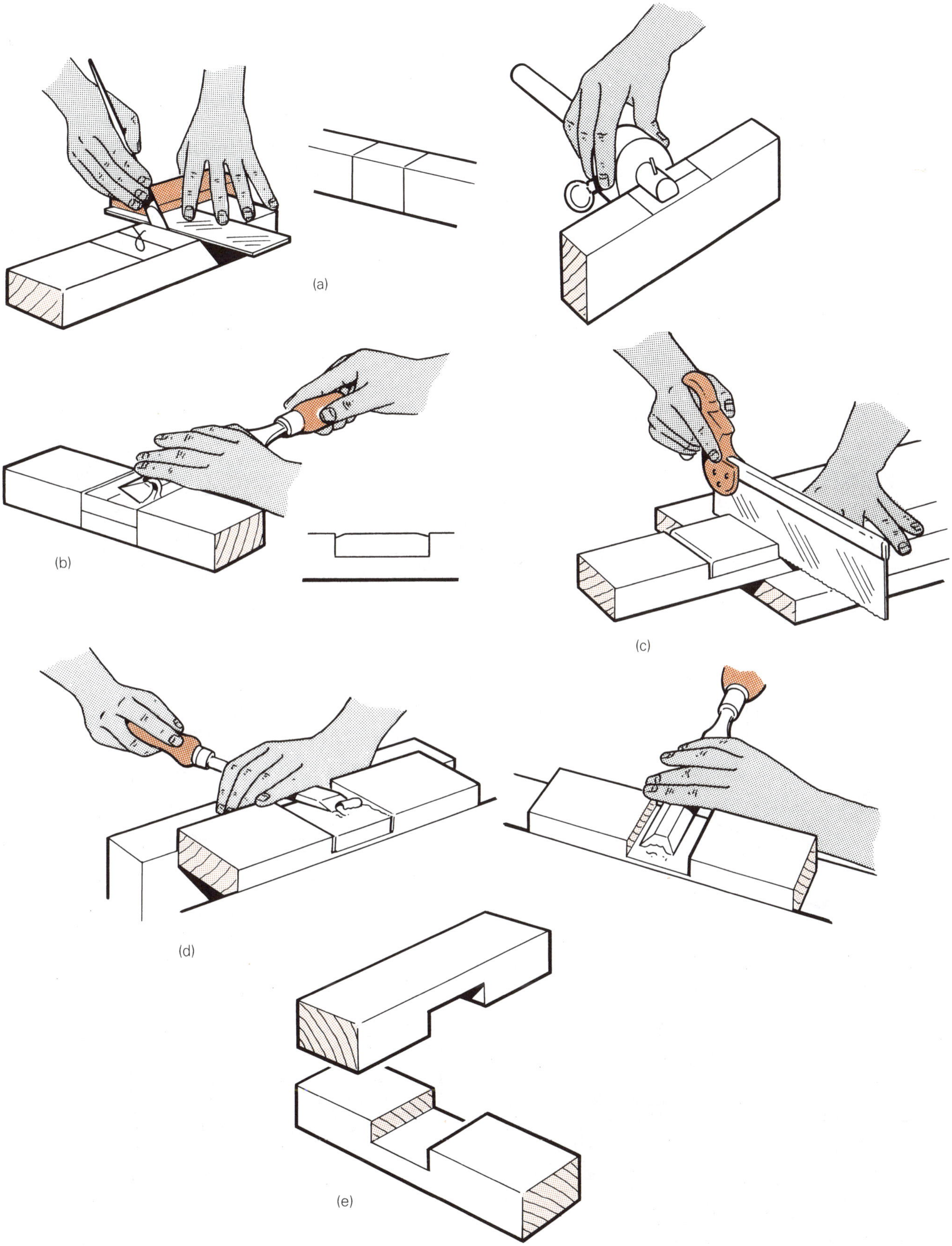

Figure 5.24 Preparing a cross-halving joint: (a) mark out, (b) make a sloping groove, (c) cut the sides, (d) remove the waste, (e) assemble.

constructions where the pieces meet end to end, with their edges at right angles. Those chiefly used are: dowelled joints, mitred joints, halving joints, bridled joints, and mortise and tenon joints.

The **dowelled joint** (Figure 5.23a) is a butt joint strengthened by dowels. It is used in cheap table and chair construction for joining rails to legs.

The **mitred joint** (Figure 5.23b) is an improved butt joint. The ends of the meeting pieces are cut at 45° to one another (mitred) to prevent the end grains showing. It is chiefly used for picture frames and architrave.

Halving joints (Figure 5.23c) as the name indicates, are halved in thickness the ends where the pieces meet or cross each other, so that when assembled the faces are flush. Types of halving joint are: the **corner halving joint**, used for frames of cheap flush doors, and other light frames; the **mitred corner halving joint**, used for picture frames; the **tee halving joint**, used for frames where a rail is joined to a stile away from the end; the **dovetailed tee halving joint**, used for connecting cross rails of frames to resist outside strain; and the **cross halving joint**, (Figure 5.24) used at places where cross or diagonal rails must meet and cross each other with their faces flush.

Bridled joints (Figure 5.23d and Figure 5.25) are similar to mortise and tenon joints except that

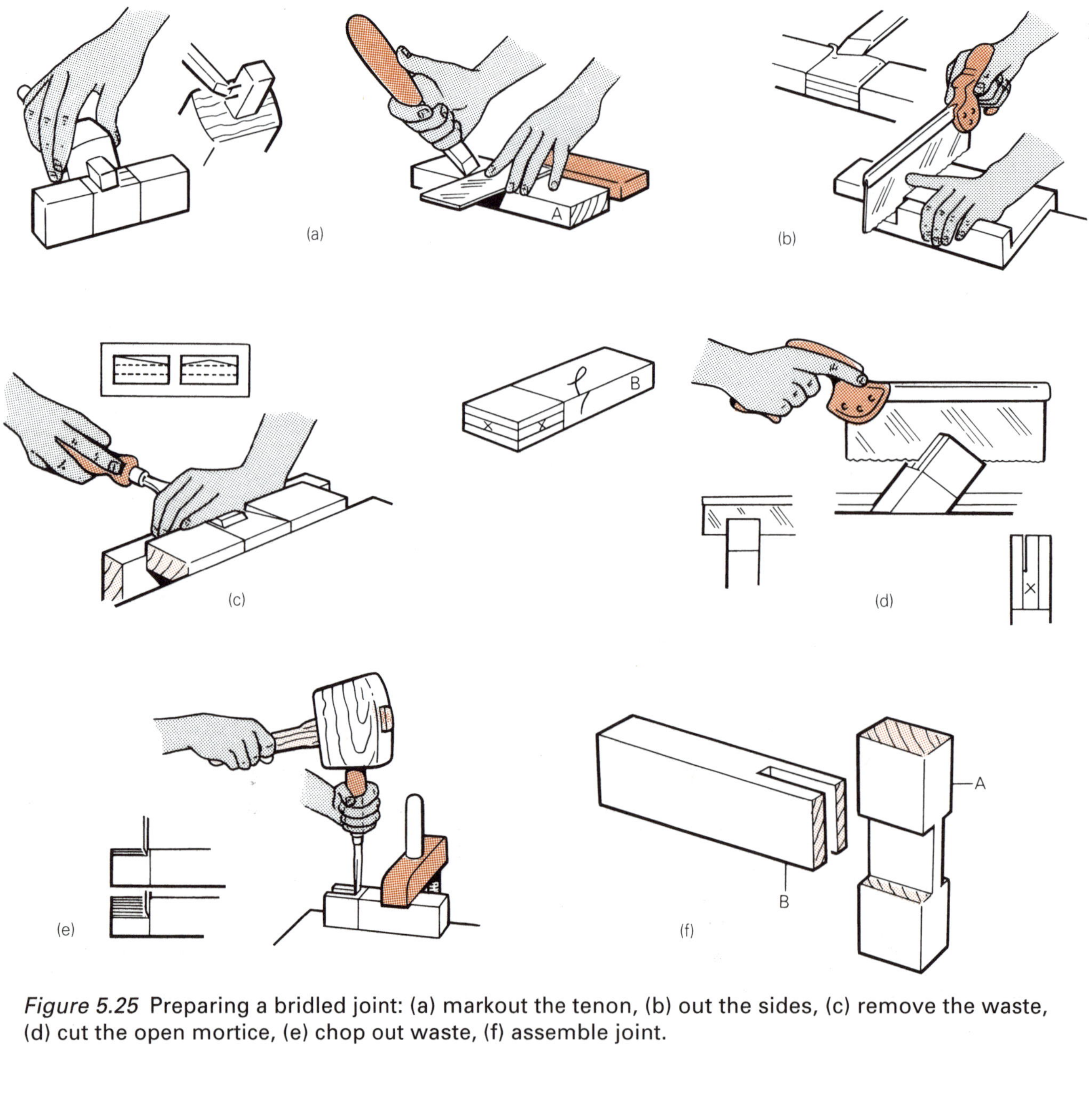

Figure 5.25 Preparing a bridled joint: (a) markout the tenon, (b) out the sides, (c) remove the waste, (d) cut the open mortice, (e) chop out waste, (f) assemble joint.

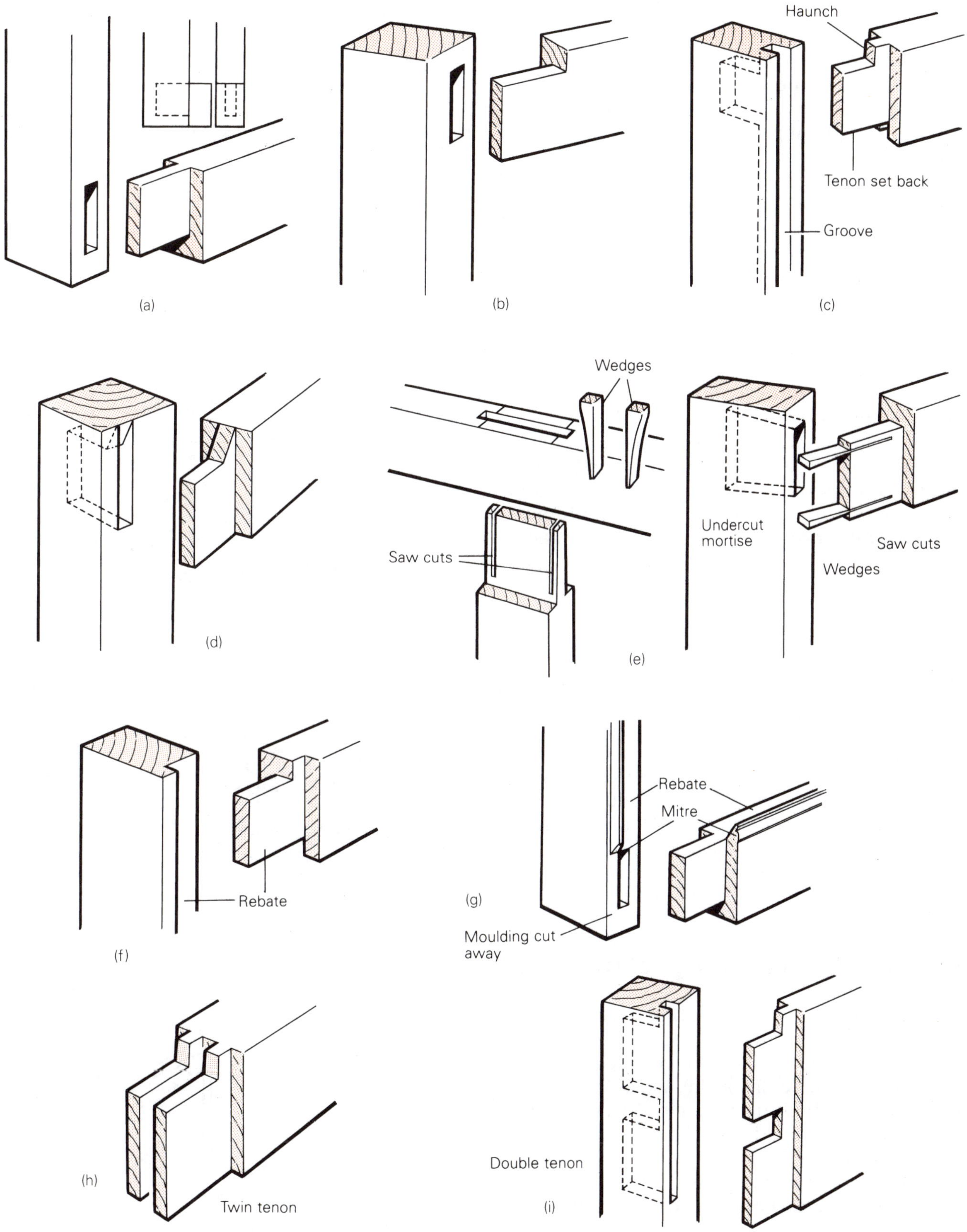

Figure 5.26 Types of mortise and tenon joint: (a) common, (b) barefaced, (c) haunched, (d) concealed haunch, (e) wedged, (f) long and short shouldered, (g) moulded, (h) twin, (i) double.

the supposed mortise is cut opened at the top thereby exposing the tenon and pins at the top of the joint. For this reason, a bridle joint is sometimes called an **open mortise joint**. The three common types are the **tee bridled joint**, used where one meeting piece connects with the other some distance away from the end; the **corner bridled joint**, used at the corners of frames; and the **mitred corner bridled joint**, used in place of a mitred halving joint where a stronger joint is desirable, as in mirror frames.

Mortise and tenon joints (Figure 5.26) are the strongest of all the frame joints and are the most commonly used in panelled frames for doors, framed carcases for cabinets, and for joining rails to table and chair legs. The most common forms are: the **through** or **common mortise and tenon joint**, used where one of the meeting pieces connects with the other some distance away from the end; the **haunched mortise and tenon joint**, used where the meeting pieces join at the ends; the **long and short shouldered mortise and tenon joint**, used where the pieces are rebated to receive panels, glasses, etc.; the **barefaced mortise and tenon joint**, used when thinner rails are to be joined to thicker stiles or legs; the **stub mortise and tenon joint**, used where it is undesirable for the end grain of the tenon to be seen as in high-class cabinet work; the **double tenon**, used for wide rails where a wide tenon would shrink and weaken the stile; the **twin tenon**, suitable for thick rails where a very strong joint is needed; and the **compound double tenon**, applicable for wide and thick rails as in the middle rails of some doors where a mortise lock is fixed into the stile.

The procedure for constructing a mortise and tenon has been described on page 49.

Forms of carcase construction

Carcase is a term used for the box-like portion of a cabinet or chest of drawers when the drawers, doors and other fittings are not yet fixed. It is the framework or skeleton of the actual cabinet. A carcase may be built up in three common forms; framed carcase, solid end carcase and solid carcase (Figure 5.27).

The **framed carcase** consists of two forms: the **simple framed carcase** and **the framed panelled carcase.** The simple framed carcase consists of a frame-like structure of narrow stiles and rails, covered by a sheet of plywood (6 mm ply) or hardboard. In the case of the framed panelled carcase, the sides (ends) are framed panels (grooved-in), joined at the top and bottom with cross rails lap dovetailed into the ends of the stiles. The top may be of solid wood, multi-ply or particle boards, and secured with screws through the top rails. The **solid end carcase** as the name implies, has ends which may be a solid wood, multi-ply, or particles board with cross rails that are lap dovetailed at the top (both front and back) and bottom (again, both front and back). The top is covered with plywood (6 mm) by screwed onto the top rails.

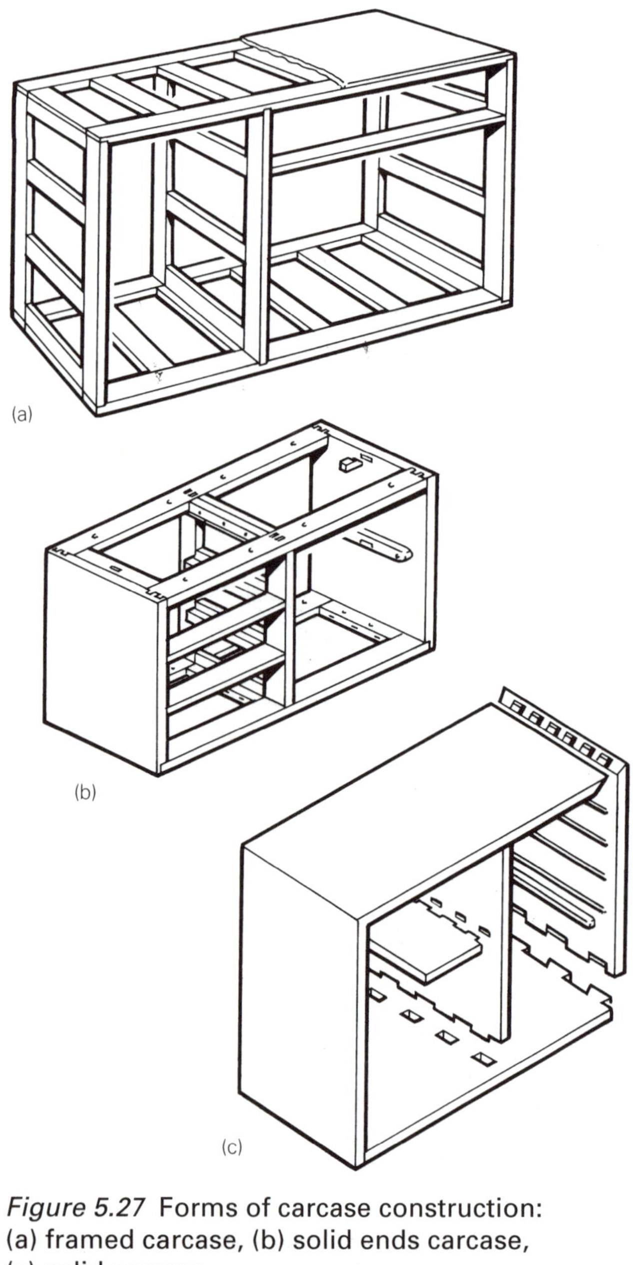
(a)
(b)
(c)

Figure 5.27 Forms of carcase construction: (a) framed carcase, (b) solid ends carcase, (c) solid carcase.

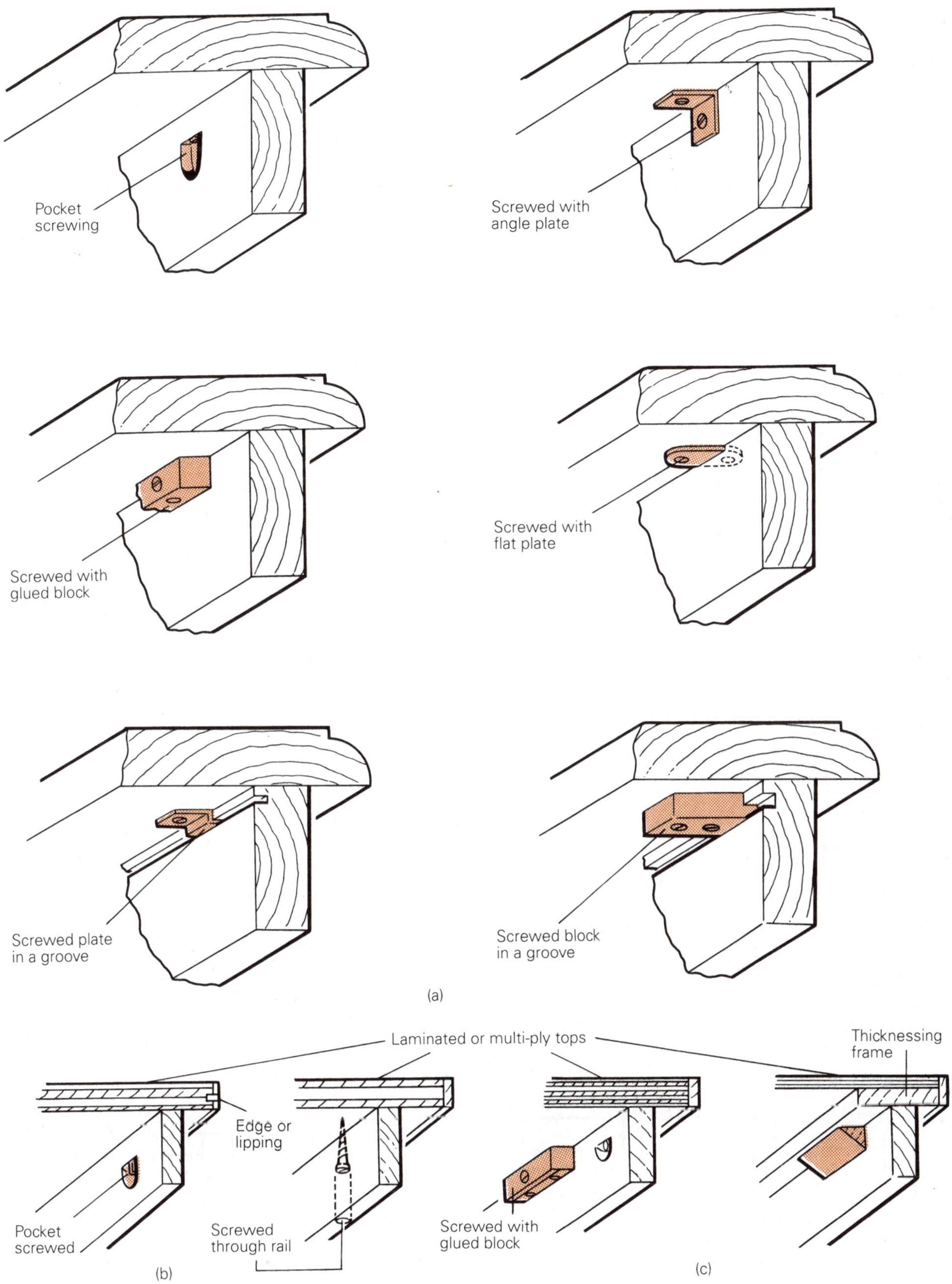

Figure 5.28 Methods for fixing a table top: (a) solid top, (b) laminated or multi-ply top, (c) 3-ply top.

A **solid carcase** consists of solid ends, top and bottom. The shelves and partitions may either be solid or framed.

Joints that are predominantly used for carcase construction include the angle joints and frame joints and they have been discussed earlier in the chapter. When you use manufactured boards to construct a carcase you should conceal the unattractive and porous end grains. You can use wood edgings (strip of wood), flexible extruded plastic edgings, metal edgings, veneer or formica. The method of concealing the end grains of a carcase is known as edging and the process of edging is called **lipping**. Figure 5.28 indicates methods of securing table tops. Figure 5.29 shows the appropriate joints for manufactured boards, Figure 5.30 shows the methods for concealing the ends or edges of manufactured boards, The method used will depend upon the construction of the frame of the table and the material used for the top, i.e., either solid wood or manufactured board. If solid board is used provision should be made to allow for swelling and shrinkage due to changings in the humidity of the atmosphere. Figure 5.31 illustrates the methods for fixing shelves while Figure 5.32 deals with drawer construction and fitting.

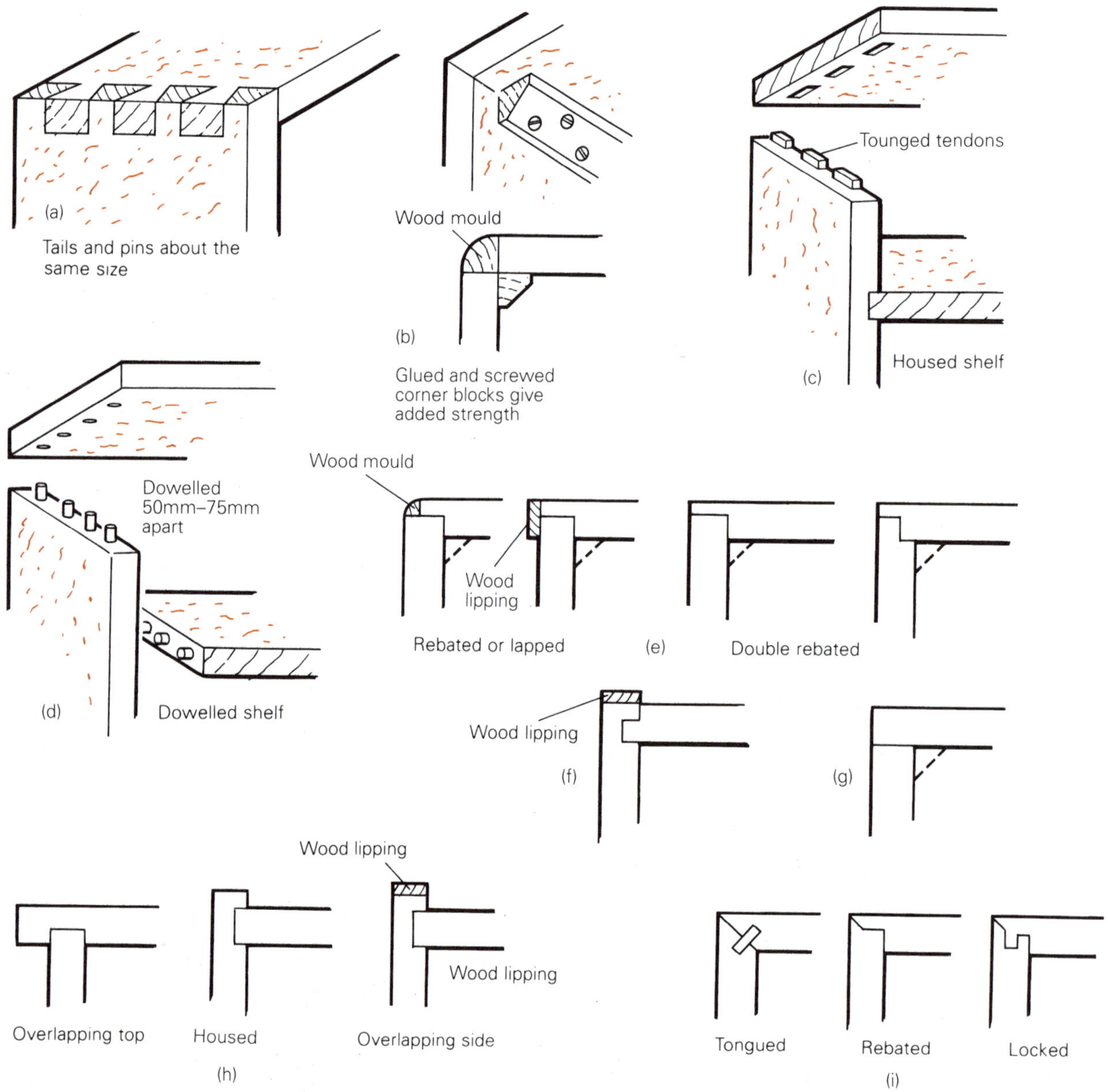

Figure 5.29 Suitable joints for manufactured boards: (a) dovetailed, (b) glued and screwed, (c) tongued tenons, (d) dowelled, (e) rebated, (f) tongued, (g) butt, (h) housed, (i) mitred.

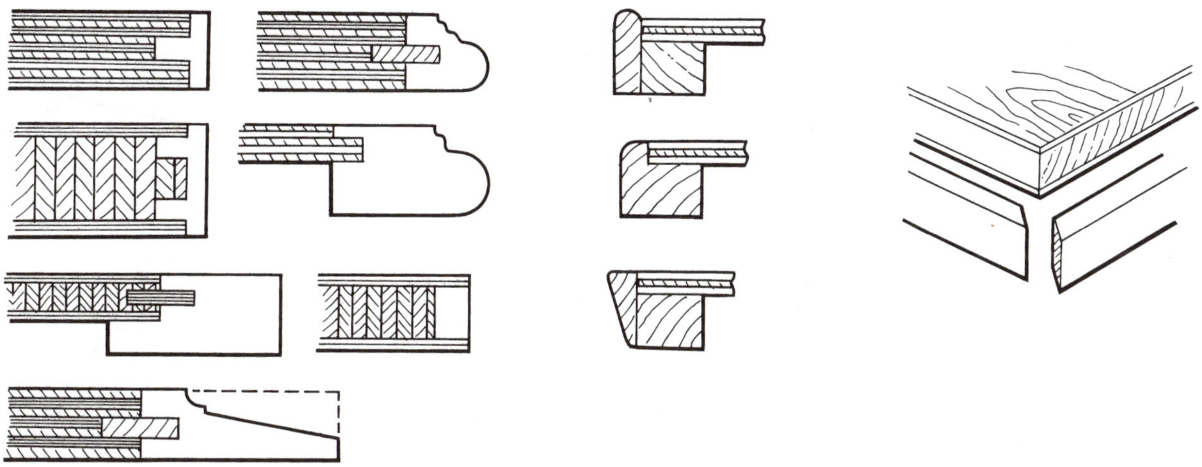
(a)

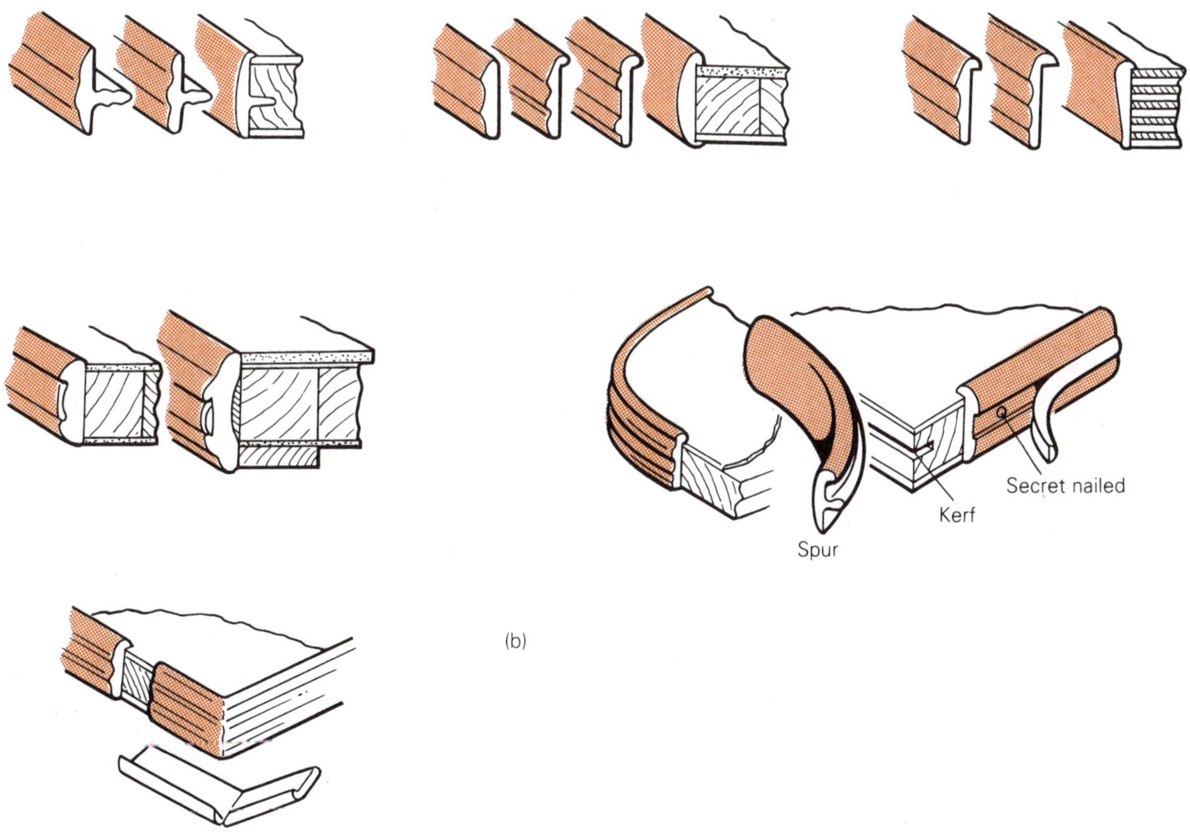

(b)

Figure 5.30 Types of edging: (a) solid wood edgings of manufactured boards, (b) flexible extruded plastic and metal edgings.

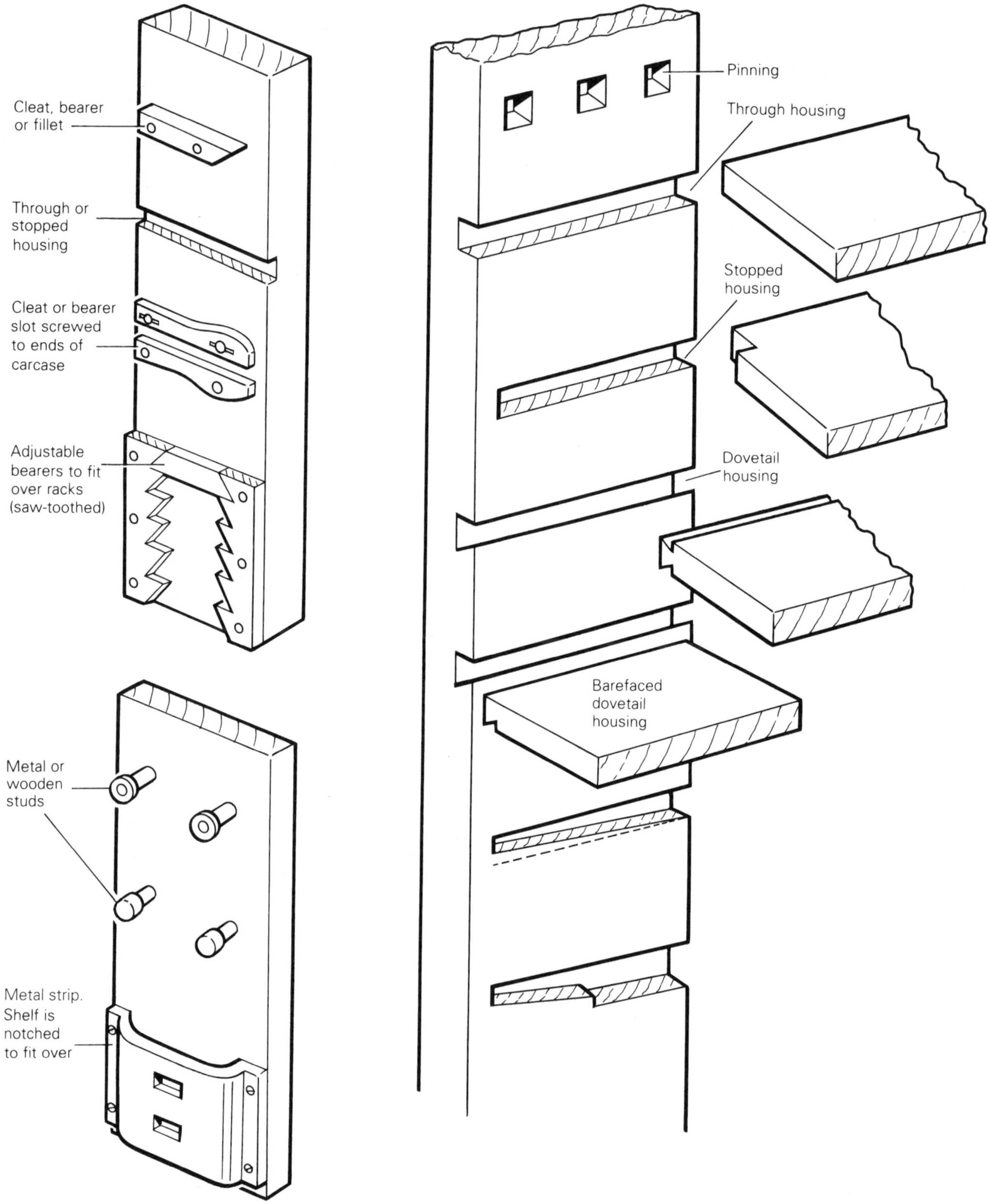

Figure 5.31 Methods of fixing shelves.

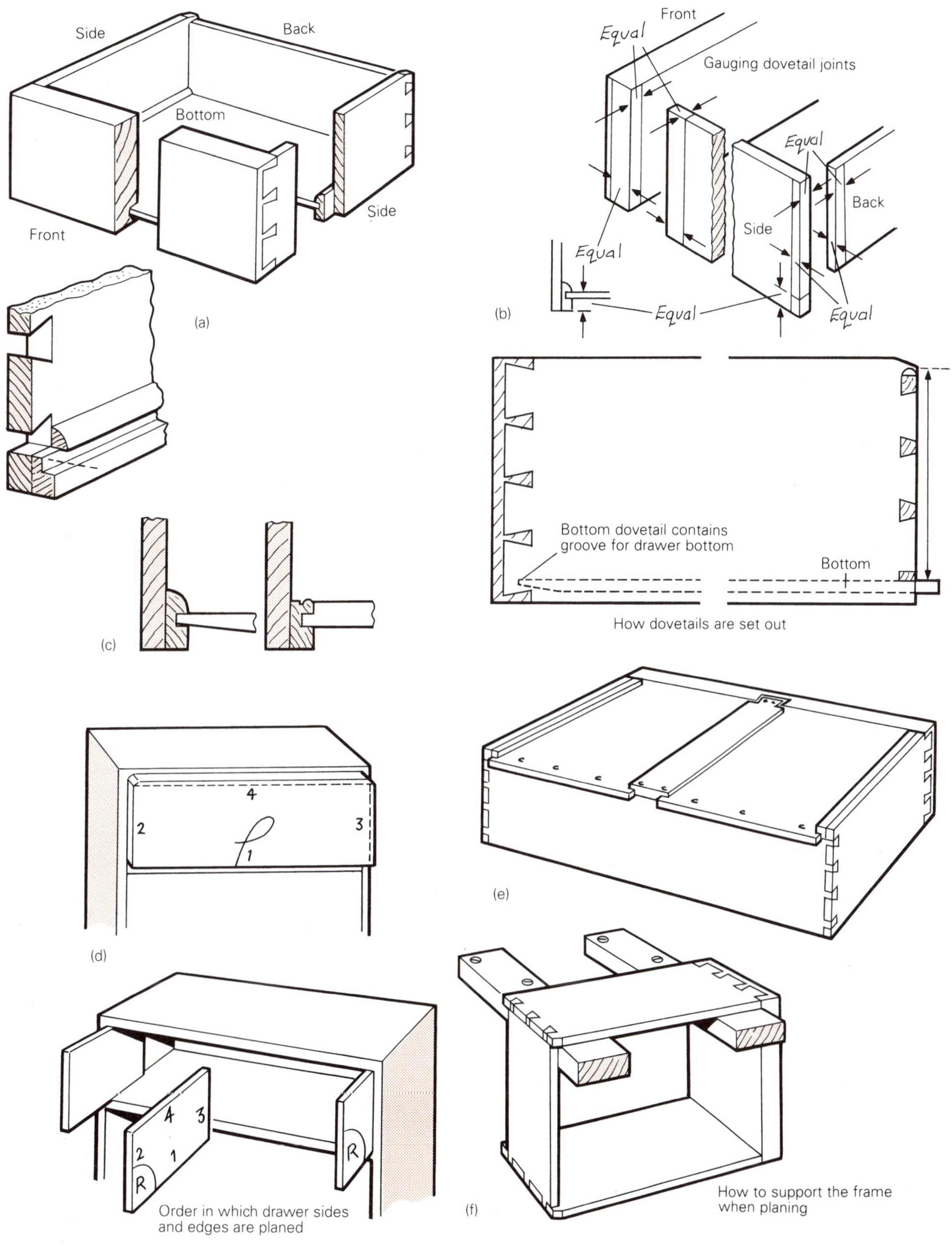

Figure 5.32 Drawer construction: (a) form of drawer construction, (b) gauging and setting out dovetail joints, (c) slips used for drawer bottom, (d) fitting the front, (e) how to fit the bottom of a long drawer, (f) planing the sides.

CHECK YOUR UNDERSTANDING

- Planing a piece of wood involves taking off shaving from the wood, using either a hand plane or a machine plane.
- A planed piece of board is known as a dressed board.
- A recess made across the grain of a piece of board is called a trench whilst a recess along the grain is known as groove. The processes of making these recesses are known as trenching and grooving respectively.
- A rebate is a rectangular recess along the edge or across the end of a piece of board. The act of doing so is known as rebating.
- When the end of a piece of board (either the full thickness or rebated) is fitted into a trench in another board, the joint formed is called a housing joint. But when pins are formed at the end of board and fitted in corresponding sockets on the other board, then the joint so formed is called a pinned joint.
- The terms setting out and marking out refer to different processes but they are quite often confused to mean the same. Setting out is the process of laying out, or drawing to full size, the joints to be formed on a piece of board or plywood. This serves as a reference for marking out. The board or plywood on which the joints are set out is called a working rod. Marking out, on the other hand, is the process of locating or transferring the correct positions of the joints from the working rod on to the workpieces.
- The three classes of woodwork joints are widening joints, angle joints and framing joints.
- A carcase is the box-like structure of a cabinet with or without the drawers, doors and other fittings.
- A carcase can be constructed by using a framed structure for the sides, top and bottom; solid boards/plywood for the sides and framed top and bottom or solid boards/plywood for the sides, top and bottom.
- The end grains of solid or manufactured boards used for constructing a carcase may be concealed by using solid wood edgings or flexible plastic or metal edging.

REVISION EXERCISES AND QUESTIONS

1. Explain the term 'dressed' board and list out briefly the steps for dressing a board.
2. Explain the significance of face markings when marking out.
3. How will you avoid splitting when you are to rebate across the grain of a workpiece?
4. Name four forms of mortise and tenon joints and give one example where each is used.
5. If you are given two pieces of board to be joined edge to edge, describe six methods by which you can join them and name the term given to each method.
6. Supposing you are to construct a carcase, given solid boards, list the joints you will use for constructing: the top (if the end grains are not to be shown), the bottom, the shelves, and the partitions.

6 Fasteners, adhesives and hardware

Introduction

The chapter deals with materials used for securing and bonding together wooden structures such as frame-works and members of a carcase. The methods of securing or fastening members of a structure may be considered as *temporary* if the members that are secured together can be taken apart without causing damage to them. Such methods include nailing and screwing. The use of adhesives for bonding members of a structure or carcase is *permanent*, as it is likely damage will be caused to the members if they are separated. The process of bonding is known as gluing. The various hardware fastenings or fittings used in woodwork are also discussed here.

Common fasteners

Nails (Figure 6.1) are generally made from mild steel wire or plate and consist of three parts: a **head**, **shank** and **point**. Some nails are coated with zinc or cadmium to prevent them from rusting, particularly when used on structures that are exposed to the weather. Other metals, such as brass, copper or aluminium, may be used for making special purpose nails, such as are used in boat building where they are liable to be in contact with water most of the time. Nails are classified according to shape of the head, shape of the cross section, and the uses to which they are put.

Roofing nails, sometimes called **springheads** have dome-shaped heads with either round or square twisted shanks. The length ranges from 44 mm to 46 mm. Because of their exposure to the weather, they are often **galvanised** (coated with zinc) to prevent rusting. As the name implies, these nails are chiefly used in roofing buildings.

Flat head nails are of two types: the **round shank** (sometimes called a **French nail**), with a large flat head and the **square shank**, that is twisted sometimes to provide greater holding power. It has the same shape of head as the round shank type. The head is large and flat and that prevents it from being pulled through the piece of board. These nails are generally used for packing cases, and box construction.

The **diamond head nail** has a round shank and a diamond-shaped head, and its length ranges from 18 mm to 150 mm. It is used for fixing flooring boards, lining boards, packing case or box construction and general constructional work.

The **jolt head nail** or **bullet head**, has a round shank and a round head that enables it to be punched into the board neatly. Due to its ability to enter cleanly into the board is sometimes called the **lost head**. The length ranges from 18 mm to 150 mm and it is used for the same purposes as the diamond head.

The **panel pin** is a thin round wire nail with an inverted cone-shaped head that enters the board easily and cleanly. The length ranges from 12 mm to 50 mm. Its uses include fixing plywood to frames, cabinet backs, and mouldings.

The **veneer pin** is a very thin wire nail, sometimes without a head, used to hold veneers in position during setting out and cutting or gluing. Its other uses include fixing very small mouldings and the backs of picture frames.

The **upholstery nail** is of a very thin round shank with a dome-shaped head. It may be made from brass or plated steel, according to the work for which it is required. It is used in upholstery

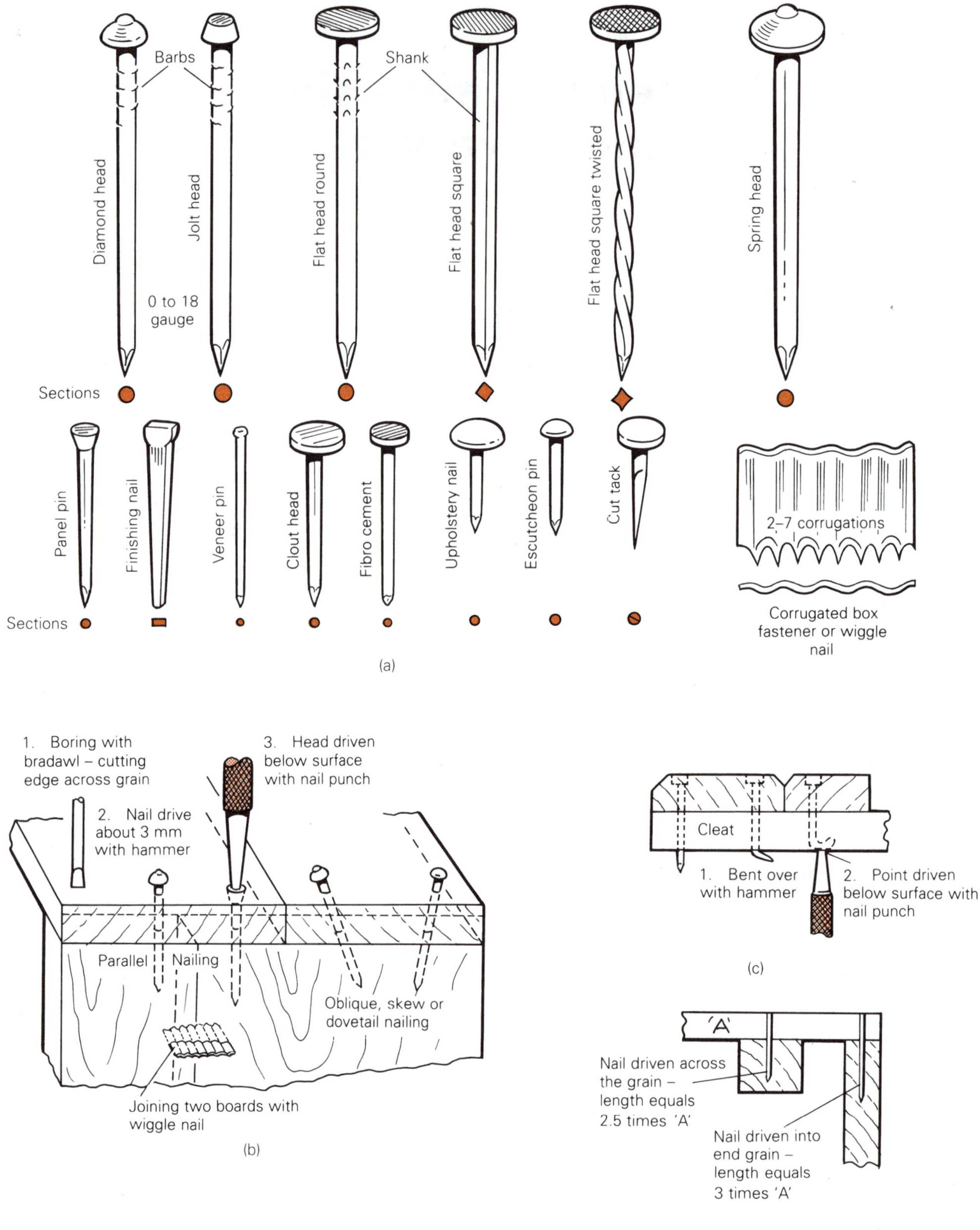

Figure 6.1 Nails and nailing: (a) common types of nails, (b) methods of inserting nails, (c) clinching nails.

work generally for fixing tapestry, leather and other materials to wood.

The **escutcheon pin** is a small brass or nickel-plated nail, having a small dome-shaped head. Its general use is for fixing small or ornamental plates, such as the escutcheon plates that protect keyholes from being enlarged through usage.

The **wiggle nail**, or **corrugated box fastener**, is often made from corrugated steel plate. The size varies from 6 mm to 25 mm. It is chiefly used in the construction of widening joints to strengthen the edge-to-edge joint, as may be seen in the sides of packing cases. It is also used for joining framed cores of cheap flush panel doors, and even in framed carcase construction.

In order to use nails satisfactorily, it is essential that you procure the right type, taking into account the factors that determine the holding power, and then apply the correct technique to drive in the nail for the job on hand.

> When you are purchasing nails, particularly when ordering, state the weight required (nails are sold by weight); the length of the nails (this is the length that goes into the board); the gauge (thickness); the type; the metal of which it is made; and the type of finish.

The ability of the nail to hold the pieces together depends upon the following factors:

- how firmly the fibres of the board can grip against the shank of the nail – closed-grained hardwood has greater holding power than open-grained softwood;
- whether driven across the grain or along the grain (you should use a length of 2.5 times the thickness of the top piece for nailing across the grain and 3.5 times the thickness for nailing along the grain);
- the type of shank – for example, square nails with twisted shanks have greater holding power than round nails;
- the angle of nailing – oblique or dovetail nailing has a better hold;
- the depth of penetration – a nail that projects through both pieces and is clinched along the grain has greater strength;
- the size of the head – a large flathead nail that is unlikely to pull through the top board gives a stronger hold.

Consider the following hints when you are nailing:

1. To avoid dents on the surface of the top piece, hammer the nail until the head is about 2 mm above the surface, drive it just below with a nail punch and fill the hole later with putty (this permits the surface to be planed off).
2. Strike the head of the nail directly, as a blow that slips off the head may cause the nail to bend. Always ensure that any traces of glue, oil or dirt that are likely to cause the hammer to slip off the head are removed before you begin the nailing operation.
3. Drill or bore a pilot hole (a hole slightly less than the gauge of the nail) in hardwood boards to prevent them from splitting or the nail from bending.
4. When nailing in chain (in other words when using a series of nails in a row), *stagger* them (so they are not in a straight line) to prevent the boards from splitting.
5. When you are removing a nail from a board place a waste piece of board under the jaw of the pincers or the claw hammer to protect the surface of the top piece and provide a greater leverage (see Figure 6.2).

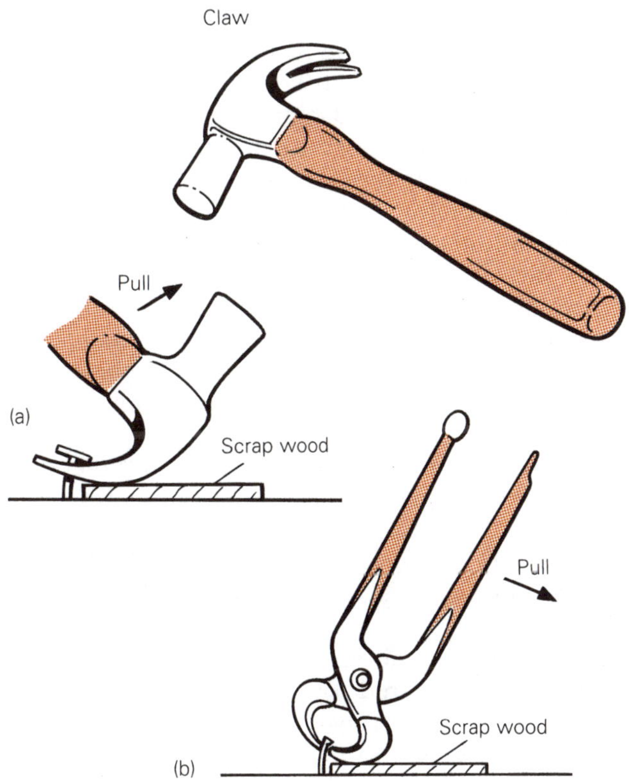

Figure 6.2 Extracting a nail: (a) using a claw hammer, (b) using pincers.

A **screw** (Figure 6.3) is a fastening device with a threaded shank that cuts into the piece of wood being screwed into. It is commonly made from mild steel, brass or copper. The essential parts are: the head, shank, core, thread and point. The **head** is of various shapes that determines the name of the screw. It is either straight slotted or cross slotted to take a screw driver. The **shank** is the portion below the head and about two-thirds of it contains the thread. The **thread** is the spiral profile around the core that draws the screw into the piece of wood. Because the thread cuts itself into the piece of wood, its holding power is greater than that of a nail. Generally, screws are used for securing members of wooden pieces together, strengthening joints and fixing fittings to carcases. The commonly used woodwork screws are: the countersunk head, round head, raised head and square head.

The **countersunk screw** has a tapered head with a flat top that fits flush with the surface of the work-piece or fitting. It is chiefly used for fixing chair seats, table tops, locks, hinges and other metal fittings.

The **round head screw**, as the name implies, has a rounded head that projects over the surface of the work-piece. It is used for securing parts of a job together, and also for fixing metal fittings at places where the visible heads serves as an ornament. It is quite often used for holding together pieces that are thin and cannot be countersunk.

The **raised head screw** also known as an **oval head**, is a combination of countersunk and round head. The combination enables it to combine the strong holding power of the countersunk screw with the decorative quality of the round head screw. It is used for fixing thick metal fittings to carcases where both strength and ornamentation are needed. Another area of use is in the motor body building, for fixing metal panels, cover strips and mouldings.

The **square head screw**, known as the **coach screw**, has a square head, and is a much stronger screw. Its chief use is in structural construction, such as bridge construction, coach and carriage constructions. It is also used for fixing gate hinges.

In order that the screw must hold firmly the pieces being secured together, it is essential that the head of the screw should always be larger than the size of the clearance hole in the top piece and the thread must cut its way deep into the fibres in the bottom piece. Care must therefore be taken when drilling or boring a pilot hole for the screw thread in the bottom piece to leave enough material for the thread to cut sufficiently into the fibres. For this reason, it is not necessary to provide a pilot hole in a softwood as the thread on the screw can easily cut its way into the fibres. It is also important that it turns freely in the clearance hole in the top piece so that the two pieces are drawn together tightly. Always use a screw-driver that fits well down in the slot in the head of the screw and make sure it is not wider than the screw head. A loose fit may damage the slot.

> When you are buying (ordering) screws you give the following information; the number (quantity) required; the length (the amount of the screw inserted into the board); the gauge number (thickness or diameter of the shank, indicated by a standard wire gauge number); the type (for example, round head); the metal; and the finish. For example, when you are ordering a No. 8 gauge, 100 mm brass countersunk screw, you should state: "100 × 8 brass countersunk".

> Never drive in a screw with a hammer, as this may break off the fibres to the thread diameter of the screw thereby preventing the thread from cutting its way in the fibres, reducing the holding power.

Wood adhesives

Wood adhesives are bonding substances used for fixing wooden joints or pieces permanently together. They are commonly known as **glues**. They can be grouped into three varieties: protein adhesives, synthetic resin adhesives and contact adhesives. The choice for the type to use depends greatly on their individual properties.

The common **protein adhesives** are animal glues, casein glue, vegetable glue and blood albumin glue.

Animal glue, as the name implies, is made from parts of an animal such as the hide trimmings, fleshing of the skin, bones, sinews, intestines and cartilage. It is non-toxic; non-

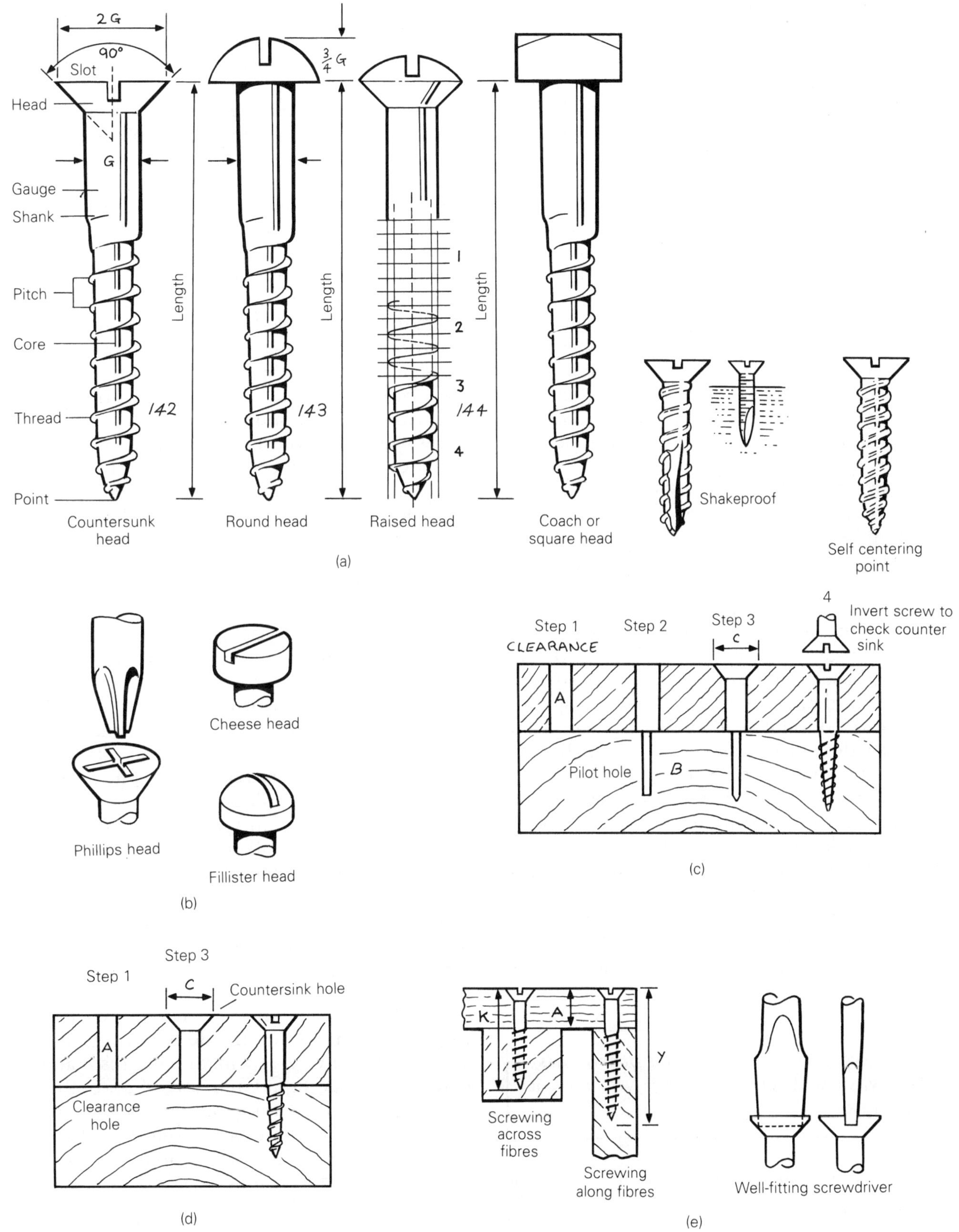

Figure 6.3 Screws and screwing: (a) screws, (b) screwheads, (c) screwing in hardwood, (d) screwing in softwood, (e) screwing in fibres.

staining and has good gap filling ability, but its water resistance and heat resistance are low. For satisfactory application, it must be used hot and the work must be assembled and cramped immediately, before it sets or chills. In countries where the weather can be cold, it is advisable to warm the pieces to be glued before the glue is applied.

Prepare the right quantity for the job, as freshly made glue has a stronger holding power. Make sure the glue has the right consistency (thickness): when you dip a brush into the glue it should run down freely from the brush without breaking into drops. A boiled glue may weaken the consistency, thereby causing it to lose its holding power. Keep the parts to be glued clean and dry. Apply glue to both parts. Note that a thick glue cannot penetrate into the fibres or pores of the wood and will produce a weak joint. Likewise a thin glue may not form enough bond and the joint will become 'starved' (dry). Animal glue is mostly supplied and sold by weight in the form of sheets or cakes, or else powdered or granulated. Those manufactured from fish offal are in liquid form and sold in tubes. The other types can be treated with acids so that they remain in liquid form and are kept in sealed containers, ready for use.

Casein glue is manufactured in powder form from the curds of skimmed milk and **alkalis** (chemical bases that dissolve in water) such as sodium fluoride, borax and sodium phosphate. They are added to make it soluble in water. Lime (calcium hydroxide) is also added to enable the glued joint to withstand water. Casein glue is mixed and used cold, permitting up to 40 minutes assembly time; it can be applied either by hand or machine; is fairly water and heat resistant but not waterproof; stains hardwoods; dulls cutting edges, has good gap-filling properties; dries and sets by evaporation of water due to chemical action. The stain may be removed by washing the surface immediately with a solution of oxalic acid in water.

> ▲ **Take care**: this solution is ***poisonous***. Wash with clean water if you spilt it on your skin.

When you are preparing casein glue ready for use, add the powder (casein) to cold water in the correct proportion as stated on the label; stir the mixture continually until it thickens; leave it for about 10 to 20 minutes to enable the chemicals to react with the water; then stir again for a few minutes before you apply. Casein glue is chiefly used for manufacturing plywood, laminated boards, and caul veneering.

Vegetable protein glues are manufactured from the residue obtained after oil is extracted from vegetables such as soya beans, peanuts, cotton seed etc. Chemicals are added to improve qualities such as water resistance. These glues are mainly used for manufacturing water resistant plywoods.

Blood albumin glues are manufactured from albumin obtained from animal blood. Certain chemicals are added. They are used in the plywood industry, mainly for manufacturing water resistant plywoods. Special machinery is required for heating and pressing the glued work.

Synthetic resin adhesives are the commonest glues used at the present time. There are different varieties and the widely used ones for bonding wooden pieces are: urea formaldehyde, melamine formaldehyde, phenol formaldehyde, resorcinol formaldehyde, epoxy and polyvinyl acetate (P.V.A.). All these varieties, except the epoxy and polyvinyl acetate, set by a chemical reaction called polycondensation.

Urea formaldehyde resin (U.F.) glues are supplied in three forms: **viscous syrup resin** with either powder or liquid hardener, having a storage time or **shelf life** of three to six months (the time it can be stored between manufacture and mixing for use before it loses its holding power); **white powder resin** glues with a powder or liquid hardener, having a shelf-life of one to two years; and **powder resin** with a powder hardener mixed ready for use, and having a shelf life time of one to two years. U.F. glues have the following characteristics: they are insoluble in the common organic solvents, resistant to mild acids, alkalis and water (but not completely waterproof); non-staining, fairly heat resistant, and resistant to fungi. Applied cold, they can set by the addition of an acid known as a **catalyst**, **accelerator** or **hardener**, They may be applied in two ways: by mixing the resin and the hardener as one liquid and spread on both parts of the work; or by applying the resin (liquid) to one part and the hardener (liquid) to the other part, then assembling and cramping to set. They are mainly used for caul or cold press veneering, plywood and particle board manufacture. They are also suitable for any assembly gluing.

Melamine formaldehyde resin (M.F.) glues are supplied in two parts; white powdered resin and a liquid or powder hardener. They have a strong waterproof and heatproof bond. They set by polycondensation and are chiefly used for waterproof plywood, veneering, and waterproof joints. They are also used in the manufacture of plastic laminates.

Phenol formaldehyde resin (P.F.) glues are manufactured and supplied in two parts; powder (brownish) or liquid resin and a powder filler. They are mixed with water before use. Phenol formaldehyde resin glues set by polycondensation at a high temperature (210–280°F) under pressure (clamping). They are chiefly used for bonding waterproof plywood, particle board, fibre board and metals.

Resorcinol formaldehyde and **resorcinol phenol-formaldehyde** glues are manufactured and supplied in two parts; reddish-brown liquid resin and liquid or powder hardener. They are non-staining; waterproof; resistant to solvents, heat, moulds; set cold or hot under light pressure. A temperature between 70°F and 150°F may be enough for proper setting. They are mainly used for bonding waterproof plywood and laminates (formica) to wood.

Epoxy resin glue is manufactured and supplied in two parts: an amber-coloured viscous liquid resin and a viscous liquid hardener. It is waterproof, heatproof, non-staining, shock resistant, gap filling and has a high chemical resistance. It gives a very strong bond between most materials, except thermo-plastics. It sets at a room temperature of 70°F by a process known as **polymerization** (cross-linking of molecules). It allows a period of 1 hour for assembling and requires up to 48 hours clamping. For a maximum strength, a job may remain in a cramp for 1 week. It is used for bonding wood to wood, wood to metal, and metal to metal. It is also used for bonding glass, china, tiles, cement rubber, thermo-set plastics and any non-porous material.

Polyvinyl acetate (P.V.A.) glues are manufactured and supplied as a white or cream liquid contained in plastic tubes, bottles, jars, etc. They are non-staining, non-toxic, and resistant to mould and fungi. They have a low water resistance, but are fairly heat resistant up to about 200°F, and have good gap filling qualities. They are used cold, applied cold and set (dry) by evaporation of the water contained in them. They have very strong holding power but low creep strength and so are not suitable for high permanent loads. They permit an assembly time of 10–20 minutes, require light clamping for 1–2 hours, and for maximum strength, the work should be clamped up to 12 hours. They are mainly used for internal work, such as gluing of carcases, frames, cold or caul veneering and for fixing wood and plastics to porous surfaces.

Contact glues are 'instant bonding' – that is, adhesion takes place as soon as the parts are put together, without clamping. They are manufactured by dissolving synthetic rubber, particularly neoprene, in solvents. Fillers such as phenol are added to improve upon their gap filling qualities. They are viscous liquids supplied in small tubes and tins up to 4.5 litres. They set or dry by evaporation of the solvent and may have a shelf life of up to 1 year. When applying them you should coat the surfaces of parts being joined and allow about 10 to 20 minutes drying time (the coated surfaces must no longer be sticky to the touch) before you bring them together. Press the parts together by tapping with a hammer or running a roller over the top piece. Full strength is attained after several hours. Contact glues are mostly used for bonding dissimilar (different) materials such as formica, fabrics, leather, glass, plywood, and particle boards. Nevertheless, they are used for bonding pieces of the same material as well (i.e., bonding wood to wood or glass to glass).

> ▲ Always be careful when using glues. Read the manufacturer's safety instructions carefully *before* you use them. Avoid contact with your skin and make sure that you use them only in a well-ventilated place. The fumes from many glues, and especially the synthetic ones, are ***poisonous***.

Hardware (fittings)

Hardware items are fittings that are used for finishing a piece of cabinet work. They play an important part in the final finish and appearance of the cabinet. The quality of the fittings depends largely on their material, the method used for construction and the kind of finish applied to them. The types of fitting covered in this text are: locks, catches, handles or pulls, stays and castors.

Locks (Figure 6.4) are fittings used for locking doors and drawers of cabinet work. The three

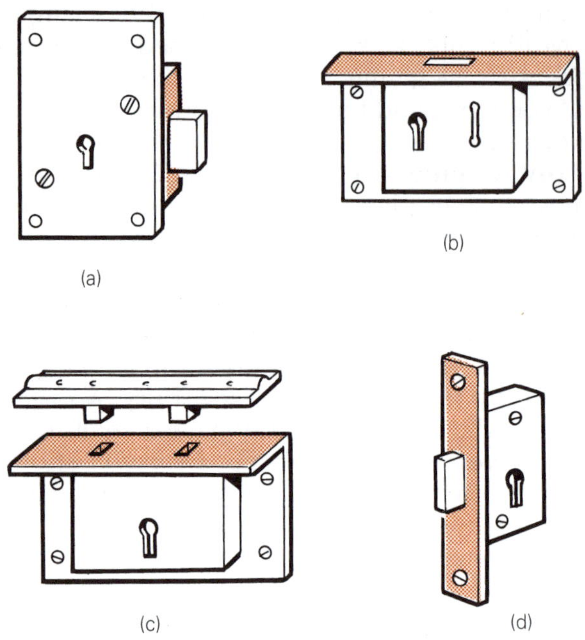

Figure 6.4 Locks: (a) straight lock, (b) cut lock, (c) box lock, (d) mortise lock.

common locks in woodwork are: the straight lock, the cut lock and the mortise lock.

The **straight locks** are used for doors and drawer. They are usually screwed flat onto the inside face of the door or drawer without a recess.

The **cut locks**, also known as **till locks**, are fitted into a recess made in the inside face of a door or in drawer front. They fit flush with the face of the door or drawer front, and are therefore neater and stronger than straight locks. The **horizontal pattern** is used for drawers and the **upright pattern** is for doors. The latter is made *right* or *left* (i.e., right lock fits on a right-hand stile and a left lock fits on a left-hand stile. Another type of cut lock, the **box lock**, is used for lids of boxes and desk top flaps. This has a catch plate carrying two projecting pins. The catch plate is fixed to the lid while the lock is let in flush on the inside of the box.

Mortise locks are normally used for larger and thicker doors. A mortise (recess) is made in the front edge of the door to receive the body of the lock.

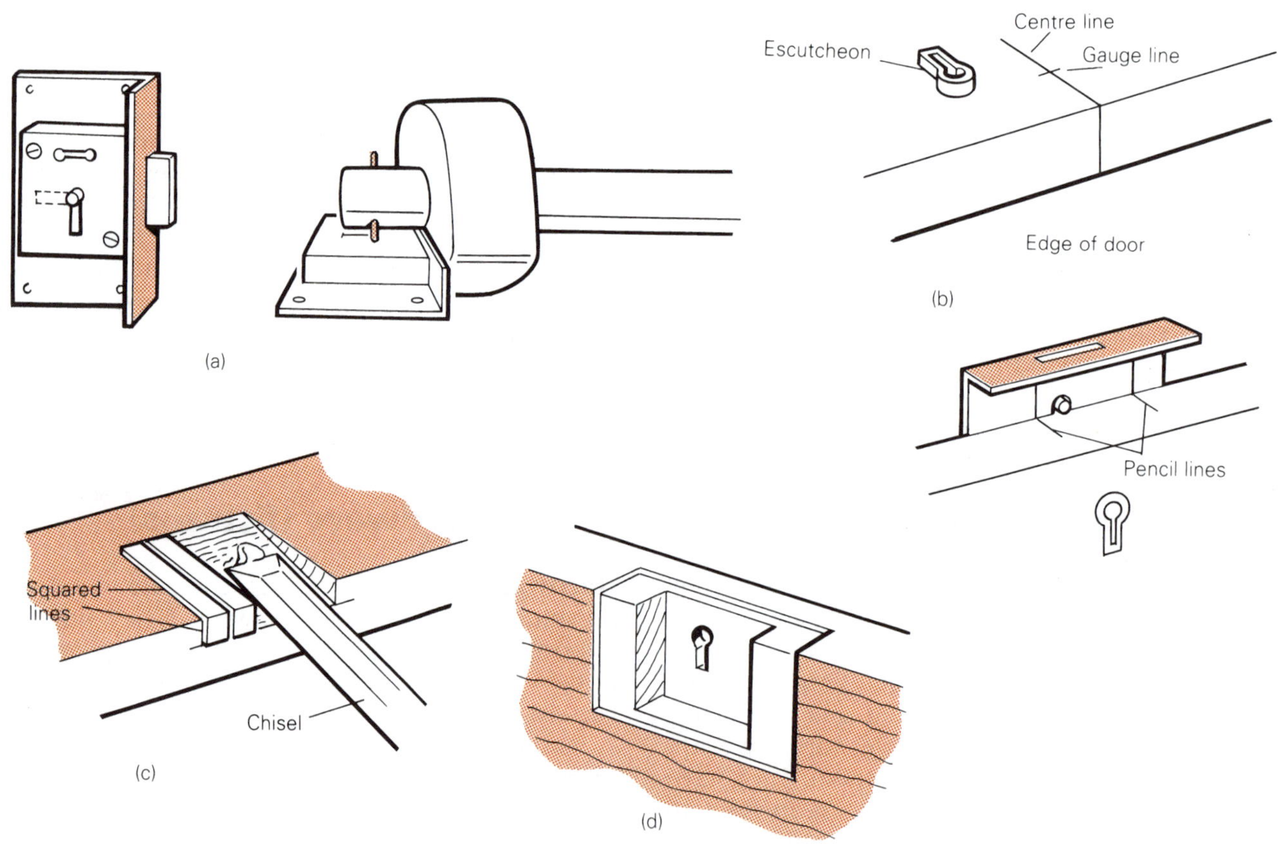

Figure 6.5 Procedure for fitting a lock: (a) setting the gauge (b) marking out, (c) paring out waste, (d) recess for the plate.

The **link-plate lock** is used for doors that fit over the face of the cupboard.

The procedure for fitting locks is as follows (Figure 6.5):

1. Mark the centre (height or length) of the door or drawer front and square on the face and edge (use pencil).
2. Set the marking gauge to centre of the lock pin from the top plate (Figure 6.5a).
3. Gauge for the keyhole from the edge of the door or the top edge of drawer.
4. Bore the keyhole and, using a keyhole saw, cut the sides and pare out waste with a chisel. Where an escutcheon is used, mark the shape of the escutcheon to guide the saw around sides of the keyhole recess (Figure 6.5b).
5. Mark the length of the body of the lock, ensuring that the pin is on the pencil line. Square the lines on the edge and on in the inside (Figure 6.5c).
6. Gauge the lines for the body and the plate.
7. Saw the recess for the body of the lock.
8. Chisel and pare out the waste (Figure 6.5c).
9. Check for length and width of the plate and cut recess for the plate (Figure 6.5d).
10. Screw the lock in position.
11. Mark the position where the bolt will strike the frame or on the inside of the drawer rail by putting chalk, paint or dirty oil on the bolt, closing the door and turning the key.
12. Chisel out waste.
13. Close the door or drawer and turn in the key to lock it. Make any necessary correction.

Catches (Figure 6.6) are metal devices used for securing doors and falls in position. There are many varieties and two main categories are discussed in this text: the spring action type, that has automatic closing and opening and the non-spring type, having automatic closing only.

The **spring action type** requires a handle or knob for pulling the door open. Among this type are: the ball catch, magnetic catch, roller mortise catch and French catch.

The **ball catch** is made in two parts: the **ball**, fitted into the edge of the door, and the **strike plate**, screwed to the side of the cabinet.

The **magnetic catch** has a strong permanent magnet housed in a plastic case that is fixed to the side of the cabinet and a metal (steel) plate that is fixed to the door. The door is held in position by the force between the magnet and the steel plate.

The **roller mortise catch** has a roller bolt, housed in a metal case in the form of a mortise lock, mortised into the edge or end of the door, and a metal strike plate fixed to the side of the cabinet.

The **French catch** is similar to the roller mortise catch except that the bolt is vee shaped and does not roll. The case is also mortised into the edge of the door and the strike plate is fixed to the side of the cabinet.

The **non-spring catches** need a button or knob on the outside of the door to operate the opening device of the catch. The two commonly used members of this category are the surface cupboard catch and the push button cupboard catch.

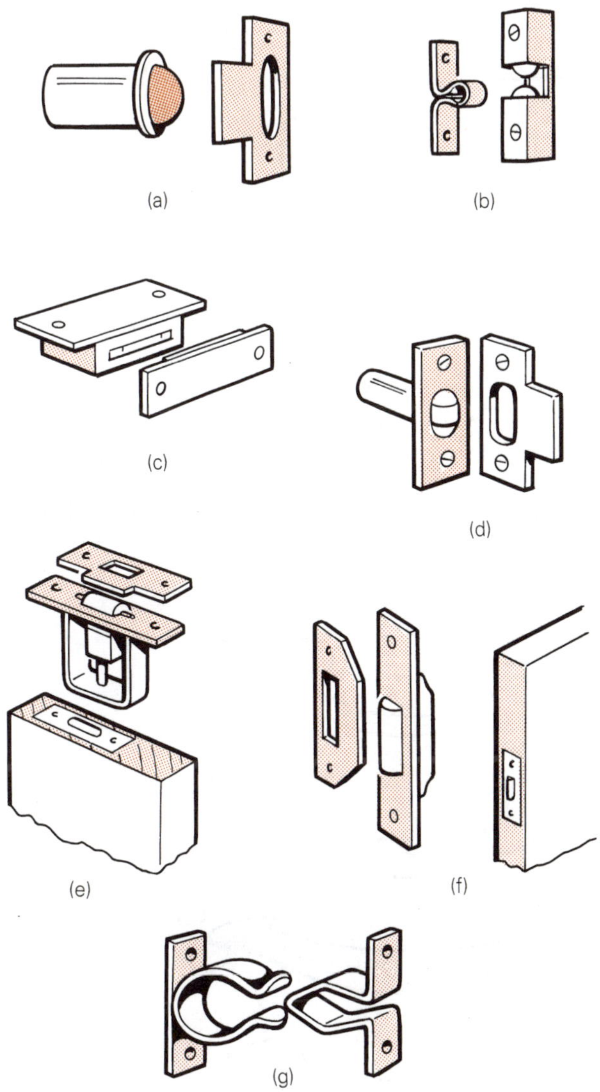

Figure 6.6 Types of catch: (a) ball, (b) double ball, (c) magnetic, (d) roller mortise, (e) French, (f) spring.

The **surface cupboard catch** is similar to the straight lock. It is often called an **earl lock**. Both the body and the strike plate are fixed to the surface of the doors if it is a double door cabinet. When the strike plate is fixed on the inside of the cupboard, then a key may be necessary for opening.

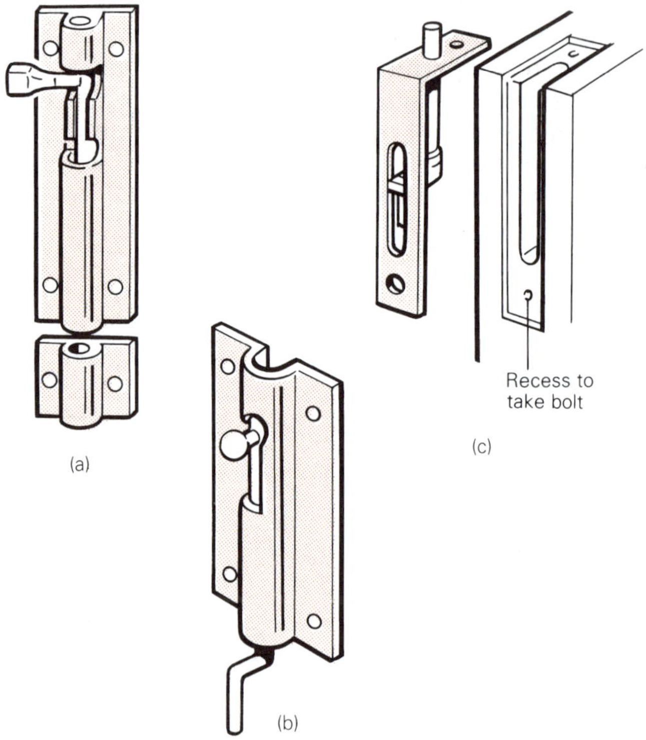

Figure 6.7 Types of bolt: (a) barrel bolt, (b) offset bolt, (c) flush bolt.

The **push button cupboard catch** is a knob type, fixed into a hole through the doors to hook the strike that is fixed on the shelf in the cupboard. It has a button in front of the knob which, when pushed, disengages the strike for opening.

Bolts (Figure 6.7) are devices used for hooking one of the pairs of door in position so that the other pair can be catched or locked in place. They are fixed at the top and bottom in the inside of one of a pair of doors. The common types are the **barrel bolt**, **offset bolt** and **flush bolt**.

Handles (Figure 6.8) are fixed to the outside of cabinet doors and drawers and also serve as ornaments. Various materials are used, including brass, copper, chrome plate, satin nickel, antique brass and plastic. They are often die-cast. The three main types are **handles** having a wide opening to accommodate the whole hand; **knobs**, which are round or circular in shape; and **pulls** (ring or finger) which have either a hole (eye) or else a recess into which the fingers are inserted.

Stays (Figure 6.9) are metal fittings of different designs that are fixed to lids of boxes, folding or sliding table tops so that the lids or tops are supported when opened. They are made in left-hand and right hand forms.

Castors (Figure 6.10) are wheels fitted to legs and bases of furniture to enable the furniture to be moved about easily. There are a range of designs and sizes. They are supplied in sets of four.

Hinges (Figure 6.11) are devices used for securing doors and lids to cabinets and boxes. The common types are: butt hinges, back-

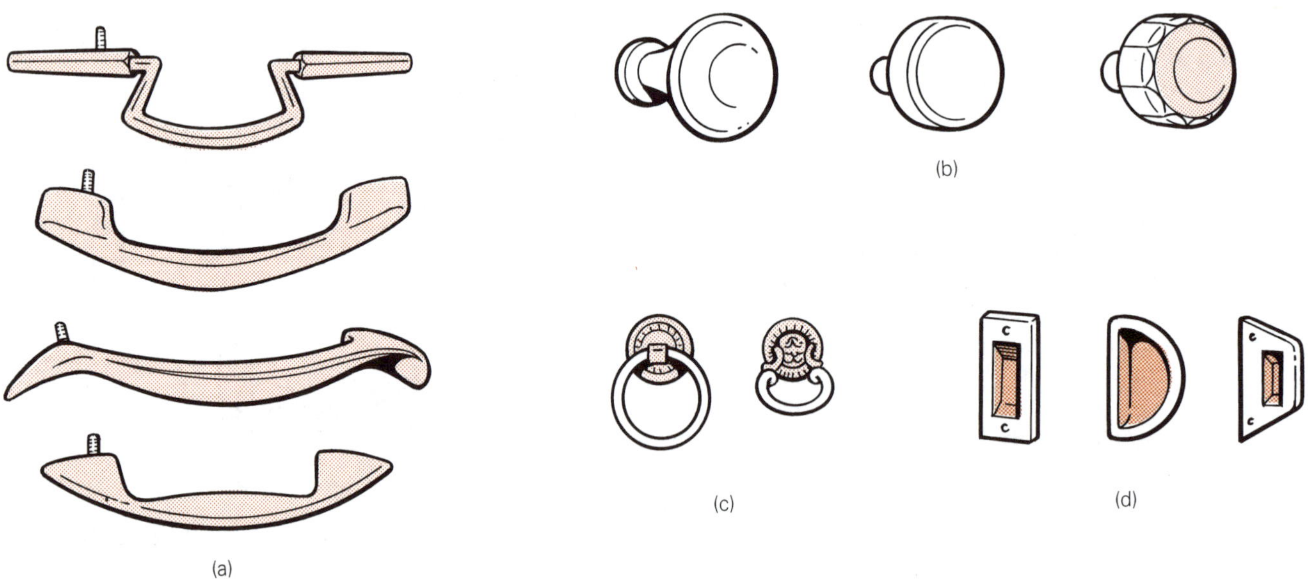

Figure 6.8 Types of handle: (a) handles, (b) knobs, (c) ring pulls, (d) finger pulls.

flap hinges, strap hinges and hook and band hinges.

Butt hinges are the most commonly used hinges when fixing lids and doors to boxes and cabinets respectively. They are available in lengths from 20 mm to 150 mm and are made of brass, steel, plastic or nylon.

Back-flap hinges are widely used on the fronts of writing bureaux and on the leaves of drop-leaf tables. They are either recessed into the surface or simply screwed directly onto it. They are made of brass or steel and are in available in lengths from 20 mm to 75 mm.

Strap hinges, also called **tee hinges** are mainly used for securing garage doors, shed doors and light gates. They are usually made of mild steel. When used outside, they may be coated black to prevent rusting. The common sizes range from 50 mm to 300 mm.

Hook and band hinges are used for securing heavy gates and doors. These are made of mild steel and to individual specifications.

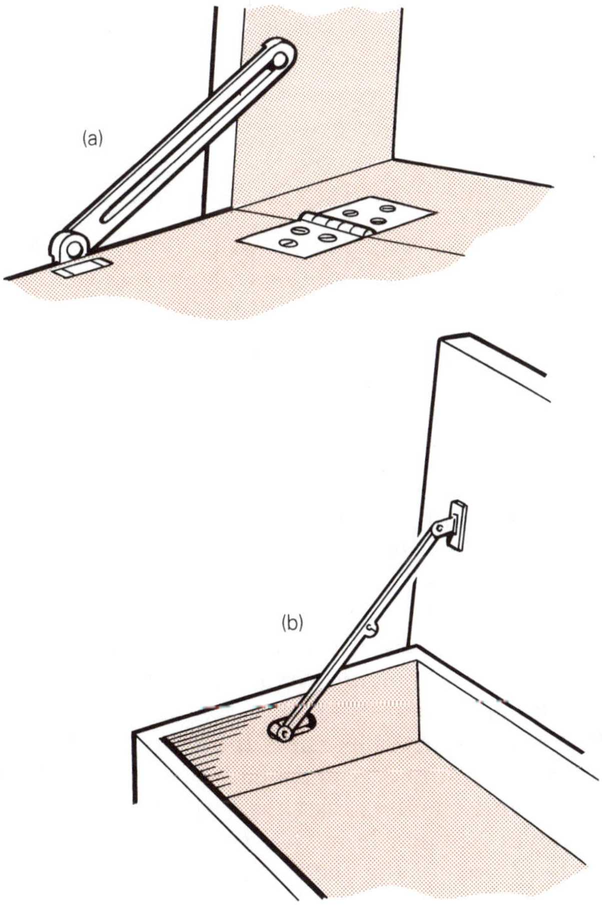

Figure 6.9 Types of stay: (a) sliding stay, (b) folding stay.

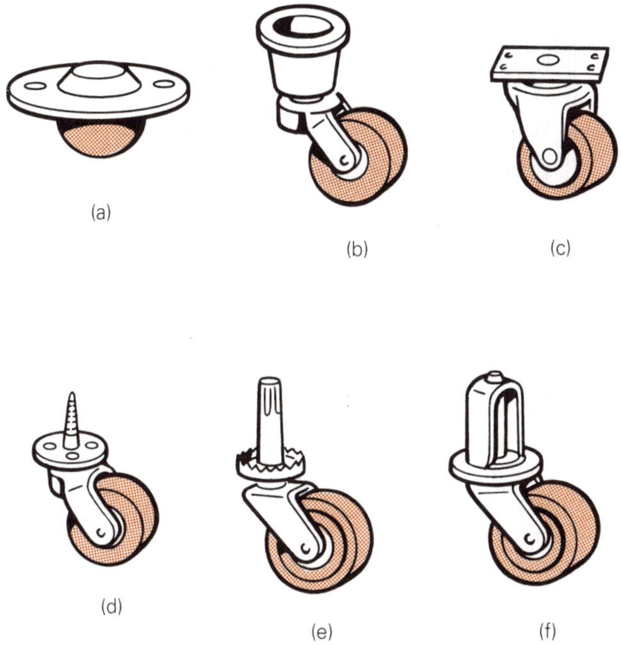

Figure 6.10 Types of castor: (a) ball, (b) round socket, (c) plate, (d) screw, (e) grip-neck socket, (f) tubular.

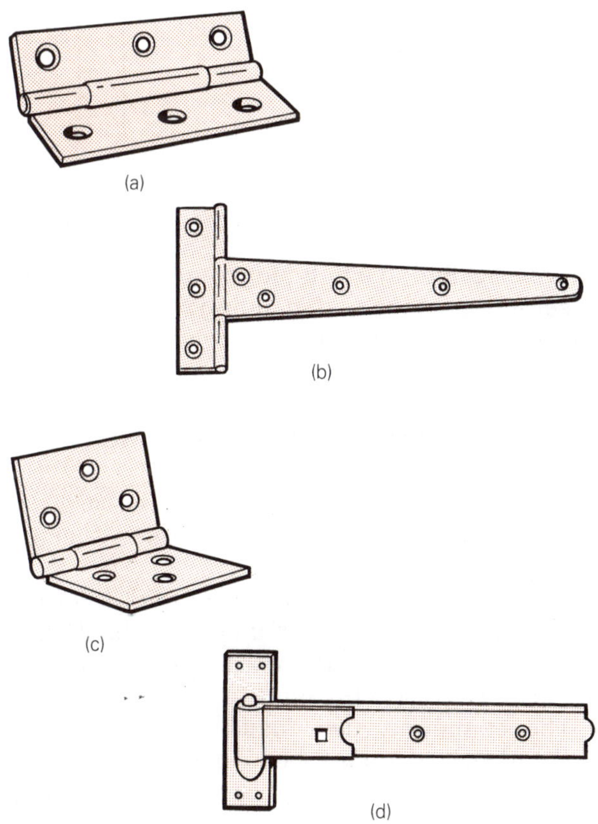

Figure 6.11 Types of hinge: (a) butt, (b) strap, (c) back-flap, (d) hook and band.

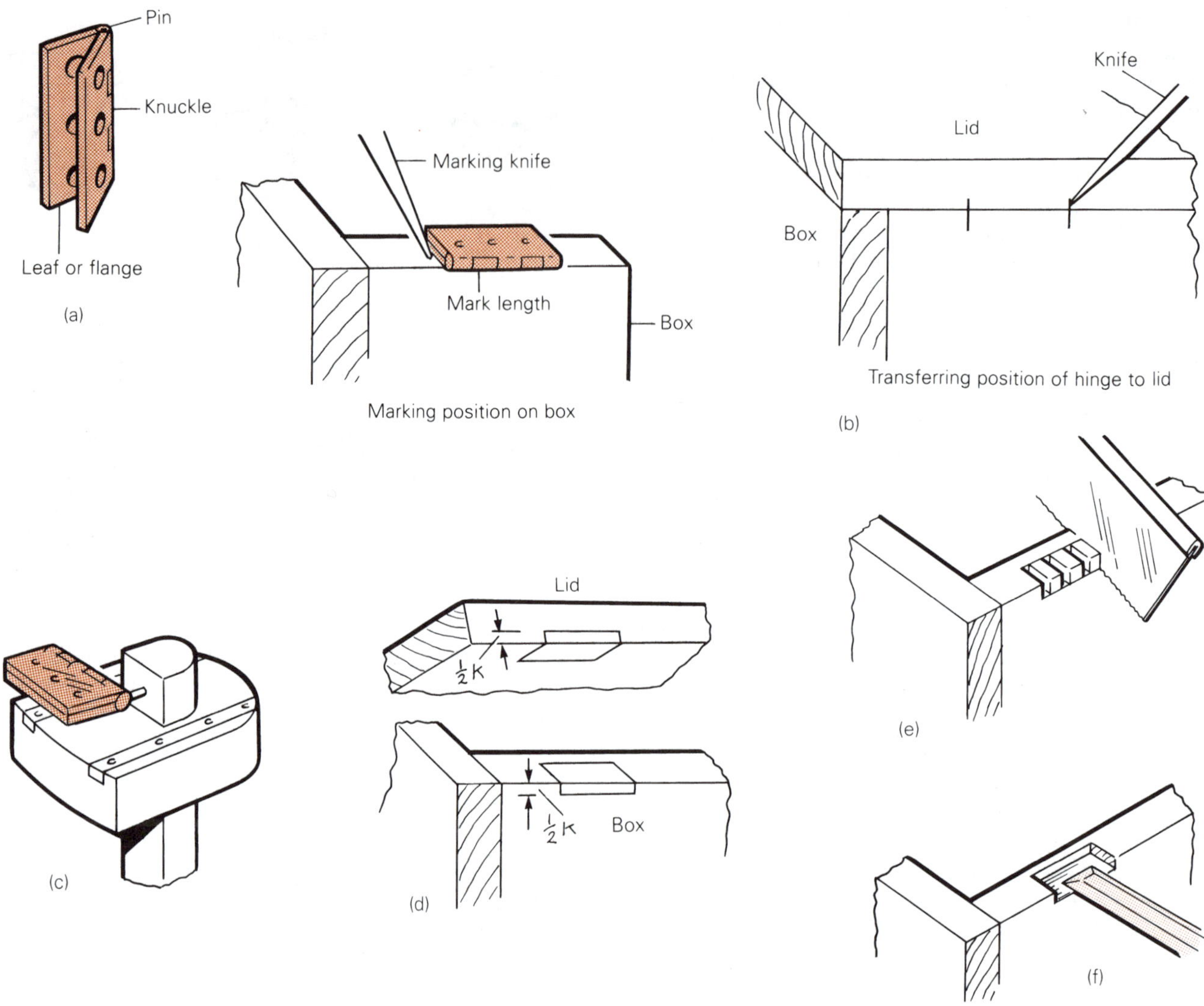

Figure 6.12 Fitting a butt hinge: (a) parts of a butt hinge, (b) marking the position of the hinge, (c) setting the gauge, (d) gauging width of leaf and half thickness of knuckle on box and lid, (e) sawing waste, (f) chopping and paring waste.

CHECK YOUR UNDERSTANDING

- Fasteners are materials or devices used for securing wooden pieces or structures together. They include: nails and screws.
- Nails are generally made from mild steel wire or plate and they consist of a head, a shank and a point, and are classified according to the shape of the head, cross-section and use.
- The commonly used nails are: roofing nail, flat head nail, diamond head nail, jolt head nail, panel pin, veneer pin, upholstery nail and wiggle nail.
- The holding power of a nail depends upon the following factors: the ability of the wood fibres to grip the shank, the type of shank, the angle of nailing, the depth of penetration, and the size of the head of the nail.
- A screw consists of a head, shank, core, thread and point. Screws are used for fastening wooden members together and fixing fittings to carcases or cabinets.
- A screw derives its name from the shape of the head; for example, round head screw, raised head screw, and countersunk (head) screw.
- Never hammer home a screw as this will break off the fibres of the piece and prevent the thread from screwing itself into the fibres, thereby reducing the holding power.
- Wood adhesives are bonding substances that

hold pieces together permanently. The three varieties of wood adhesives are: the protein types, synthetic resin types and the contact adhesives.

- The term hardware refers to fittings attached to cabinet works as part of the final finish.
- These include locks, catches, bolts, handles, stays and castors.
- Locks are devices fitted to doors and drawers of a cabinet work such a cupboard or a sideboard for locking them up. The common types are straight locks, cut locks and mortise locks.
- Catches are metal devices used for holding doors and falls in position when they are closed.
- Bolts are metal devices fixed at the top and bottom of one of the pairs of a door for hooking it in position so that the other pair is locked or catched in place.
- Stays are metal fittings of various designs fixed to lids of boxes or falls for supporting them when opened.
- Handles are fittings fastened to cabinet door and drawers, mainly for pulling them open.
- Castors are wheels fitted to legs and bases of furniture to enable it to be moved about easily.

REVISION EXERCISES AND QUESTIONS

1 Name any four types of nail and describe the features of each. State where each is used.
2 What factors are essential to state when ordering (purchasing) nails?
3 Explain why a screw has a stronger holding power than a nail.
4 What precaution needs to be taken when drilling or boring a pilot hole for screwing.
5 Name the three classes of wood adhesive, give an example of each class and state the applications of each example.
6 Name the common types of lock and state where each is used.
7 Explain the main difference between the spring action catch and the non-spring action type.
8 When would you use a bolt?
9 State the three main types of handle and briefly describe each.
10 What is the main function of (a) a stay and (b) a set of castors?

7 Woodworking machine tools

Introduction

Machines tools are those tools that are driven by an electrical motor – that is, they are power driven. Their use saves labour and time. They provide a wide range of operations accurately. We discuss here the basic woodworking machines that you are likely to use in the workshop, so that you can acquaint yourself with their functions and operations. Most of the machines are dangerous and you should read the general safety precautions to be exercised when working in the machine shop (see chapter 1, page 2). You must acquaint yourself with the factory safety regulations prevalent in your own country.

Circular saw

The **circular saw** is one of the woodwork machines that can be found in most workshops. The most versatile variation of the circular saw is the **dimension saw** (Figure 7.1). Its primary function is to re-saw timber. It can be used for a range of operations, such as ripping, cross cutting, bevelling, chamfering, mitring, trenching and grooving. It has a heavy cast iron **base** upon which the spindle or saw arbor, table and electric motor are mounted. The base, for some makes is mounted on the floor while others are mounted upon a strong steel stand.

The **table** is the working surface on top of the base. The **spindle** carries the circular blade and runs on ball bearings. The **electric motor** provides the power for driving the spindle. The motor may be directly coupled to the spindle or else indirectly coupled by a belt and pulleys. The whole assembly can be tilted for bevel sawing, and raised or lowered to set the depth of the cut.

The **blade** is circular and fixed on the spindle by a flange washer and nut. There are varying sizes and thicknesses of blades. They are made from high speed steel or chrome-molybdenum.

For safety, a **saw guard** is provided to prevent accidental contact with the blade. The guard also prevents the piece being sawn from kicking back (that is, being forced backwards) towards the operator. The **fence** is an adjustable device attached to the table for guiding boards so that a uniform width can be ripped. It is also used when grooving, rebating and trenching. Most of the fences can be tilted to enable bevelling and chamfering to be carried out. Another essential attachment is the **mitre gauge** that is fitted to slide in a groove machined along the surface of the table on each side of the blade. This can be adjusted to cut angles ranging from 45° to 135°. It is used for guiding and holding a piece of board when cross-cutting and mitring.

The circular saw with this standard equipment can be used to carry out a number of operations. Their names and meanings are the same as those discussed in chapter 5, except that a machine is being used to perform the operations.

Procedure when **ripping** (Figure 7.2):

1. Check and ensure that the ripping saw is in place and runs free.
2. Set the fence at the required distance away from the saw (blade).
3. Set the saw to project about 6 mm above the board to be ripped.
4. Adjust and set the saw safety guard in position.
5. Switch on the machine. Stand away from the saw.
6. Press the board against the fence and push it forward. Keep the board against the fence

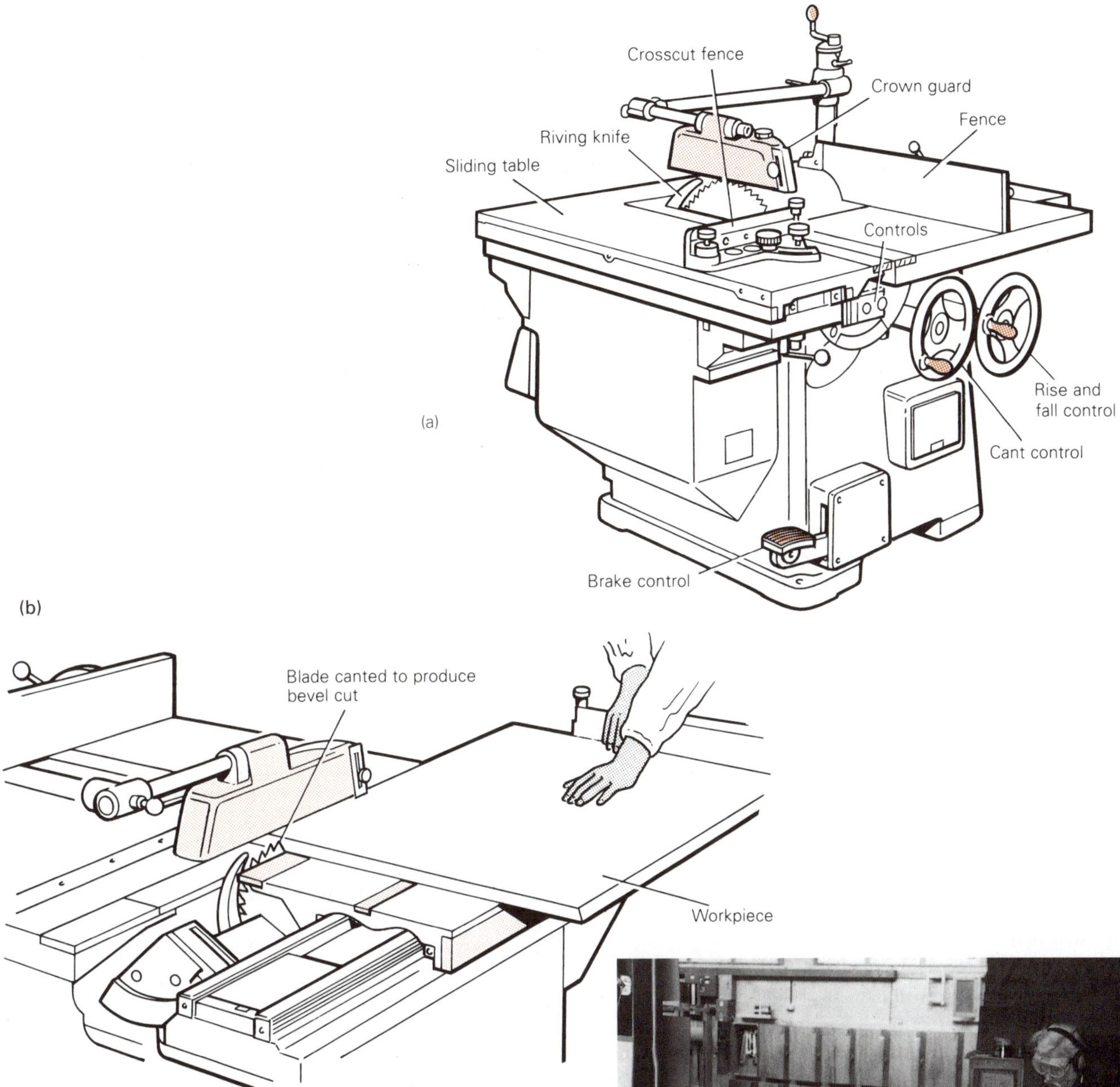

Figure 7.1 The dimension saw: (a) the saw, (b) detail of the slidingtable.

Figure 7.2 Using the dimension saw. Note the safety precautions (goggles, ear protection, guard in use).

throughout. When ripping narrow boards, use a forked push stick to push the board when the cut is getting close to the end of the board.
7. *Switch off the machine after working.*

Procedure for **cross-cutting**:

1. Check whether the blade in place is a cross-cutting saw. If it is not, remove it and mount the correct one.
2. Set the saw to project 6 mm above the surface of the board.
3. Adjust the fence to allow the material to clear it.
4. Put the mitre gauge in place.
5. Set the saw safety guard in position.
6. Switch on the machine.
7. Hold the board against the mitre gauge and push it forward against the revolving saw. *Use push sticks.*
8. When cross-cutting a number of pieces of the same length cramp a small wooden block to the fence (this provides clearance between the end of the board and the fence when sawing). Set the fence with the block at the required distance; support the board on the mitre gauge and push it against the block; then push the board together with the mitre gauge forward against the revolving saw.
9. *Switch off the machine after working.*

Procedure for **rebating**:

1. Mount the ripping saw.
2. Set the fence to the required distance.
3. Set the saw to the required depth of cut. Ensure it is adequately guarded.
4. Switch on the machine.
5. Hold the work against the fence and push it forward against the revolving saw. *Use push sticks.* If the size of the rebate is wider than the thickness of the saw you will need to re-set the fence a number of times and repeat the cutting until the required width of the rebate is attained.
6. *Switch off the machine after working.*

Procedure for **grooving**:

1. Mount the ripping saw.
2. Set the fence to the required distance. Note that the setting must be altered several times until the required width of the groove is obtained.
3. Adjust the saw to the required depth of cut.
4. Switch on the machine.
5. Hold the work-piece against the fence and push it forward to cut. *Use push sticks.* Re-set the fence and cut until the required width is obtained.
6. *Switch off the machine after working.*

Procedure for **trenching**:

1. Mount the cross-cutting saw.
2. Adjust the fence.
3. Fix the mitre gauge.
4. Adjust the saw to the required depth of cut. Ensure it is adequately guarded.
5. Switch on the machine.
6. Support the work-piece against the mitre gauge and cut. Adjust the piece and repeat the cutting until the required width of the trench is obtained. *Use push sticks.*
7. *Switch off the machine after working.*

Mitring on the circular saw involves similar steps to cross-cutting except that the mitre gauge is off-set to the required angle, usually 45°. Hold the work-piece very firmly for accurate cutting.

Chamfering and **bevelling** are performed following similar steps for *ripping* (if along the length) or for *cross-cutting* (if at the end of the piece). The saw must however be tilted to the desired angle and the fence or mitre gauge set on the lower side of the saw or table.

Band saw

A **band saw** (Figure 7.3, 7.4) is one of the useful machines in the woodwork shop that is generally used for cutting external curved shapes due to its endless narrow flexible blade. It can also be used for straight cutting or a combination of straight and curved work. It has a range of industrial applications. It is useful in the general woodwork shop, the small cabinet shop, the pattern maker's shop, the sample maker's shop, and the school workshop. In the timber industry, the band saw is used for cutting logs into commercial sizes, such as planks, boards and scantlings (timbers of small cross sections). Generally, all woodworkers should possess skill in the operation of the band saw.

It has a heavy cast iron **frame** on which all the other parts are mounted. In order to prevent vibration, it must be firmly fixed, either on a strong and rigid stand, or else directly to the floor. It has two **pulleys** or **wheels** (lower pulley and upper pulley) that run on ball bearings and are

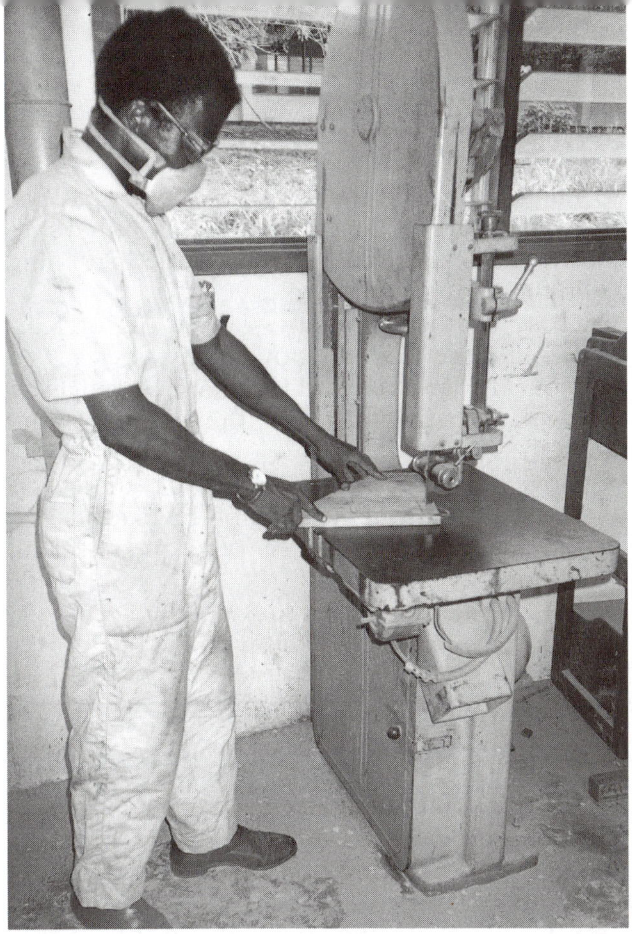

Figure 7.3 The band saw: (a) the band saw, (b) using the band saw.

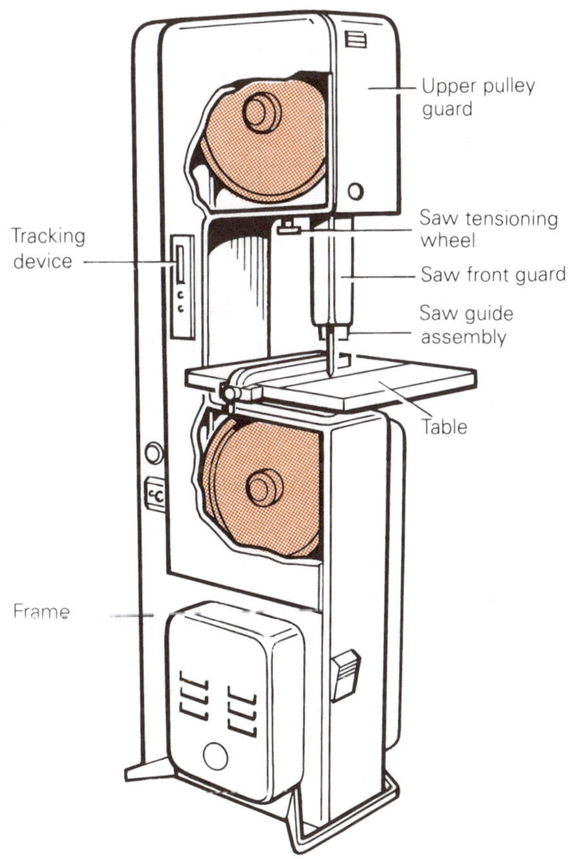

Figure 7.4 Parts of the band saw.

covered by strong guards. The **lower pulley** is fixed and is either connected directly to the motor, or else coupled by a belt and pulleys. The **upper pulley** is carried on an adjustable slide that enables it to be raised or lowered vertically to allow new blades to be fitted and to provide the correct tension. It can also be tilted forwards or backwards for adjustment of the blade. In order to preserve the set on the blade and also to provide adequate friction for driving the saw blade, each pulley is fitted with a rubber band or **tyre**. The working **table** is also made of cast iron and is ribbed for strength. It can be tilted and locked at any angle up to 45°. To prevent the teeth of a broken blade running against the hard cast iron table and causing damage, a soft aluminium throat is fitted into where the blade passes through the table. The **blades** are of two types: **plain** and **bevel back**. The bevel back type is thinner on the back edge and has the advantage of providing more clearance for fine curve cutting. The thickness of the saw blade largely depends upon the diameter of the lower and upper pulleys over which it runs. The width of the blade can also affect the thickness required. The smaller the diameter of the pulleys or the width of the blade, the thinner the blade must be. The width of blade also affects the radius of curvature of the cut; narrow blades can cut smaller or sharper curves than wide blades. Two **saw guides**, one above the table, the other below, are provided to enable the blade to run true. Each guide has two bronze jaws that prevent the blade from wobbling or twisting in the cut. The guide also has a ball-bearing thrust wheel behind the jaw to support the back of the blade when sawing. The jaws are adjustable so that enough space can be permitted for the blade to run freely without necessarily rubbing against them. The upper guide is adjusted by means of the guide post and it should be about 3 mm to 6 mm above the work-piece. **Guards** are provided to cover the blade almost completely, except the cutting portion. The **sliding guard** attached to the sliding guide post covers the blade on the cutting side and this must always be in place before the machine is started.

Follow the following instructions when you want to fix a new or mended band saw blade:

1. Switch off all electrical power to the machine at the mains.
2. Open up the guide jaws.
3. Slide the thrust wheels back.
4. Put the blade on the upper and lower pulleys.
5. Apply tension and revolve the pulleys to check tracking (i.e., that the blade is running in the correct path).
6. Close up the guide jaws, allow slight clearance for the blade to move freely.
7. Move the jaws forward for their front edges to level with the edges of the gullets of the teeth.
8. Move the thrust wheels forward until it is about 2 mm from the back of the blade.
9. Pull the blade by hand to rotate and check all settings.
10. Put in place all the guards.
11. Adjust the table to square with the blade.

▲ When you are using a band saw:

- Never wear loose sleeves or a flying tie. This is extremely dangerous.
- Always wear ear muffs and safety goggles (Figure 7.5).
- Before switching on, ensure that all guards are in position and are secured.
- Ensure that the blade is properly positioned between the jaws and can run freely just close to the thrust wheels.
- It is unsafe to stand in front of the machine when switching it on. *Listen* to the sound of the machine: a *clicking sound* indicates that the blade is *about to break* or the blade is not properly tracking. *Switch off at once* and report immediately to the teacher if you hear a clicking sound.
- Concentrate on the work and don't talk to other people.
- Do not leave the machine after switching off *until it has stopped.*

Radial arm saw

The **radial arm saw** (Figure 7.5), originally was designed for crosscutting boards to length, but with improvements it has become a versatile machine in the woodwork shop. It can be used for other operations such as grooving, trenching, mitring, bevelling and tenoning. For special operations it can take other attachments such as: trench heads, moulding heads, shaper heads, router, rotary planes, drill chuck, saber saw, wood

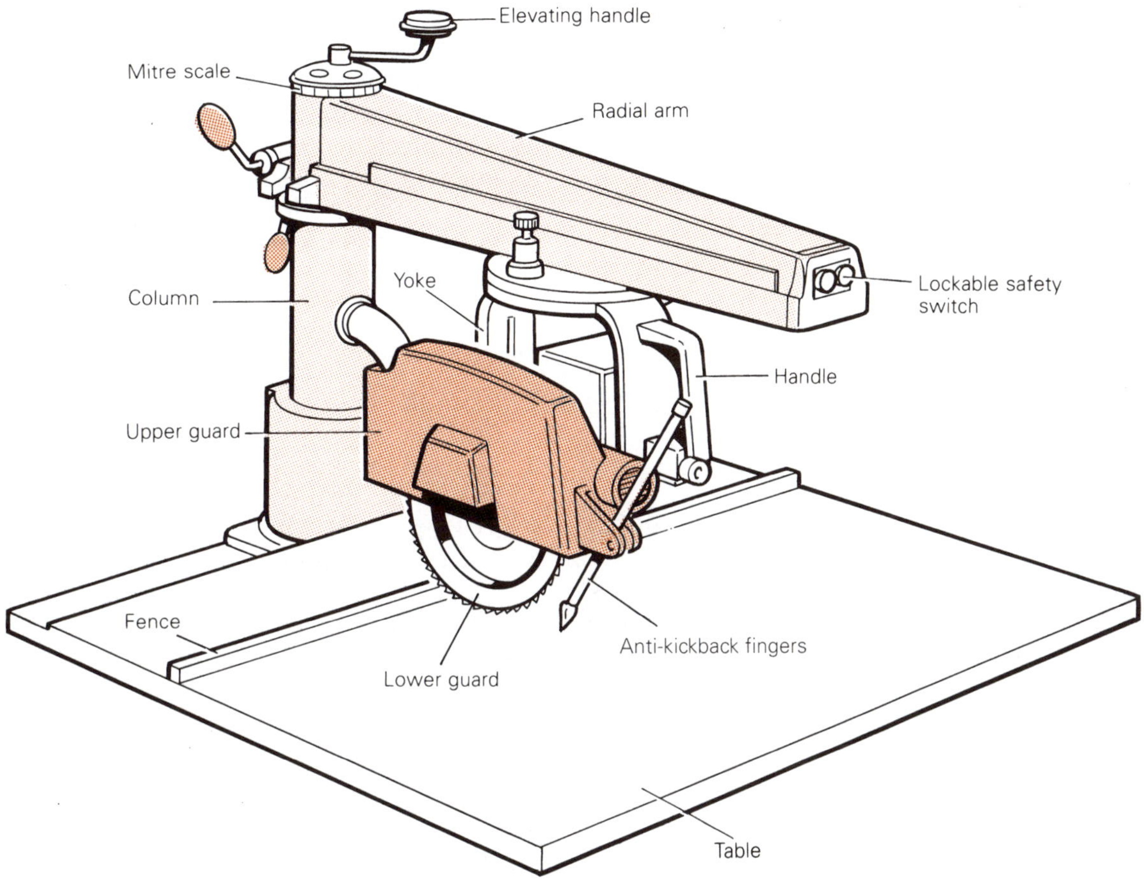

Figure 7.5 The radial arm saw.

lathe, belt sander, disc sander and grinder.

It has a metal **base** that supports the **table** and the column. The **column**, made of hollow steel supports the **radial arm** that can be raised and lowered to the correct operating position. The **yoke**, supporting the motor is attached to the arm and can be made to slide forwards or backwards by pulling or pushing the handle. The yoke is fastened in position by the **yoke lock**. The **motor** provides the power for driving the saw and its attachments which are directly fixed on to the motor spindle. The **saw blade** is covered with cast iron **guards** that direct the sawdust away from the cut. **Anti-kickback fingers** keep the board that has been cut from being thrown back at the operator during a ripping operation. The **table** is made of wood and has a **fence** (wooden guide strip) against which the board is held.

Crosscutting, ripping, grooving, bevelling and trenching are among the common operations carried out on the radial arm saw and the procedures for carrying them out are described here.

Procedure for **crosscutting**:

1. Ensure that the radial arm is at right angles to the fence on the saw table. The mitre scale will then read zero ('0'). You may make a trial cut and then check the squareness with a try square.
2. Set the saw blade to cut slightly below the surface of the table.
3. Mark the position where you intend to cut the board and place the board against the fence on the table.
4. Adjust the saw guards near to the position of the cut you will be making. The guards must not be placed so that they block your view.
5. Raise the anti-kickback fingers just above the surface of the board to be cut.
6. Set the saw blade to the back against the stop and make sure it is not in contact with any wood.
7. Switch on the motor and wait until it has reached maximum speed.

8. Hold the board firmly against the guide strip with one hand and hold the handle of the motor yoke with the other hand. Pull the handle forward to cut the board.
9. Push the saw back to the starting position and *switch off the motor.*

> It may be necessary to clamp a stop to the guide strip at the desired position when you want to cut several boards of the same length.

Procedure for **ripping**:

1. Set the radial arm at right angles to the guide strip and lock it in position.
2. Raise the column so that the saw blade is above the table. Bring the saw blade to the front of the radial arm and lock it in position.
3. Rotate the motor yoke so that the saw blade is parallel with the guide strip and lock it in position.
4. Move the motor assembly to the desired width of cut and lock is position.
5. Adjust the safety guard on the in-feed end until it clears the surface of the board to be sawn.
6. Lower the saw blade so it comes into contact with the table.
7. Put the anti-kickback fingers in place.
8. Start the motor and slowly feed the board against the rotating saw blade. Ensure the straight edge of the board is always is contact with the guide strip. *Use push sticks* as the cut gets close to the end of the board.

> **Note**: The capacity for the width of ripping is determined by the arm length and the table width.

Procedure for **cutting grooves**:

1. Remove the saw blade and replace it with trench (dado) head of the desired width.
2. Release the yoke clamp and turn the motor parallel to the guide strip, locking in position.
3. Move the motor assembly on the radial arm to the desired position and lock it in place.
4. Lower the column for the desired depth of cut and lock it in position.
5. Adjust the guard to the correct position and lock it in place.
6. Start the motor and make a trial cut on a waste piece of wood of the same thickness as the workpiece.
7. Push the workpiece against the guide strip and slowly feed it into the trench heads. **Keep your hands clear of the trench heads**. **Use a push stick** to push the workpiece as the cut gets close to the end of the piece.

Procedure for cutting **bevels**:

1. Raise the column so that the saw blade clears the table when the motor is tilted.
2. Release the bevel latch and the bevel clamp lever and tilt the motor to the desired angle, locking it in position.
3. Lower the column and set the saw blade to cut slightly below the surface of the table.
4. Adjust the saw guard correctly.
5. Start the motor and make a trial cut on a waste piece of the same thickness.
6. If satisfied, proceed as in crosscutting.

> ▲ When you are using a radial arm saw:
>
> ● All boards or work-pieces must be firmly held on the table and against the guide strip (fence) for all operations.
> ● Ensure that all cramps and locking devices are well tightened and the depth of cut is correctly set before you start the motor.
> ● Always keep the saw guards and the anti-kickback device in position. Never take them off without the knowledge of your teacher or instructor.
> ● Ensure the saw blade is returned to its position at the back of the table before you remove the workpiece from the table.
> ● Remember to switch off the motor and wait for the saw blade to stop before you make any adjustments or leave the machine after an operation.
> ● Always keep the surface of the table clean, free of waste pieces and excessive amounts of sawdust.
> ● Always seek help from your teacher/instructor if you are not sure what to do or how to set up or perform an operation.

Procedure for cutting **trenches**:

1. Take out the saw blade and replace it with a trench head of the desired width.
2. Lower the head to the desired depth.
3. Put the saw guard in place. Make sure it cannot block your view.
4. Start the motor and make a trial cut using a waste piece of the same thickness as the workpiece.
5. Make any necessary adjustments.
6. Position the workpiece correctly and hold it against the guide strip.
7. Start the motor, hold the yoke and slowly feed the head into the work. Wider trenches may be made by repositioning the workpiece and taking additional cuts.
8. Angled trenches can be cut by setting the motor to the desired angle and locking it firmly in position.

Jigsaw

The **jigsaw** (Figure 7.7) is a power-driven, reciprocating, toothed cutting tool used for cutting curves and regular shapes. Its narrow blade is an advantage for cutting internal shapes and curves as well as for very fine and detailed cutting. There are three types of jigsaw: the plunger type, the vibrating type, and the rocker-arm type.

The **plunger type** of jigsaw is the commonest in most workshops. It has a **belt-driven pulley** connected to a **cam-and-pitman mechanism**. As the arm rotates, the pitman pushes and pulls the **lower chuck**, holding the lower end of the blade, up and down. The upper end of the blade is held in the **upper chuck** attached to the **plunger-and-spring tension control**.

The **vibrating type of jigsaw** is operated by a **circuit breaker**. In the power stoke, an electromagnet pulls down the **lower chuck**, which holds one end of the blade, when the chuck reaches the lower limit of its stroke, the circuit is broken and a spring in the tension sleeve returns the blade to its upper limit.

The **rocker arm type of jigsaw** operates on the principle of the wheel and axle, similar to the plunger type. The **driven pulley** is connected to a cam that drives a **rocker arm** to produce an up and down motion.

The capacity of a jigsaw is governed by the distance between the overarm and the blade (i.e., throat capacity) and also by the maximum thickness of board permitted by the guidepost. The most common sizes of jigsaw are 450 mm and

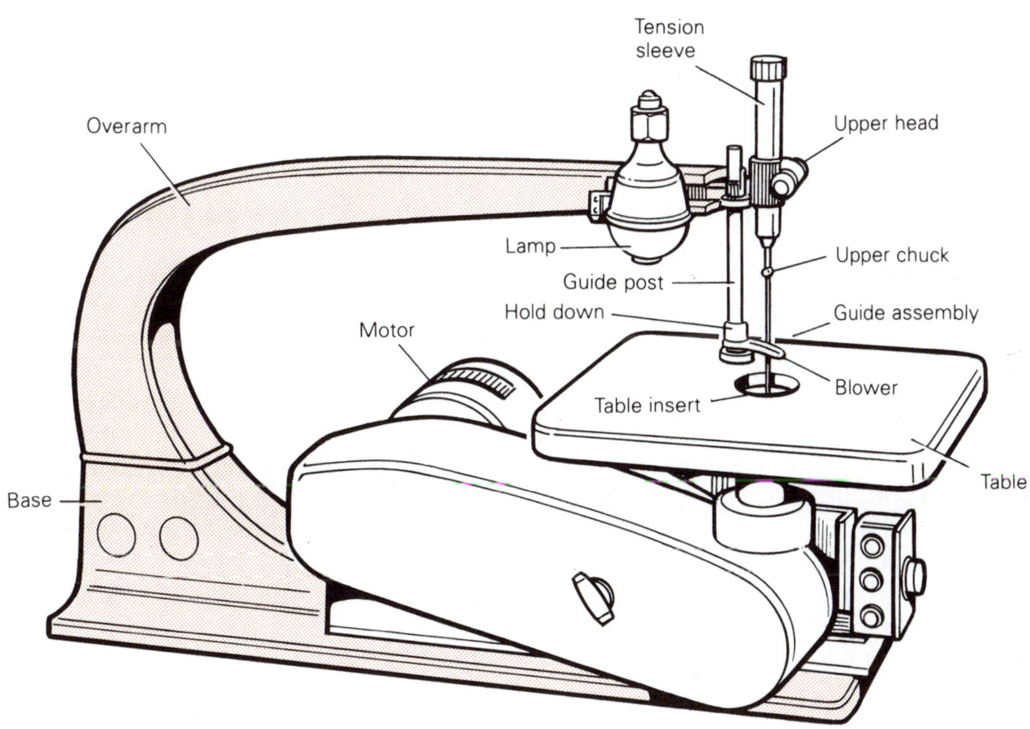

Figure 7.6 The jigsaw.

600 mm and the maximum thickness of cut varies from 18 mm to 100 mm depending on the make.

The essential components of the jigsaw include:

- a **cast iron frame** made up of two sections bolted together: the upper deep part-throated yoke called the **overarm**, to which the guidepost is fixed, the lower part, the base that is attached to the overarm houses the pitman and cam;
- a machined cast iron **table** that supports the workpiece and can be tilted 45° to the left or to the right of the horizontal;
- a **tension sleeve**, made of pressed steel, which contains a **spring** that regulates the tension on the **blade**;
- a **throat plate**, inserted in the table above the lower chuck supports the workpiece close to the blade and also it keeps the blade from being damaged if it run out of line;
- the **upper and lower chucks**, used for holding the blade at the ends, the upper chuck also holds the tension sleeve;
- a **hardened-steel blade guide**, that keeps the blade from twisting during cutting operation;
- a **guidepost** that adjusts to the thickness of the board;
- a **hold-down foot** (depressor foot), fitted to the bottom of the guidepost, resting on the board to keep it from moving up and down with the motion of the saw;
- the **cam-and-pitman mechanism** that converts the rotary motion into reciprocating motion; a **plunger** that serves as a blower, keeping the sawdust away from the saw;
- a **cone pulley** for varying the speed;
- a belt in order to protect the operator;
- an electric motor that provides the power;
- heat-treated **steel blades** that cut the material.

Guidelines for using the jigsaw correctly and safely for regular outside cutting:

1. Select the appropriate blade that will suit the work you intend to do.
2. Fix the blade into the upper and lower chucks with the teeth pointing toward the table (downward).
3. Adjust the tension on the upper chuck to keep the blade taut.
4. Set the blade guide in position.
5. Select the speed appropriate to the work to be done.
6. Adjust the hold-down foot so that the spring tension holds the workpiece down on the table.
7. Check all adjustments by turning the drive shaft one revolution by hand to ensure they have all been properly secured.
8. Start the machine and feed the workpiece evenly and gently forward against the moving blade. Press down on the workpiece toprevent it from being lifted by the motion of the blade.
9. Stop the machine, take away the work and clean the table and the work area.

For internal cutting follow these instructions:

1. Drill a small hole in the waste portion of the work at any points where there is a sharp change of direction in the curve.
2. Insert the blade through one of the drilled holes and fix the ends in the chucks tightly.
3. Follow steps 3 to 8 as for regular outside cutting described earlier on this page.
4. Raise the hold-down foot and take out the blade from the upper chuck. Remove the work.

Always remember:

- Maintain the oil level of the driving mechanism.
- Oil the motor periodically, unless it has sealed bearings.
- Release the tension on the blade when the machine is not being used.
- Apply a thin coat of oil on all machined parts when the machine will not be in use for a long period of time.

Planing machines

Planing is the second essential operation in the preparation of a work-piece after sawing. When workpieces are well planed, square and true, they produce a better job. The most efficient way of planing workpieces is using planing machines. They are designed for the high speed planing of true surfaces and edges (flat, smooth and square) for subsequent operations. The size of a planing machine is determined by the widest board it can plane. The common types are the jointer or surfacer and the thicknesser. The jointer (Figures 7.7a, 7.8a) allows planing to be carried out only on the surface of the table and is not self-feeding. It is a multi-purpose machine. Beside its principal function of planing the surfaces and edges of boards (for butt jointing, hence its name), it can be used for other operations such as rebating,

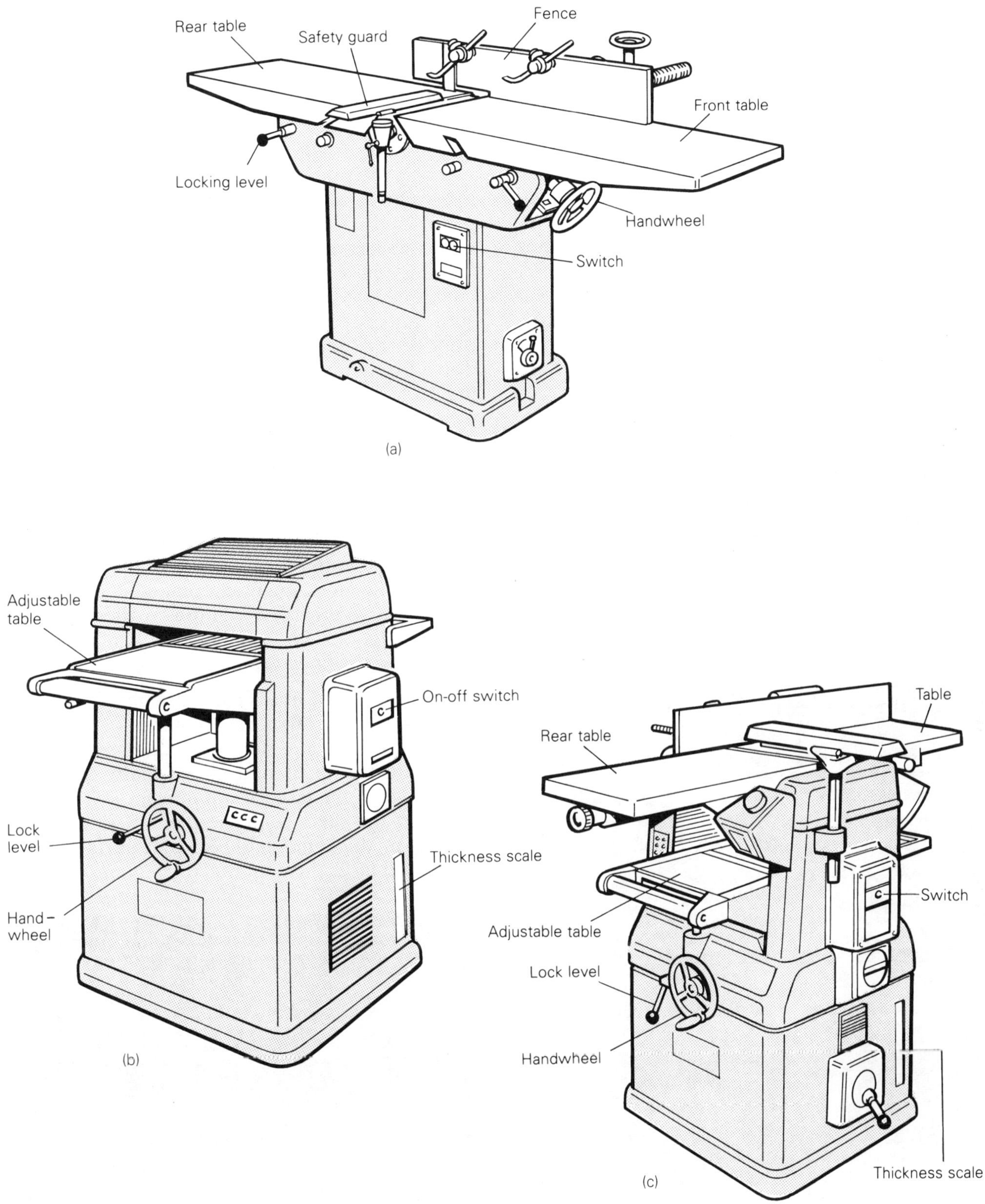

Figure 7.7 Planers: (a) jointer, (b) thicknesser, (c) combined jointer and thicknesser.

chamfering, and bevelling. It consists of a heavy cast iron base that supports the front and rear tables; a circular cutter block on to which the cutter is fitted; a fence that guides the workpiece; safety guards that cover the cutters to prevent injury to the operator's hands; a hand wheel for adjusting the tables; locking levers for locking the tables; on/off push button switches for starting and stopping the machine.

The thicknesser on the other hand, is a single purpose machine that is used mainly for planing boards to parallel and uniform thickness (Figure 7.7b). It is self-feeding and can have either a single-cutterblock or double-cutterblocks. The single-cutterblock has one cutterblock (cutterhead) located above the board to be planed. The table is mounted on vertical slides below the cutterblock and can be adjusted vertically to a required distance or height according to the thickness of the board being planed. The double-cutterblock type has two cutterblocks, one above the board being planed and the other below it to enable both surfaces of the board to be planed simultaneously. It is the most convenient for quick reducing or planing a work-piece to the exact and accurate thickness.

Both the single and the double-cutterblocks work on the principle that, as the board is fed into the machine, its top surface is gripped by a corrugated feed roller that moves it forward under the chip breaker and into the cutterblock. The chip breaker presses the board down in front of the cutterblock to prevent excessive chipping. A pressure bar is located behind the cutterblock, and runs on the planed surface, holding the board down firmly on the bed to prevent it from vibrating. Behind the pressure bar is the power outfeed roller that moves the board out of the machine. There are two (anti-friction) rollers directly under the infeed and outfeed rollers that are set slightly above the surface of the table to help carry the board forward. The table is adjustable to enable various thicknesses of board to be planed.

There is another combination of the jointer and the thicknesser, known as combined jointer and thicknesser, (sometimes simply called a thicknesser (Figure 7.7b, 7.8c)). To use a combined jointer and thicknesser:

1. Plane one of the faces straight (using the surface of the table as a jointer);
2. Plane one of the adjacent edges straight and square to the planed face, (again, using the surface of the table as a jointer);
3. Insert the piece with the planed surface lying flat on the inner table so that surface to be planed faces the cutterhead;

Figure 7.8 Planers: (a) jointer, (b) combined jointer and thicknesser.

4. Adjust the table upward to feed the depth of cut and the board travels over the cutterhead by itself;
5. Adjust the depth of cut again and repeat the process until the desired thickness is obtained.

When you are using a planer:

- Ensure that the cutters are always sharp, for dull cutters will cause vibration that may allow the board to kick back.
- Examine the board to be planed to check that there are no loose knots or splits that might break off in the cause of the planing.
- When planing, ensure that your fingers are at least 100 mm away from the cutters and never allow them to be above the cutters even if the guard is in place.
- Always put the guard in place, to cover the cutters.
- Before you start the machine, ensure that no one is standing in line with the table. You must also stand to side of the table yourself.
- It is dangerous to plane pieces of short length. They can only be done by using specially prepared push sticks.
- You need to use *both* of your hands to press the board down when feeding. When the board has moved about 150 mm passed the cutter, carefully move your left hand over to press the board down on the rear table.
- Remember to switch-off after you have finished using the machine and do not go away until the cutters stop rotating. *Never leave the machine running unattended.*

Drilling machine

The **drilling machine** (Figure 7.9) is used in the woodwork shop for drilling holes. It provides the rotary power for rotating twist drills, bits and plug cutters and many auxiliary attachments, such as grinder, sander, shaper, router and mortiser. These attachments enable various special operations to be carried out on the drilling machine.

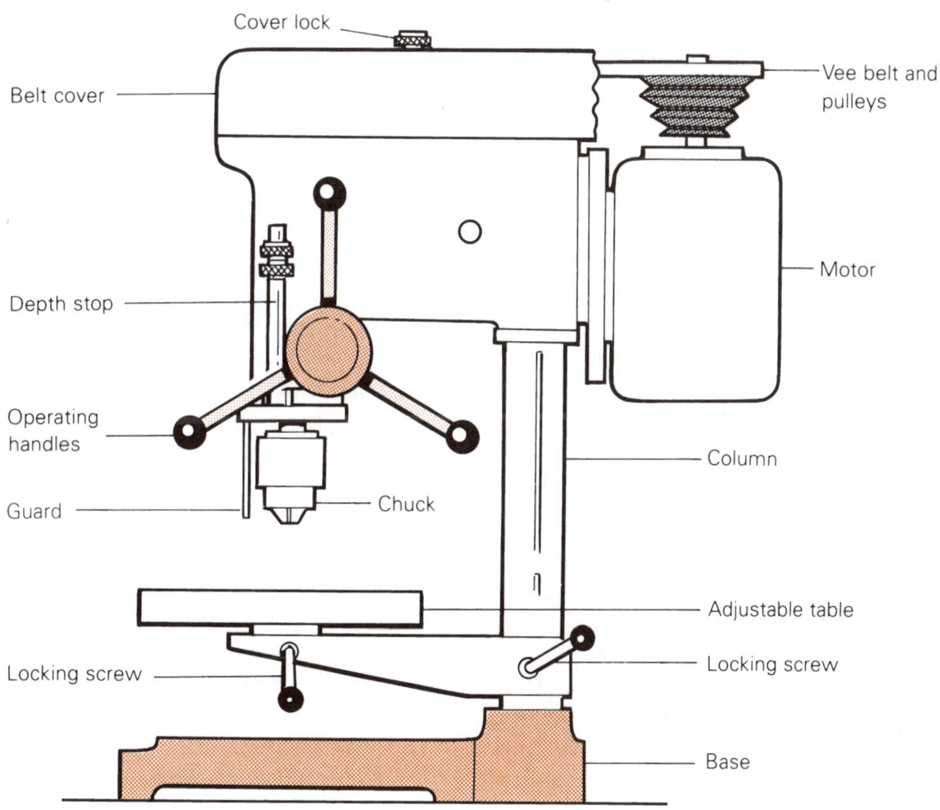

Figure 7.9 The drilling machine.

The principal components are:

- a cast iron **base** that supports the column and the other parts;
- the **column**, made of machined steel, on which the table, head and motor are fitted;
- a cast iron **table** that holds the workpiece (adjustable for height: some can be tilted to angles other than 90° to enable inclined holes to be drilled);
- a **three jaw chuck** for holding the drills and bits and tightened by a key;
- an **electrical motor** that provides the power for rotating the spindle, chuck and drill;
- a **vee-belt** that connects and transmits power through the motor pulleys to the spindle pulleys;
- the **spindle**, connected to the front pulley and into which the chuck is fitted;
- the **feed lever** which, working through a rack and pinion, is used for raising and lowering the spindle during the drilling operation;
- the **quill** that permits vertical motion without interfering with the rotary drive;
- the **head** that houses the pulleys, the bearings, the spindle and quill;
- the **start and stop switch**, used for starting and stopping the motor.

The drilling machine when equipped with attachments can be used for a range of operations other than the traditional use for drilling holes. A few selected operations are described here to guide you when performing them.

When drilling a hole:

1. Mark out the position of the hole.
2. Hold the workpiece in a vice or fasten it to the table firmly. Use waste blocks to protect the table.
3. Start the motor and bring down the rotating drill to the spot you marked on the workpiece to commence the cutting.
4. Apply a constant feed so that the shavings will be carried out of the hole and the drill will operate efficiently. Too slow feeding will create friction and overheat the drill, while too fast feeding may cause the drill to break due to excessive pressure.
5. When drilling a through hole, release the pressure on the feed as the drill is about to break through to avoid a split on the under surface.
6. If the hole is deep, withdraw the drill from the hole several times to clear the shavings from the flutes of the drill.
7. When you are drilling a series of parallel holes, use a fence as a guide to keep them in correct alignment.
8. Use a hole saw or circle cutter when making large holes.

Using the mortising attachment:

1. Fix the attachment firmly to the appropriate position.
2. Select the correct size bit and chisel and fit them in their respective holders. Allow the bit to project 3 mm below the chisel.
3. Fix the fence and hold-down fingers in position on the table to align the work with the sides of the chisel.
4. Adjust the table to the required height.
5. Adjust the depth gauge to the correct depth of the mortise.
6. Select the appropriate speed for the size of the bit being used (the larger the hole, the slower the speed).
7. Start the motor and begin to cut from one end of the mortise, making a series of parallel cuts toward the other end until all the waste material is removed.
8. Stop the machine and remove the work and the attachment. Clean the attachment and put it away safely.

Using the shaper attachment:
(The shaper attachment is used for cutting various forms of mouldings)

1. Insert the special adapter in the spindle to hold the shaper cutter. Do not use the normal chuck.
2. Select the appropriate shaper cutter and collar and fasten them in place on the spindle adapter. Lock the spindle in position.
3. In place of a collar, fasten the shaper fence in the desired position on the table to produce the correct depth of cut in relation to the vertical axis of the cutter,
4. Adjust the table to the required height and lock it in position.
5. Adjust the spring-tension guides to hold the work against both the fence and the table.
6. Select a high speed and make a trial cut on a waste piece.
7. Feed the work slowly into the rotating cutter from the left side of the table.
8. For a deep cut, make several light cuts to prevent splintering.
9. Stop the machine and remove the attachment. Clean it and put it away safely.

Using the router attachment:

1. Remove the chuck and insert the special spindle adapter in its place.
2. Select the appropriate router bit and insert it in the special spindle adapter.
3. Adjust and lock the table at the desired height.
4. Put in place the special fence on the table, or clamp a straight-edged piece on the table to serve as a guide.
5. Select the highest possible speed (5000 rpm) and start the machine.
6. Make several light cuts to remove the waste material in the groove, feeding the work from left side of the table.
7. When you are cutting a stopped groove or slot, use the stop blocks or make marks on the face of the work to show the limit of the travel.
8. Stop the machine and remove the bit and the adapter. Clean them and put them away safely.

Using the sanding attachment:

1. Remove the normal drill chuck and fix the special chuck.
2. Insert the sanding drum in the special chuck.
3. Secure a wood table plate on the machine table.
4. Adjust and lock the table at the desired height.
5. Lock the spindle at the correct position.
6. Select the appropriate speed (about 1200 rpm) and start the machine.
7. Hold the work flat on the table and start to sand, using a light pressure. Maintain even pressure and steady movement.
8. Stop the machine and remove the sanding drum and the adapter.

> When you are using a drill for any operation:
> - Wear safety glasses or goggles for all operations on the drilling machine.
> - Stop the motor before you change the position of the vee-belt to vary the speed.
> - Use the correct speed for each operation.
> - The table must always be firmly locked in position after any adjustment.
> - The work-piece must be firmly secured either on the table or in a vice.
> - When using the drilling machine, other than in a drilling operation, use a fence to guide the movement of the work in the correct position.
> - Use the guards at all times.

Mortising machine

Mortising is a process of making a recess half-way or completely through a piece of wood. It can be carried out by hand using a chisel, as described in chapter 5, or by the use of a power-driven cutting bit. The machine used for mortising is called a **mortiser** (Figure 7.10, 11). The two common types are the hollow-chisel mortiser and the chain mortiser.

The **hollow-chisel mortiser** is basically used for cutting a mortise when preparing a mortise-and-tenon joint, principally in furniture factories or door production factories. Nonetheless, it is also used in the school workshops. It is similar to

Figure 7.10 The mortising machine.

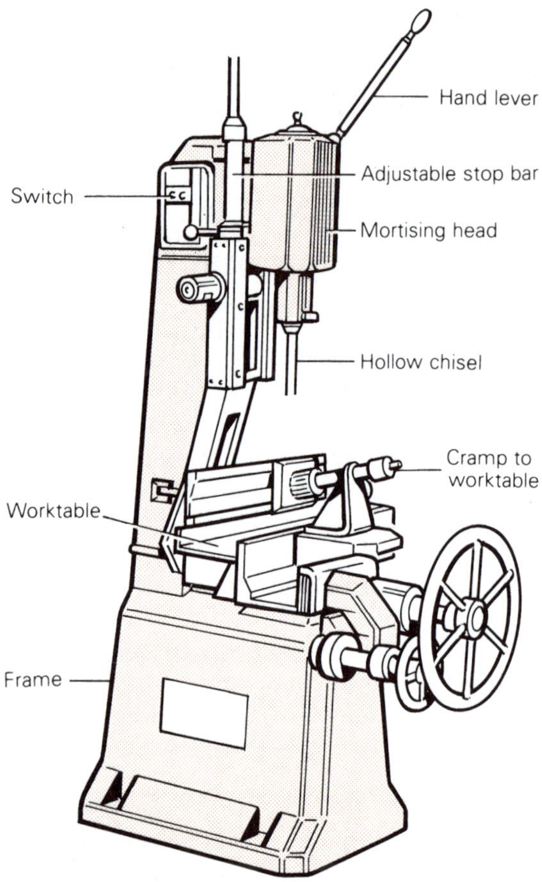

Figure 7.11 Parts of the mortising machine.

the drilling machine and is widely used because it is compact, easily adapted for light work and usable in multiple head assemblies. It consists of the following parts:

- a **frame** made of steel, supporting the table and motor;
- the **table**, made of machined cast iron, on which the work-piece is supported, with a mechanism that enables the table to be tilted to any desired angle and moved up and down;
- a **fence** and **cramp** fitted on top of the table for holding fast the work-piece;
- an **electrical motor** fixed at the top of the frame to provide power to drive the bit;
- a **chuck** that holds the bit and is attached directly to the motor spindle;
- **handwheels**, made of machined cast iron, fitted to the table and tilting mechanism for adjusting them;
- a high-carbon steel **bit** fitted into the chuck to bore the hole and carry the shavings away through the **hollow chisel** that surrounds the bit;
- a **treadle** fitted at the bottom part of the frame to lower the chisel and the bit. The treadle has a spring steel attached to it to return it to the up position.

To use the hollow-chisel mortiser:

1. Select the correct size bit and chisel for the mortise being cut.
2. Fix the bit in the chuck and the chisel in the holder. Allow the bit to project about 1.5 mm below the chisel to enable it bore a round hole before the chisel cuts the square hole. This also helps to avoid overheating and excessive pressure on the chisel.
3. Adjust the depth gauge to a height such that the depth of the mortise (stopped mortise) will be 3 mm deeper than the length of the tenon.
4. Hold fast the workpiece with the table cramp.
5. Set the table in such a way that the mortise will be cut in the correct position.
6. When cutting a number of mortises all having the same size and depth, adjust the stops so that each mortise will be the correct length.
7. Begin the cuts at one end of the mortise. To avoid the chisel being jammed in it, the first cut should not be too deep. Make the second cut adjacent to the first and a little deeper. Come back to the first position and cut to the required depth (bottom). Go to the second position and cut to full depth. Where the mortise is reasonably long, you may have to make more than two initial cuts, but you must follow the same procedure.

The **chain mortiser** is a variation of the hollow-chisel mortiser and it has similar parts as the latter, except that it has a chain saw type of cutter instead of the hollow chisel and bit. It cuts more quickly, but tends to cause splitting at the ends of the mortise when the chain is cutting. The chain consists of cutters made of special tool steel and ground with a hook of 25° on their outside. It is driven by a **sprocket** in a clockwise direction around a guide bar that can be adjusted for the correct tension of the chain. A metal **guard** covers the chains to protect the operator. Some guards are provided with a glass window to enable the operator to have a clear view of the work. A **chip breaker** is attached to the guard to enable the guard to rest on the workpiece and break the chips as the chain cuts the mortise.

The procedure for operating the chain mortiser is similar to that of the hollow-chisel mortiser already described.

Horizontal slot boring machine

The **horizontal slot boring machine** is used for boring holes, slots, grooves and mortises. It has a vertical cast iron **base,** carrying the electrical motor which moves in a vee-slide forwards and backward, at right angles to the work table. The **motor** is moved by means of a lever connected to a link attached to a sliding plate on which the motor is mounted. In front of the base is attached a cast iron **frame (kneel)** that supports the table. The frame or kneel can be adjusted to move in a vee-slide upward or downward by means of a handwheel. The **table** is made of a machined cast iron and is mounted on the top of the kneel. It can be made to traverse at 90° to the cutter. The **cutter**, in a form of a chisel bit is inserted into a **spindle** connected directly to the electrical motor. This means that the spindle and the chisel bit revolve at the same speed and direction as the motor. The workpiece is fastened to the table by a cramp attached to the table. The table can be held stationary by means of two **pins** (steel bars) fixed to the kneel under the table.

To use the hollow slot boring machine;

1. Select the correct size of bit and fix it firmly in the spindle.
2. Hold down the workpiece firmly on the table by means of the table cramp. Ensure the piece is parallel with the side of the table and perpendicular to the bit.
3. Adjust the table to bring the workpiece to the desired height. Move the bit forward to touch the piece to ensure that is will cut exactly at the position marked.
4. Start the motor, bring the marked area on the workpiece in line with the revolving bit and feed in to the required depth.
5. Withdraw the bit and take another adjacent cut. Repeat this operation until the required length and depth of the slot, groove or mortise is obtained.
6. To bore a slot to the exact length, let the first cut be close to one end of the slot and the second cut close to the other end of the slot. The subsequent cuts can be taken in between them. Move the table from side to side to clear out the wastes between the bored holes.
7. Switch-off the motor, remove the workpiece and clean the table and work area.

Tenoners

Tenoners are woodworking machines used for cutting plain tenons or shaped shouldered tenons such as those for making panel doors or sash windows. They can also be set up to cut various corner joints. Tenoners are of two models: single-end and double-end. The single-end tenoner consists of two tenoning heads, two scribing heads, a cut-off saw, and a movable carriage. The workpiece is clamped to the **carriage** that moves forward through the **tenoning heads** which cut the cheek and shoulder of the tenon. From here the piece may be passed through the scribing heads, mounted on **vertical arbors**, for contours to be formed on the shoulders if the tenons are to be fitted over moulded edges. The piece is then moved to the **cut-off saw** where the tenon is cut to length. The cutting heads have individual motors to provide power for the cutting and can be adjusted to various vertical and horizontal positions. The **scribing heads** can be tilted for cutting at angles.

The **double-end tenoner** consists of two cutter assemblies, mounted on a heavy base and one of these can be moved along the base to provide the setting for various lengths or widths of work. The workpiece is carried through the machine by two feed chains. It is held down on the feed chain by pressure beams, while a rubber padded chain moves on top of it.

Spindle moulder

The **spindle moulder** (Figure 7.12) is principally designed to cut mouldings and to form various decorative shapes on the edges of the work. It may be used to cut rebate joints or tongue and groove joints. It consists of a **base**, made of either cast iron or pressed steel that supports the table and the motor. The **table** is made of machined cast iron and supports the cast iron fence with wooden (maple) facing that serves as a guide for the workpiece. It has a vertical spindle projecting through the horizontal table. The **spindle** is a machined, high-grade steel shaft on to which the cutters are mounted. It can be raised and lowered to provide various positions for the cutter. The spindle on small machines is normally driven by a belt at speeds of 5000 to 10 000 rpm. The power

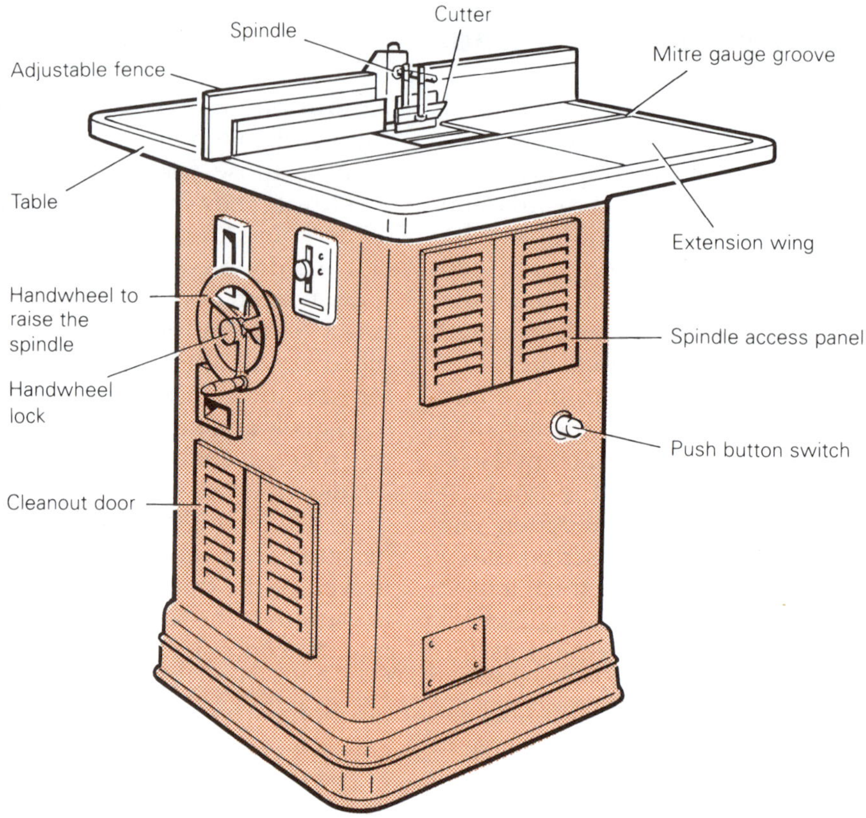

Figure 7.12 The spindle moulder.

is provided by an **electric motor** that has a reversing switch enabling the spindle to be driven round in either direction, allowing a variety of set-ups to be made. The **cutters** are made of alloy steel and are of various shapes. The two general types of cutter are the **assembled cutter** and **solid wing cutter**. The former, also called an **open-face cutter** is mostly used on larger machines for mass production work. The solid cutters are ideal for work in the school workshop because they are safer to use.

The choice of method of operation may depend largely on the shape of form of the work-piece and the number of duplicated parts required. Straight workpieces can be machined by simply running them against the fence while curved workpieces are shaped by using collars and a pin. Where the same shaping or moulding is required repeatedly, the workpiece is fastened to a template.

When shaping straight workpiece using a fence:

1. Select the correct cutter block and mount it on the spindle securely.
2. Set the work to the desired height by raising or lowering the spindle. Lock it firmly.
3. Adjust the fence to produce a cut of the desired depth and lock it firmly in position.
4. Put the guard in place.
5. Switch on the motor (without the work being fed) to ensure the correct direction of rotation of the cutter if the motor is reversible.
6. Make a trial run on a waste piece.
7. Feed the workpiece, pushing it against the fence throughout, using the pressure springs to hold the workpiece. *Keep your fingers well away from the cutter.*
8. Switch-off the motor after the operation and clean up when the cutter has completely stopped.

To shape a curved workpiece:

1. Select the appropriate cutter block.
2. Set the work to the desired height by raising or lowering the spindle. Lock it firmly.
3. Insert the pin at the appropriate place. If the cutter is to rotate clockwise, insert the pin in the hole on the left hand side of the spindle

and *vice-versa*. The pin must always be on the infeed side of the table.

4. Put the guard in place.
5. Ensure the motor is revolving in a direction opposite to that of the feed.
6. Make a trial cut on a waste piece of the same thickness as the workpiece and check for the depth of cut.
7. Feed the workpiece for the full length of cut, ensuring that it is in contact with the collar at all times. A continuous feeding will prevent the cut surface from burning. Keep your fingers well away from the cutter.
8. Switch-off the motor.

Shaping with template is a process very similar to the method of using a collar and pin – except that in this case it is the template that is run against the collar instead of the workpiece. You must however take note of the following points:

1. Prepare the required template. It must either be slightly larger or smaller than the workpiece. The size may be determined by the size of the collar, the size of the cutter and the required depth of cut.
2. Fix the workpiece to the template by means of a clamp or pins, allowing an equal overhang on all sides.
3. Follow the procedure described for shaping a curved workpiece with collar and pins.

▲ The spindle moulder (shaper) is considered the most dangerous machine in the school workshop and you need to observe the following safety precautions.

- Wear safety goggles.
- Keep your hands away from the cutter at all times.
- Ensure that the floor area is clear of debris and substances such as oil that might cause you to slip.
- Always ensure the spindle nut is tight.
- Check that all attachments and guards are firmly secured.
- Never try to shape pieces that are only of short length.
- Always feed the work into the cutter in opposite direction to the cutter rotation.

Wood turning lathe

The wood turning lathe (Figure 7.13) is mainly used for holding and rotating a piece of wood against hand-held turning tools to produce a variety of cylindrical and circular shapes. Wooden items such as tool and drawer handles, turned table and chair legs, bowls, lamp holders, flower

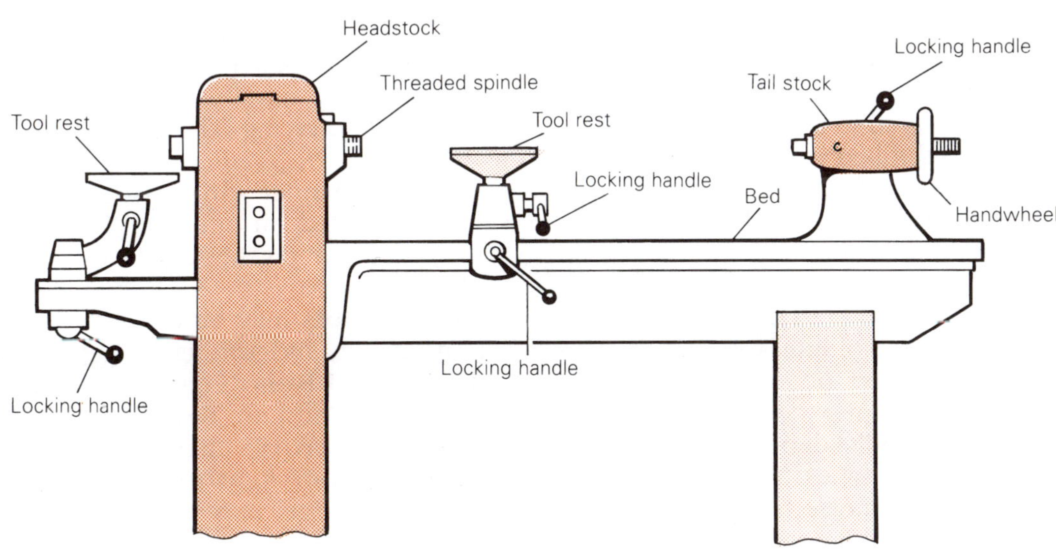

Figure 7.13 The wood turning lathe.

vases and other turned items are produced on the lathe. It consists of four main parts: the bed, the headstock, the tailstock and the tool rest, together with a number of other attachments.

The **bed** forms the main structure supporting the headstock, the tailstock, tool rest and the other attachments. A gap is provided on some lathe beds to enable large diameter discs, bases, platters and bowls to be turned.

The **headstock**, mounted on the left-hand end of the bed, contains a set of cone pulleys fixed to a **spindle** that revolves on ball or roller bearings. A morse taper bore is provided at the right-hand end of the spindle to hold spur centres and drill chucks whilst it is threaded externally to receive face plates and chucks. The left-hand end of the spindle also has a left-hand thread so that face plates can be attached for turning work of very large diameter. This end can also take a disc sander. The height of the spindle centre from the top of the bed determines the size of the lathe. This distance, however, does not refer to the maximum diameter of work to be carried out on the lathe. That is always twice the size of the lathe. For example, a 200 mm lathe will turn 400 mm diameter work and such a lathe is said to have a 400 mm swing.

The **tailstock** that is fixed on the right hand side of the bed is adjustable and can be moved along the bed and clamped at any required distance form the headstock. The end of the tailstock spindle also has a morse taper bore to take a centre or a drill chuck. The tailstock is generally used to support the end of a workpiece when turning 'between centres' (i.e., when the work is held between the headstock centre – the live centre – and the tailstock centre – the dead centre).

The **tool rest** is fitted in an adjustable bracket that is clamped to the bed in any required position between the headstock and tailstock. It can be raised or lowered on the bracket to a convenient height to suit the work being turned.

Other attachments or accessories (Figure 7.14) include the chucks, steady rest and boring guide. **Chucks** are used for holding workpieces and they are either threaded on to the headstock spindle nose or inserted into the morse taper hole at the end of the spindle depending upon the make. There are several types. The **prong chuck** (or **spur centre**) has a morse taper shank and is used for holding and driving the workpiece. The **cup chuck** is screwed on to the headstock spindle nose and has a hollow tapered mouth into which the workpiece is fitted. It is used for holding a workpiece when turning items that have to be hollowed out from the end, such as vases, mugs and egg cups. A **screw chuck** has either a threaded hole or a morse taper shank to fit the headstock spindle nose. It has a screw point at the face that holds the workpiece. It is a suitable holding device when turning small work, such as a knob, a small bowl or an egg cup. A **face plate** is a kind of chuck with a flat circular plate that is screwed on to the headstock spindle nose. There are several holes in the plate through which the work to be turned can be screwed. A **steady rest** is an attachment used for supporting a long, small-diameter workpiece to prevent vibration when the spindle is turning. It is fixed onto the bed at the appropriate position.

Turning involves removing waste wood from the piece being turned or shaped. The two methods normally employed for removing the waste wood are scraping or cutting. In this respect, there are two categories of turning tool: scraping tools and cutting tools.

Scraping tools (Figure 7.14) are those bevelled on one side only and are of different shapes. The common shapes are the flat, square nose also known as facing chisel; the spear point chisel and the round nose chisel.

The **flat**, **square nose (facing) chisel**, ranging in size from 12 mm to 25 mm is slightly bevelled along one or both edges for side scraping. It is used for surfacing and rounding convex surfaces and beads when face turning and also for forming rounds, beads and cylinders when centre turning.

The **spear point chisel** is ground to a point of about 60°. It is generally used for scraping vee-grooves, rounds and beads, square shoulders and also for flat surfacing. The useful sizes are 12 mm to 18 mm.

The **round nose chisel** is ground round and slightly bevelled on one side. It is used for hollowing bowls and other concave shapes and also for forming coves.

Cutting tools may be divided into two types: turning chisels and turning gouges.

Turning chisels (Figure 7.14) have straight cutting edges and are bevelled on both sides of the blade to enable them to be used in both directions (from right to left and from left to right). The edge may be square or skewed. Turning chisels are used on spindle turned work for smoothing surfaces after gouging out the balk of the waste. They are also used on straight, tapered and slightly hollowed work, for marking positions of details, squaring and rounding shoulders and for forming beads and convex shapes.

Turning gouges (Figure 7.14) are half-round in shape and are ground either square across or pointed on the outside (i.e., the convex side only). They are used for quick removing of waste on both face turned and spindle turned work and also for forming shapes such as coves or hollows when spindle turning.

A variety of operations can be performed on the wood turning lathe. Their names are normally derived from the method of holding the workpiece for the operation.

Spindle turning (Figure 7.15a) refers to the process of shaping a piece of wood held between the headstock centre and the tailstock centre where the grains of the wood run parallel to the axis of the lathe. It is also known as **turning between centres**. Scraping tools may be used but you need to sand off to remove the rough

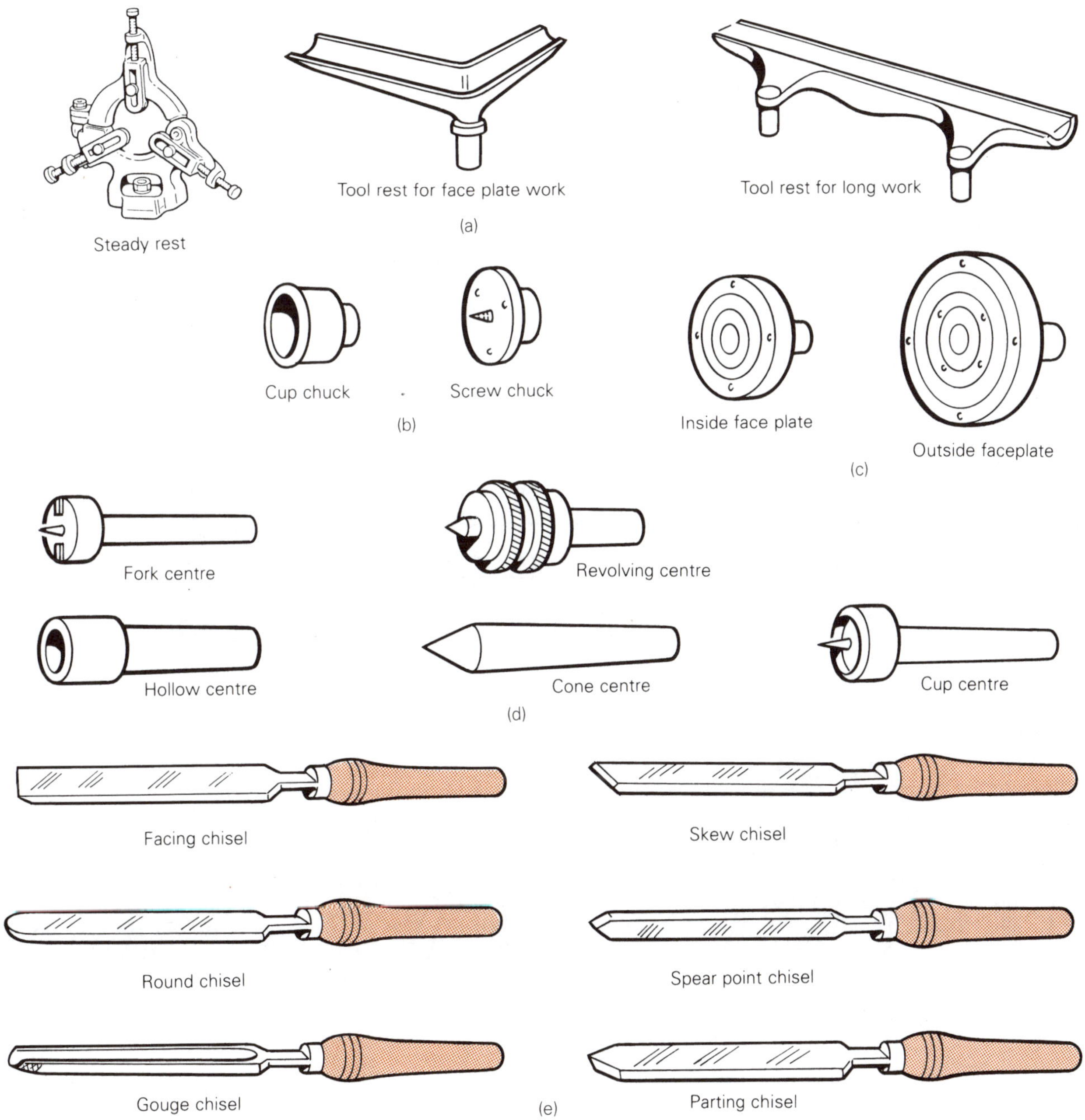

Figure 7.14 Lathe accessories.

surface. Examples of spindle-turned items are table and chain legs, furniture legs, lamp columns, tool handles and rolling pins.

Face turning (Figure 7.15b) is the term used to describe the process of producing circular edges and surfaces, and hollow ends of a workpiece mounted on a face plate. Workpieces that are face turned usually have their grains running at right angle to the axis of the lathe. Always select a speed appropriate to the diameter of the work or the type of wood you want to turn. Scraping tools are generally used for face turning. Examples of items that are face turned include bowls, bases for stands, circular stool seats and bread boards.

Cup chuck turning is the process of shaping a workpiece mounted in a cup chuck. The workpiece is first spindle turned and one end slightly tapered to be driven into the tapered mouth of the cup chuck. For internal hollowing, bore with a flat bit held in a drill chuck in the tailstock and then scrape with a suitable shaped scraping tool. Examples of cup chuck turned items are egg cups, salt and pepper shakers etc.

Boring is a process of producing a hole in or through a workpiece. When boring through spindle work, bore half way through from each end. Withdraw the bit frequently to clear the hole of chips. It is advisable to bore before you shape the work for easy manipulation and to avoid splitting – especially if you have a glued up workpiece. Plug the ends of the holes before you turn the shape. A lamp column is an example of a bored item.

The procedure for turning involves the following steps:

1. Mount the workpiece firmly.
2. Set the tool rest close to the workpiece and adjust it to the correct height.
3. Put on goggles.
4. Rotate the work piece by hand and then start the machine and allow the workpiece to gather momentum.
5. Hold the chisel or gouge flat on the tool rest with the left hand (for a right-handed person) and hold the handle in the right hand keeping it slightly above the horizontal so that the edge of the chisel pointing down is prevented from digging in.
6. For external shapes, work from large diameters to smaller diameters.
7. If you are hollowing out a bowl where the grain of the piece is across the axis of the lathe, work from the outside towards the centre.
8. If you are hollowing out on the end grain, such as an egg cup or a vase, work from the centre outwards.

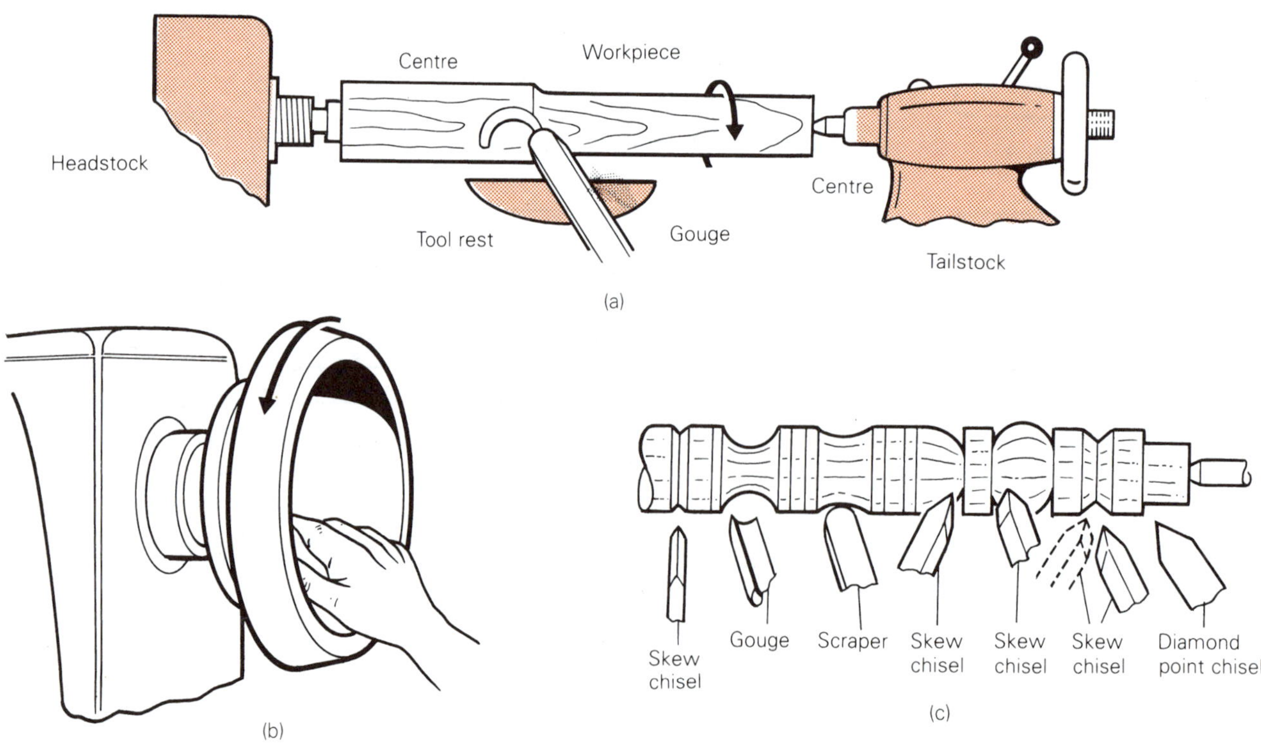

Figure 7.15 Wood turning operations: (a) spindle turning, (b) face turning, (c) applications of turning tools.

9. If you are turning the edges of a disc or a bowl, work across the piece instead of pressing directly towards the centre of the work.
10. Use a slow speed when you are working on a large diameter work, particularly when it is a glued-up piece.
11. Keep the chisel moving from side to side.
12. When you switch off the machine, allow the work to stop rotating before you remove it.

> ▲ When you are using a lathe:
>
> - Wear tight fitting clothing, an apron should be tied at the back.
> - Wear an eye shield or goggles.
> - Ensure the lathe is in a good working condition.
> - Check the workpiece to ensure it is free from splits and knots etc.
> - Take time to mount the workpiece correctly and securely. Apply grease or wax on the end of the tailstock centre if work is to be held between centres. Tighten the tailstock spindle clamp lever.
> - Ensure the tool rest bracket is clamped correctly and firmly.
> - Set the tool close to the workpiece (about 3 mm) to avoid trapping of the tool (chisel) and adjust for the correct height.
> - Select the correct tool (tools) and ensure it is correctly sharpened and in good condition.
> - While working pay full attention. Do not talk to others.
> - Stop the machine before you do any adjustment, for example, to move the tool rest closer to the work, to change the speed or to check the size of the work.
> - Never leave the lathe running unattended – for example, when searching for tools.
> - Sanding and polishing must be done with care. Remove the tool rest and bracket and change to slow speed.
> - Switch off when the work is completed, replace all fittings and tools in their proper places and clean the lathe.

Sanding machines

There are many kinds and sizes of sanding machines (Figure 7.16) ranging from small portable power tools to large multi-belt machines. The common types installed in the school workshops include, the spindle sander, disc sander and belt sander.

The spindle sander consists of base, table, spindle, motor and an oscillating mechanism. The **base** is made of either cast iron or pressed steel and supports the other parts. The **table**, made of machined cast iron serves as the working surface, supporting the work. The **spindle,** made of machined steel and rubber-coated, rotates and at the same time oscillates. The **motor** provides the power that causes the spindle to rotate. The **oscillating mechanism,** made of hardened steel, moves the spindle up and down as it rotates. The **spindle sander** (Figure 7.17a) is mainly used for

Figure 7.16 The sander.

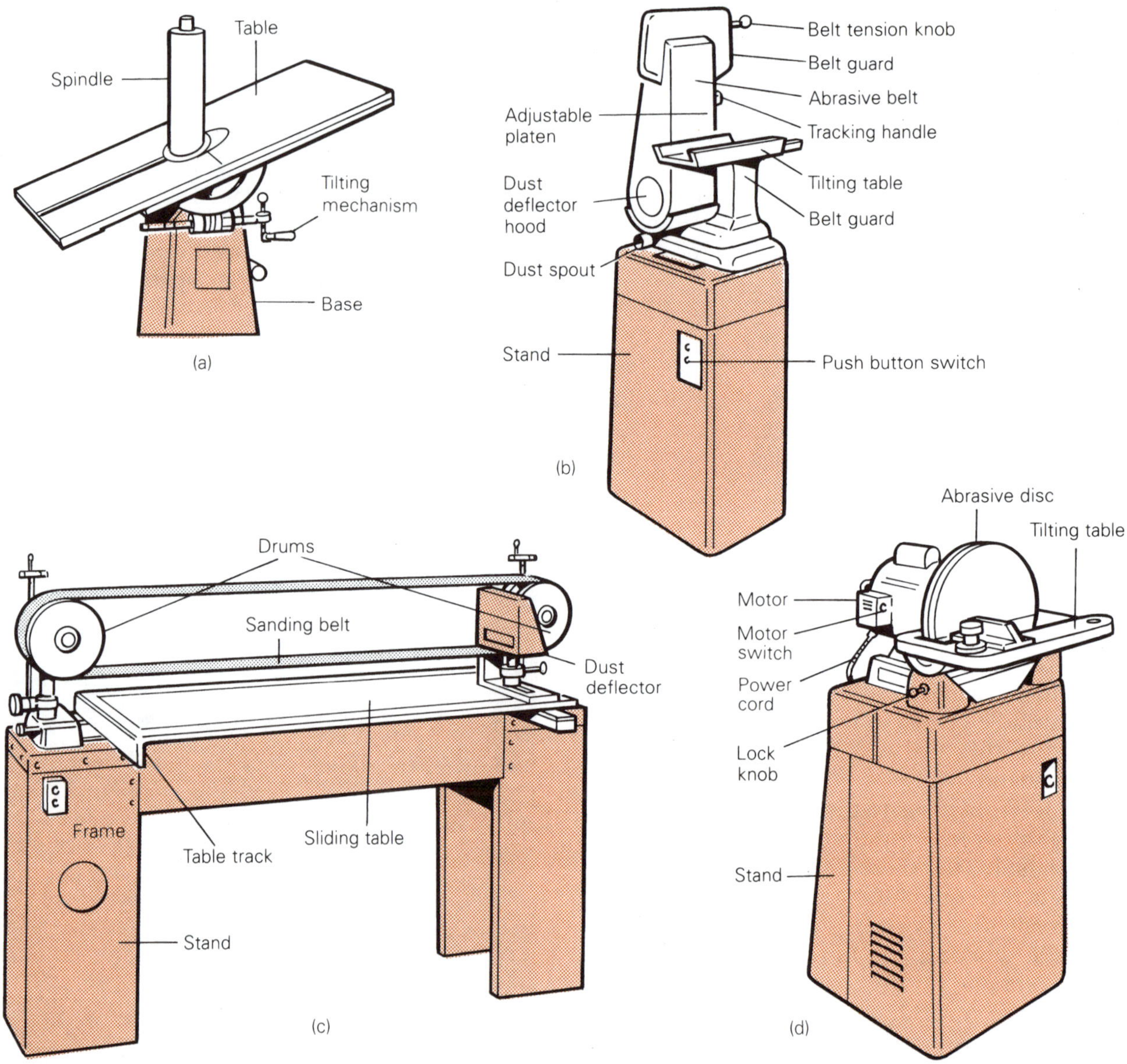

Figure 7.17 **Types of sander: (a) spindle sander, (b) disc sander, (c) floor-model belt sander, (d) horizontal belt-stroke sander.**

sanding concave curves on the edge of the workpiece.

The **disc sander** (Figure 7.17a) is used for sanding straight edges and convex curves on the edges of the workpiece. It comprises of a stand, table, motor and disc. The **stand** that supports all the other parts is made of cast iron or pressed steel and mounted to the floor. The **table** made of machined cast iron, is fixed on the top of the stand, and can tilt from 45° away from the disc to 20° toward the disc. It is used to support the work. The **motor** provides the power to rotate the disc. The disc, made from either machined cast iron or aluminium serves as a face or place to which the abrasive is fastened. The diameters of the disc range from 75 mm to 450 mm.

The **belt sanders** are of two types: the vertical type and the horizontal type (Figure 7.17c,d). They are used for sanding flat surfaces. The **vertical belt sander** is has a cast iron or pressed steel **stand** that supports the sanding unit and the motor; an **adjustable table** that supports the work and which tilts 45° away from the belt and 30° towards the belt; a **motor** that provides the driving power; and two **drums** (lower drum and upper drum), made of pressed steel or aluminium that drive the belt. The **lower drum** is connected to the motor and provides the driving force for the belt. The **upper drum** is adjustable to track and keep tension on the belt. Other parts includes the **belt-tension knob** and the **tracking handle**, used for adjusting the tension and positioning the belt on the sanding drum; an **adjustable plate** that supports the belt that keeps the belt from sagging when pressure is applied; **belt and pulley guards** that protect the operator; and the **dust deflector** and **spout** that collect and re-route the sanding dust. The **horizontal belt sander** has similar parts that performs similar functions as the vertical belt sander. The table for the horizontal belt sander slides, allowing lateral movement during sanding operations.

CHECK YOUR UNDERSTANDING

- The circular saw is one of the versatile machines in the woodwork shop because a variety of operations can be carried out on it. These include: ripping, cross-cutting, bevelling, chamfering, mitring, trenching, grooving, tenoning etc.
- The band saw, though is mainly designed for cutting external curved shapes, it can also be used for straight cutting or for cutting a combination of straight and curved work. In the timber industry, it is used for converting timber log into boards.
- The radial arm saw is traditionally designed for cross-cutting boards to length but with improvement in design it can take other attachments to enable a range of other operations to be performed. Besides the traditional cross-cutting, it can be used for grooving, trenching, mitring, bevelling, tenoning, shaping, routering, rotary planing, drilling, turning, sanding, grinding, etc.
- The jigsaw is of three types: the plunger-type, the vibrating-type and the rocker-arm type. Their main application is for cutting curves and regular shapes including internal shapes and curves.
- The planing machine is designed to plane surfaces and edges of a board at a high speed. The three main forms are the jointer or thicknesser and the combined surfacer and thicknesser.
- Modern drilling machines are equipped with attachments that enable a range of operations to be carried out. Besides the traditional application of drilling holes, the drilling machine can be used for mortising, shaping, routing, sanding, etc.
- The mortising machine is of two main types: the hollow-chisel type and the chain type.
- Tenoners are machines designed for cutting plain and shaped shouldered tenons. The two types are the single-end tenoner and the double-end tenoner.
- The shaper, also called the spindle moulder, is purposely designed for cutting various forms of moulding and to form a range of decorative shapes on the edges of a workpiece. It can also be used for rebating and for cutting tongues and grooves.
- The general wood turning lathe is purposely designed for producing a variety of turned items.
- The three types of sanding machine commonly used in a school workshop are the spindle sander, the disc sander and the belt sander.

REVISION EXERCISES AND QUESTIONS

1. Name the safety device attached to the circular saw that prevents the piece being sawn from kicking back at the operator.
2. Briefly describe how to rebate a piece of board on the circular saw.
3. In what way does the width of the band saw blade affect the radius of curvature of a cut that is possible?
4. Apart from drilling, what other operations can be performed on the drilling machine?
5. The spindle moulder (shaper) is one of the most dangerous machines in woodworking. State any four safety measures that you need to take when using the machine.
6. A variety of operations can be performed on the wood turning lathe. Name the major operations.
7. Name the types of sanding machines and state the major application of each type.

8 Portable power hand tools

Introduction

The use of portable power hand tools in the workshop and on site is ever-increasing. You must, therefore, know how to make use of these serviceable tools. They will enable you to work faster, with increased productivity, while at the same time cutting your costs and still maintaining a high and safe standard of workmanship. Since safety in the use of machines and tools is essential, the following safety precautions should be observed whenever you are using portable power hand tools:

1. Seek permission from your instructor before you use any portable power tool with which you are not familiar.
2. Do not use any power tool unless you have had proper instruction in its use.
3. A power hand tool must always be earthed while in use, unless it is double insulated.
4. Check each part of the tool and tighten any loose parts before you start to use the tool.
5. Always use the correct tool for the work to be done.
6. Disconnect from the supply point before making any adjustment.
7. Never drag the tool around by the cable.
8. Only ever use sharp cutters and saws: dull cutters and saws cause accidents.
9. Use the tool's safety guards correctly.
10. Allow the tool to obtain its maximum speed before applying it to the job.
11. Always make sure that the trigger switch is in the 'off' position before placing the plug in the socket.
12. Clamp the workpiece down firmly before you apply the tool to it.
13. Wear protective clothing and the correct protective equipment for the job. The equipment must include safety goggles, dust masks and safety helmets if you are working on site.
14. Concentrate on the work you are doing and do not allow your attention to be distracted.
15. Proper maintenance and servicing should be carried out regularly by a competent person.
16. Check the data plate, fixed to the portable power hand tool by the manufacturer, for the voltage range of the supply line. The power supply of a portable power tool can either be 240 volts mains supply or 110 volts (voltage reduced by a step-down transformer). 110 volts is usually used on site. The lower voltage decreases the effect of electric shock, should a fault occur. Most power tools manufactured now use the double insulated principle. In this case, the motor and other live parts have a second protective insulation barrier over and above the normal insulation as extra protection against electric shock. No earth wire is therefore required for these tools. All other tools that are not double insulated must be earthed.
17. All power tools which must be earthed have a three core cable which should be correctly wired up in accordance to the standard *colour code*. The colour code used for older tools is:
    - Green for the earth wire;
    - Red for the live wire;
    - Black for the neutral wire.

    Nowadays, tools have the new international colour-coded cable, which consists of:
    - Green and yellow striped for the earth wire;
    - Brown for the live wire;
    - Blue for the neutral wire.

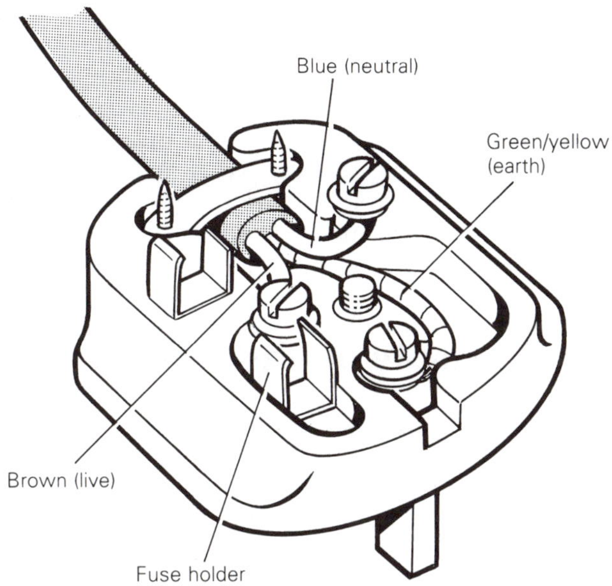

Figure 8.1 Wiring a plug.

18. The three core flexes should be connected to a three pin plug as follows (see Figure 8.1):
    - Green, or green and yellow striped wire (earth) to the largest terminal, which is marked E;
    - Red or brown wire (live) to the terminal which is marked L;
    - Black or blue wire (neutral) to the terminal which is marked N.
19. When the connection has been completed, tighten the cover firmly.

Portable electric saw

The **portable electric saw** (Figure 8.2a) is a power-driven rotary cutting tool with toothed, circular blade that revolves at an arbor speed of 3200 to 4600 revolutions per minute (rpm). It is a handy piece of equipment suitable for construction work. It is principally used for cutting boards to length and width and for making angular cuts. It is classified according to the size of the blade.

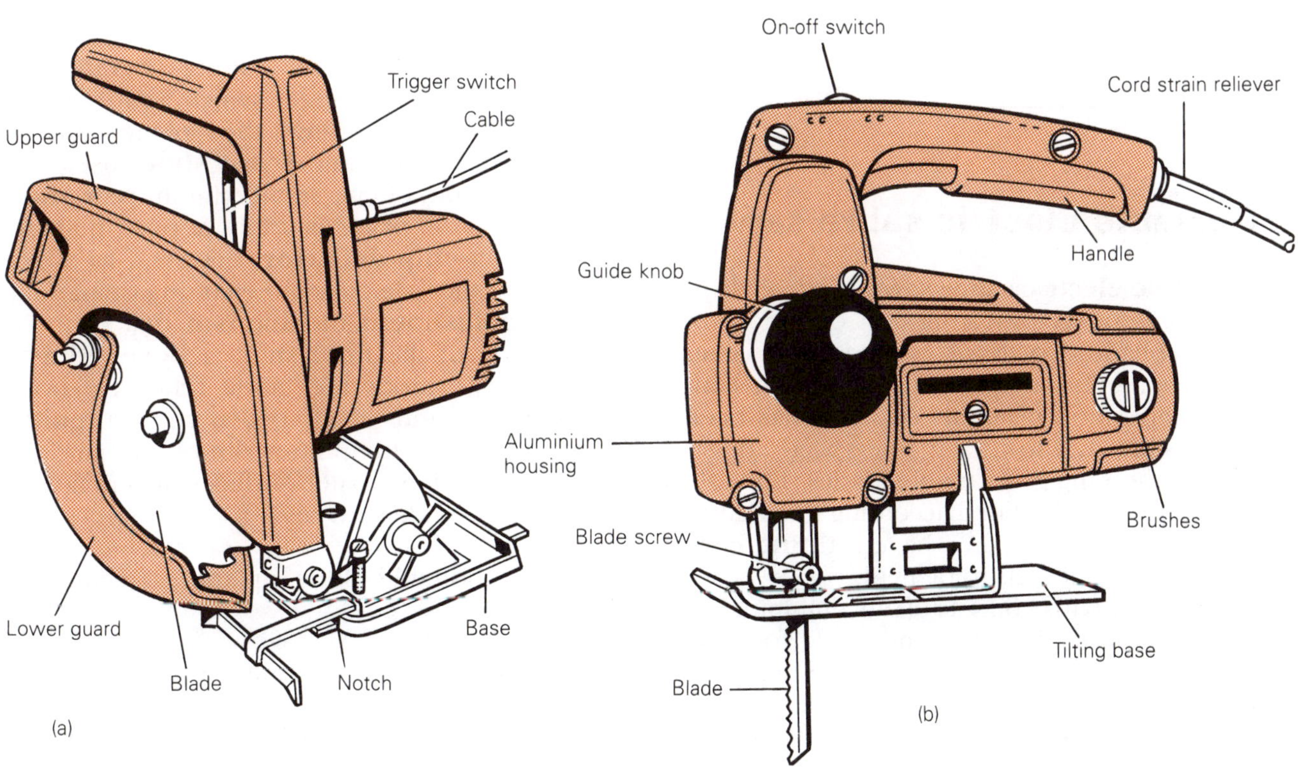

Figure 8.2 Portable circular saw and sabre saw: (a) circular saw (b) sabre saw.

The essential parts are an aluminium or plastic **base**, also called the **shoe** or **soleplate**, that supports the saw and is adjustable to control the depth of cut and the angle of the blade; a **motor** that provides the power; a **handle,** used for holding and controlling the operation; a **trigger switch,** built into the handle, used for starting and stopping the motor and also for quick control of the cutting action; a **front knob** providing additional holding and control by the other hand; an **upper saw guard,** covers the upper part of the blade; a **lower saw guard**, covers the saw at the lower part when the saw is not in use; and a high-speed steel **saw blade**, used for the cutting action.

> ▲ When using the portable electric saw:
>
> - Ensure that the saw blades are always sharp and properly set.
> - Lubricate the gear reduction unit periodically.
> - Store the saw properly when not in use to prevent rusting.
> - Use the saw only when the guard is in place.
> - Avoid cutting pieces of short length.
> - Always keep the electric cord (cable) out of the path of the saw.

Portable electric sabre saw

The **portable electric sabre saw** (Figure 8.2b) is a power-driven, reciprocating tool designed for cutting scrolls and patterns is wood and other workable materials such as metal, plastic laminates, composition materials etc. This has been made possible due to its variable-speed adjustments and the range of variable blades. Portable electric sabre saws are rated according to amperes under load (per load). They have a **tilting base** that enables angular cuts to be made.

The principal parts consists of: an aluminium or a plastic **housing**, designed and made for easy handling and it also guides the cutting action; a **motor** that provides the driving power unit; a **cam-and-pitmans unit** that changes rotary motion into reciprocating motion (up-down motion); a hardened steel **chuck**, used for holding the blade, some are adjustable to four positions; a **handle** that normally forms part of the housing is usual for carrying the machine and also for guiding the machine during operation; a **trigger switch**, housed in the handle and used as the starting control. Certain **guides** are available as attachments for ripping, crosscutting and mitre-cutting operations. Sabre saws take **blades** ranging from 56 mm to 60 mm long with a length of stoke ranging from 10.5 mm to 15 mm and speeds of 3000 to 4500 strokes per minute.

Portable electric plane

The **portable electric plane** (Figure 8.3) is power-driven rotary cutting-edge tool. The cutting is done by a continuous-motion shearing at speeds varying between 20 000 and 25 000 revolutions per minute. It is used for planing wooden surfaces and edges smooth and true in a similar manner to the hand plane, but with less effort and time. The two common sizes are the **regular power plane**, used for longer surfaces, and the **block plane** that is suitable for planing short surfaces.

The principal parts of the portable electric plane include: a cast aluminium or plastic **body** that houses the motor and the cutter; a high-speed steel cutting **blade** with a fluted, spiral cutting edge, used for the cutting operation; a **handle**, housing the trigger switch and used for holding and controlling the plane during operation; an aluminium **front shoe**, used for setting the depth of cut; an aluminium **rear shoe** that supports the plane as it moves over the path of the cut; a **chip deflector**, designed to direct the chips to the side during the cutting operation; a **depth-adjustment mechanism** that controls the position of the front shoe; a **bevel adjustment** that positions the guide fence for both outside-bevel and inside-bevel cuts; a **guide fence,** used for setting the width of cut, particularly when planing rebated surfaces. It is removed when planing wide surfaces.

To use the portable electric plane correctly and safely:

1. Hold the workpiece firmly in a convenient position on the work-bench.
2. Adjust the front shoe to set the correct depth of cut.
3. Check that the electric cable (cord) is long enough and properly connected to the power supply.

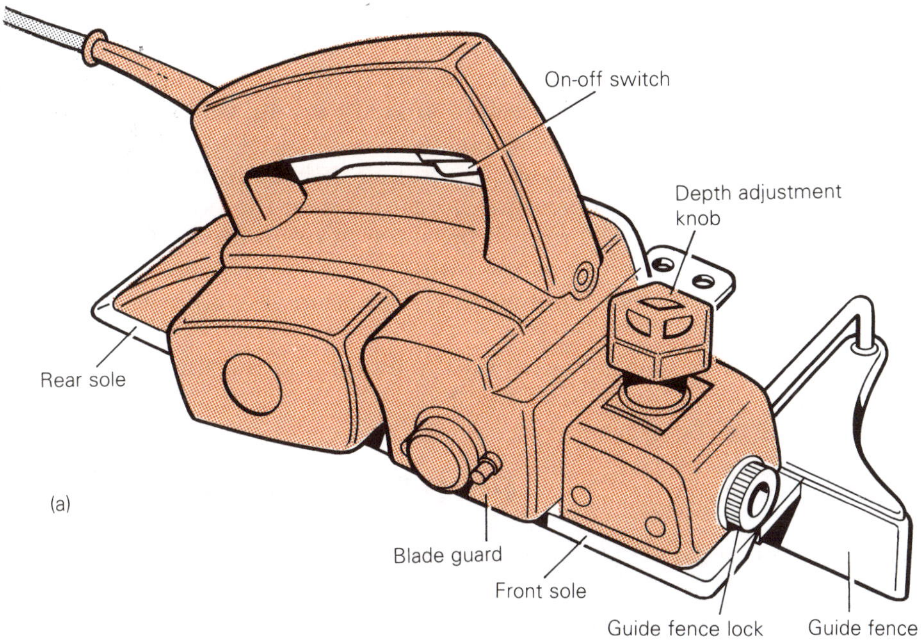

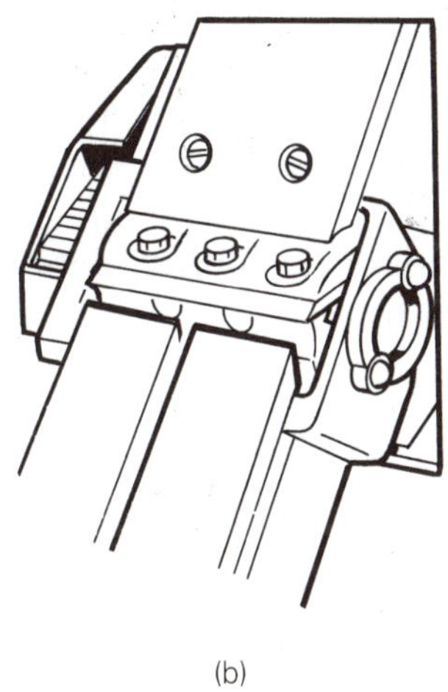

Figure 8.3 Portable electric plane: (a) portable plane, (b) plane from underneath.

4. Hold the plane over the workpiece with one of your hands on the handle trigger and the other on the front-shoe knob.
5. Switch on the motor, bring the front shoe down on the workpiece and begin the cutting operation. Maintain equal pressure on both hands. Finish the cut with pressure on the rear shoe.
6. To make a bevel cut, adjust the fence to the correct angle and follow the above procedure.
7. To plane surfaces that are wider than the width of the cutter, remove the fence and use your hands to guide the plane over the workpiece.
8. Switchoff the motor after working, clean it up and put it at a safe place.

Portable power drill

The **portable power drill** (Figure 8.4) is a motorised rotary driving tool that is operated by a small, high-speed electric motor having a gear-reduction driving unit. There are many sizes, to cater for a range of rotary cutting tools. Primarily, they are made for drilling operations, but they can be equipped with other attachments to enable them perform other operations such as sanding, polishing and grinding. The size of the power drill is determined by maximum opening of the chuck, i.e., the largest drill size that the chuck can accommodate.

The principal parts include: a **housing**, made of either plastic or aluminium, designed for holding, guiding, and applying pressure to the twist drill; an **electric motor** that provides rotary motion of the drill; **reduction gears** that produce a high-torque driving force; a **chuck** with three jaws made of high-carbon, to hold the drills firmly; and a **trigger switch**, housed in the handle, for starting and stopping the motor.

The portable power drill is supplied with the following standard attachments: **friction clutch**, used for driving screws; **flexible unit**, used for drilling in close quarters and at an angle; **drill stand**, used for supporting the portable electric drill to enable it to be used as a drill press; **drill fixture**, attached to the portable power drill for

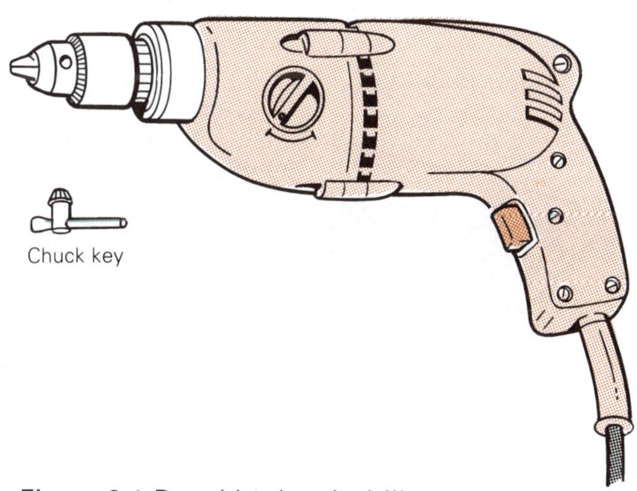

Figure 8.4 Portable electric drill.

drilling holes at 90° or at predetermined angles; **offset right-angle attachment**, used for drilling holes in close quarters. Besides these, other special purpose attachments are available that enable the portable electric drill to be used as a grinder, a buffer, a polisher, a jigsaw, a circular saw, a plane, a lathe, a sander etc.

The procedure described here shows the major steps that can be adopted when you want to use the portable electric drill for drilling holes in a piece of wood.

1. Mark and centre punch the position to be drilled.
2. Select the correct drill size and fix it firmly in the chuck.
3. Place the drill at the point to be drilled, positioning it in such a way that the hole will be drilled at the desired angle to the work surface.
4. Start the motor and apply light pressure, ensuring that the drill is square with the work or at the desired angle to it until the required depth is reached.
5. If you are drilling a through hole, relieve the pressure as you get near the bottom to prevent the drill breaking through the bottom surface, causing tearing.
6. Remove the drill from the hole while it still rotates.
7. Switch off the motor and put the portable drill in a safe place.

Portable hand router

The **power hand router** (Figure 8.5) is a power-driven shaping tool having bits that revolve at a spindle speed of between 5000 and 27 000 rpm, used for cutting mouldings, grooves for inlaying,

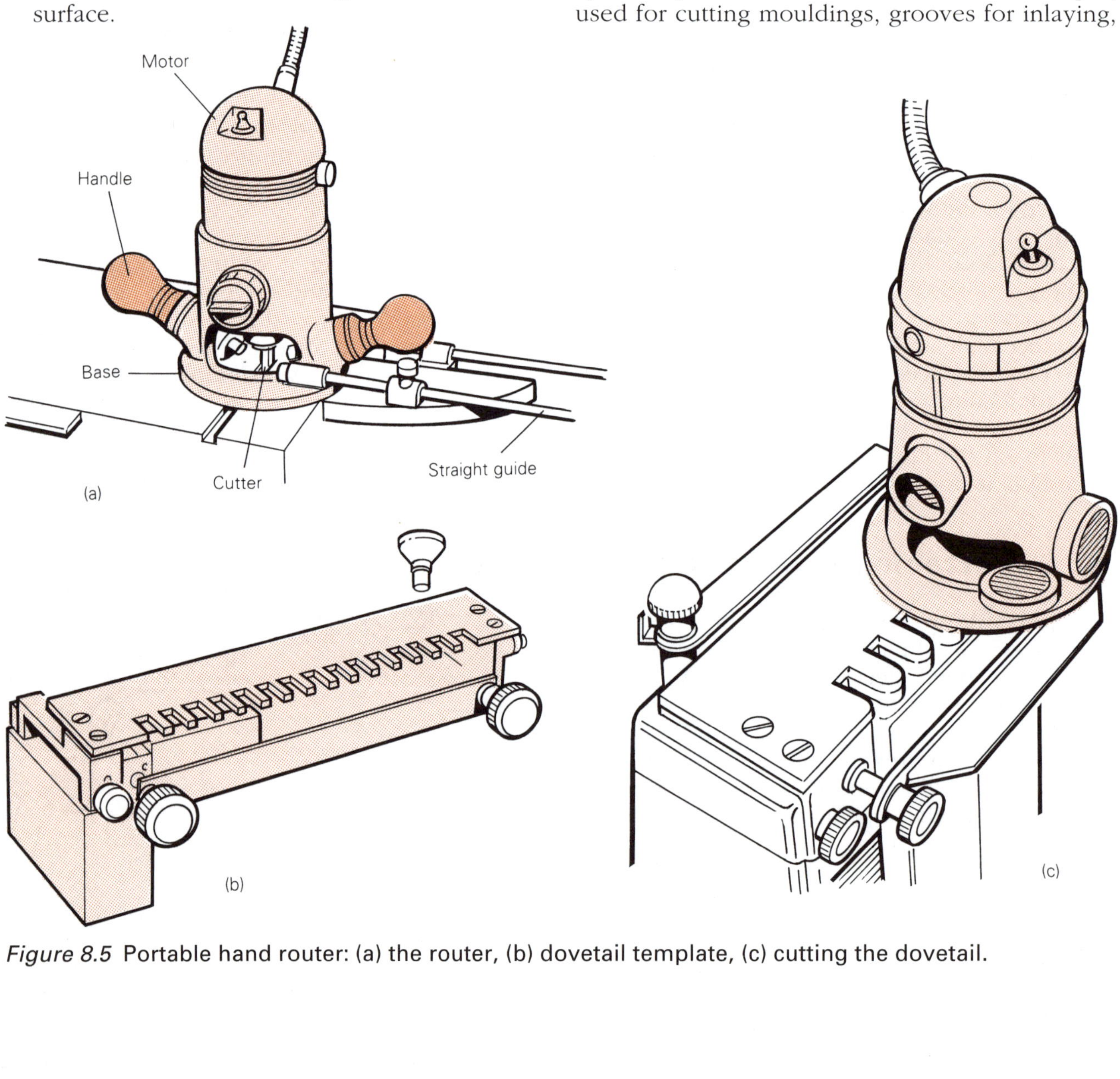

Figure 8.5 Portable hand router: (a) the router, (b) dovetail template, (c) cutting the dovetail.

dovetails, trenches, rebates, flutes, beading, decorative edges, etc. Portable hand routers are classified by the horsepower of the motor and the diameter of the cutter shaft. The capacity is determined by the depth and diameter of the bit that may be used. The main parts are: the **base**, made of machined cast aluminium, and adjustable to enable the depth of cut to be varied; the **motor** that provides the driving power; **straight** and **circular guides**, made of cast aluminium to act as a fence for routing out or shaping parallel to either a straight or a curved edge; **bits** made of either tempered, high-carbon, or tool steel, used for shaping the work; and a rosewood or plastic **knob** for holding and guiding the router.

In order that the power hand router may be used for a range of operations, certain attachments are available for use with it. These include the **dovetail template**, used for cutting dovetails; the grinding wheels; the attachment for sharpening bits; the carving attachment or pantograph attachment, used to follow a pattern when routing irregular shapes of duplicate parts; the rotary files; the circle-cutting device; the fluting and beading attachment; and the **router table**, used for positioning the router when used as a shaper.

The power hand router is used for carrying out a range of operations as described in the following procedures.

To cut a rebate, trench or groove with the router:

1. Choose a straight router bit of the correct size and fix it in the chuck.
2. Adjust the base to give you the correct depth of cut.
3. Set the straight guide to give the width of cut you desire.
4. Hold fast or clamp the workpiece on the table.
5. Start the motor and move the router over the workpiece. Hold the base of the router flat on the surface of the workpiece and the edge against the gauge. Then begin the cutting operation.
6. To cut a wider rebate, trench or groove, adjust the gauge and make additional cuts until the required width is obtained.
7. Stop the motor, remove the work and clean up.

When moulding:

1. Select the correct moulding bit, preferably the type with a pilot-tip so that the pilot tip serves to limit the depth of cut.
2. Fix the bit firmly in the machine and adjust the base to the height required.
3. When you are not using a pilot-tip bit, attach the gauge (straight or circular) and adjust it for the correct depth of cut.
4. Start the motor and move the machine slowly along the edge of the workpiece while the base of the machine is held flat on the surface of the workpiece. Ensure that the pilot-tip is at all times against the edge of the workpiece, or that the edge of the piece runs in contact with the gauge if a pilot-tip is not used.
5. Take precautions to avoid burning the moulded surface. Burning may occur when the machine is moved too slowly, the cutter is blunt, or too much pressure is applied to the pilot-tip against the edge of the piece.
6. Stop the motor after the operation, remove the work and clean the machine and work area.

Cutting dovetails with the power hand router is possible when it is used in conjunction with a dovetail jig, guide template, and dovetail bit.

The procedure is as follows (Figure 8.5):

1. Fix the guide tip to the base of the router.
2. Fix the dovetail bit in the chuck.
3. Adjust the base of the router for the dovetail bit to be set at the correct depth of cut.
4. Mount the dovetail jig (attachment) on the bench. Let the base project slightly beyond the front edge of the bench.
5. Clamp the workpiece that will take the tails (if you are making a drawer this will be the drawer side) to the side of the jig, with the face side facing you (outside). It should be in contact with the locating pin and should project above the surface of the jig a distance equal to the thickness of the second workpiece that is to take the sockets (the drawer front).
6. Place the second piece (taking the pins) across the surface of the jig, touching the locating pin and level with the top and of the first piece (the projecting end) with the face side upwards.
7. Fasten the guide template to the face of the jig lying flat over the second piece.
8. Start the motor, hold the base of the router flat on the template and slowly follow the recesses on the guide template to cut the tails and pins simultaneously.

9. Stop the machine, remove the guide template, remove the workpieces, take off the jig and clean up.
10. Fit the two pieces together.

Portable sanders

The **portable sanders** (Figure 8.6) are power-driven abrading tools used for sanding (smoothening) wooden surfaces. The three basic types are: the belt sander, the disc sander and the finishing sander. Each type is designed and made to meet a specific need. The capacity of a sander is determined by the amount and shape of the sanding surface of the machine.

The **belt sander** employs a coated-abrasive belt that revolves over a pad area by an idler and a driving drums. It is basically used for flush or regular sanding. The belts are made in various grades (grit numbers) and in a variety of widths (75–100 mm) and lengths (525–675 mm) to fit a particular make of machine. It consists of a **housing**, made from aluminium or plastic die-casting; a **motor** that provides the driving power; a **driving mechanism** that connects the motor to the driving drum through a direct gear, a rubber timing belt, a spur gear or a chain-worm gear; a **driving drum**, made of aluminium and covered by a rubber tyre on which the belt moves; a **tracking mechanism** that controls the alignment of the tracking drum to enable the belt to travel within the correct path; a **tension device**, a spring unit that maintains, automatically, the right tension of the belt by positioning the idler drum; a **striker bar**, a hardened steel bar that prevents the belt from running against the housing; a **sanding shoe**, positioned in line with the rim of the pulleys and covered by a pad and spring-steel strip, used to maintain even pressure on the belt during the sanding operation; a **vacuum system** to collect the sanding dust so that the belt does not get *clogged* and thereby promoting smooth sanding and prolonging the life of the belt; a plastic **handle**, specially designed and made to give positive control in any position; a **trigger switch**, built into the handle and used for starting and stopping the motor.

The **disc sander** has a coated-abrasive disc that rotates on a motor spindle. It is principally used where a scratch-free surface does not matter much. The disc can be fitted with a pad to enable it to be used for polishing.

It consists of: an aluminium or magnesium die-cast **housing**; a **universal motor** that provides the rotating power; a **driving unit** that connects the motor spindled (shaft); a **backing disc**, made of steel, on which the abrasive disc rotates on its surface pad; a **disc-retaining nut** that holds fast the abrasive to the centre of the disc; two plastic **handles**, built into the housing or attached to the housing, and used for holding and controlling the machine; a **trigger switch**, built into the handle and used for starting and stopping the motor.

The abrasive discs are available in diameters of 75 mm, 150 mm, 175 mm and 225 mm.

The **finishing sander** uses a coated-abrasive strip, fitted over a pressure pad that is driven either in an orbital or an in-line oscillating motion. It is classified according to the type of sanding motion it uses. The **orbital-motion type** is primarily used on flat surfaces for finish sanding. The in-line motion type with its back and fourth straight line cutting action is suitable for final sanding of wooden surfaces. Unlike the orbital type, the in-line type does not leave sanding marks on the surface.

The finishing sander consists of an aluminium or plastic **housing,** designed for easy handling; a **universal motor** that provides the driving power; a **driving mechanism** that provides the orbital or in-line motion for the sanding operation; a **sanding pad**, backed up with a felt or rubber that enables the pressure to be distributed evenly on the abrasive paper; **pad clamps**, made of spring steel and used for keeping the abrasive paper in place; **handles**, fitted into the housing to provide easy manipulation during the sanding operation (on some makes, a plastic **knob** is fitted to the front or side of the housing to provide further control); a **trigger switch** built into the handle for starting and stopping the motor.

To use the belt sander:

1. For sanding a flat surface, place the machine in a horizontal position; start the motor and hold the workpiece flat against the revolving belt; keep an even pressure on the whole surface.
2. For sanding straight or convex edges, place the machine in a vertical position, or at the desired angle if the edge is bevelled: hold the

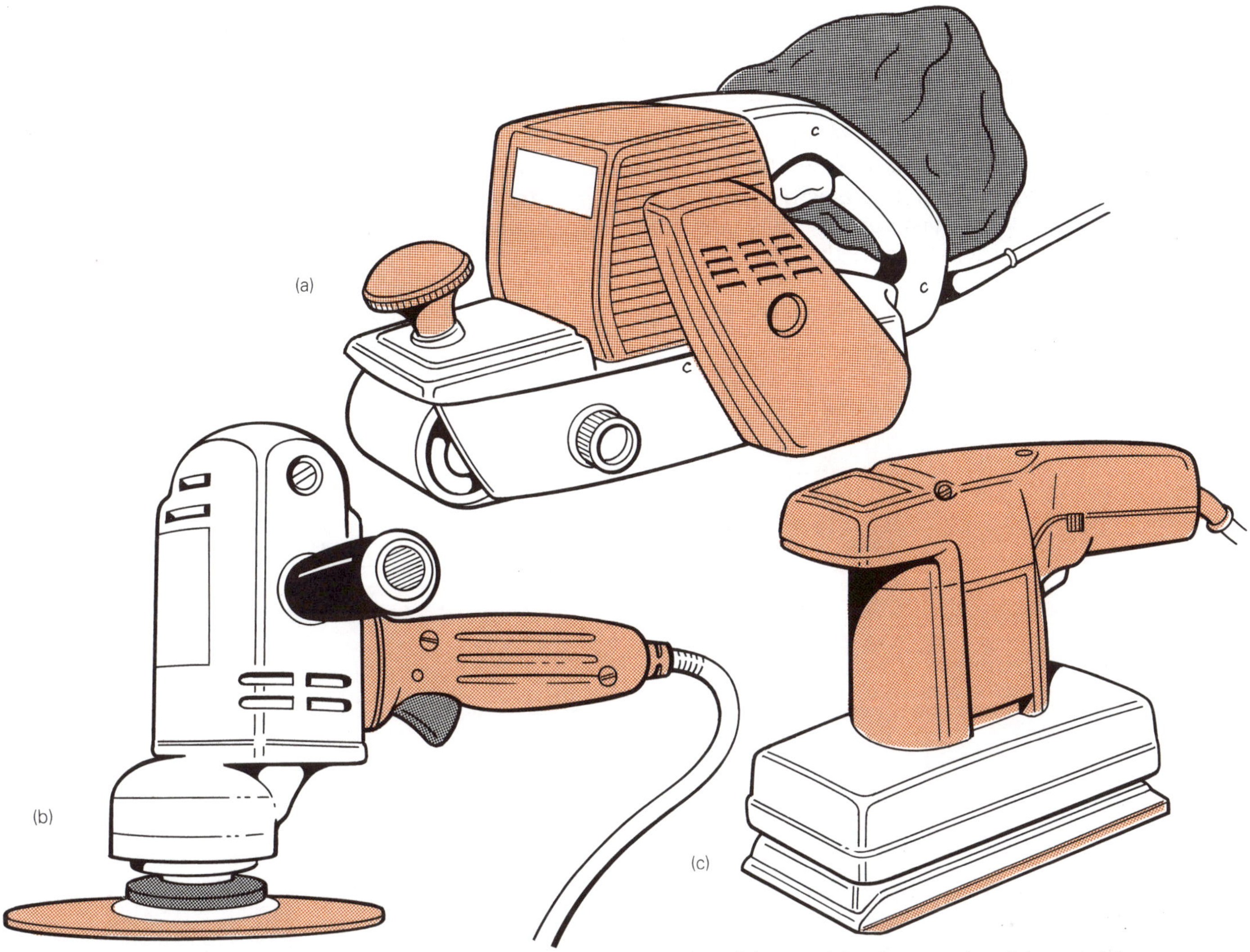

Figure 8.6 Types of portable sander: (a) portable belt sander, (b) portable disc sander, (c) portable finishing sander.

workpiece firmly against the table and gradually bring it into touch with the revolving belt, moving it along slowly until the entire edge is sanded.

3. For sanding concave surfaces, place the machine in a horizontal position and remove the top guard; bring the concave surface gently into touch with the curved top section of the machine; support the workpiece firmly with both hands during the sanding operation; replace the top guard immediately after completing the sanding.

CHECK YOUR UNDERSTANDING

- The portable electric saw is a power-driven cutting tool used mainly for cross-cutting and ripping boards.
- The portable electric sabre saw is a reciprocating, power driven cutting tool, used mainly for cutting scrolls and patterns in wood, metal, plastics and laminates, using the appropriate cutting blade for each type of material.
- The portable electric planer is a power-driven rotary cutting tool, used for planing wooden surfaces and edges smooth and true.
- The portable power drill is a motorised rotary driven tool that is primarily used for drilling holes, but which can take other attachments to enable it to be used to perform operations such as sanding, polishing and grinding.
- The power hand router is a power-driven shaping tool used for cutting mouldings, grooves, trenches, rebates, flutes, dovetails, beading, etc.
- The portable sander is a power-driven a brading tool used for sanding surfaces of a piece of cabinet work. The basic types are the belt sander, the disc sander and the finishing sander.

REVISION EXERCISES AND QUESTIONS

1 Explain how the depth of cut and the angle of the blade can be adjusted to suit the job when you are using the portable electric saw.
2 How can you distinguish between the portable electric saw and the electric sabre saw?
3 Name the two common types of the portable electric plane and state the main application of each type.
4 Name any three standard attachments that are supplied with the power drill and state the function of each.
5 Explain how the power hand routers are classified.
6 How can the capacity (size) of the power hand router be determined?

9 Timbers

African timbers

The tropical forest zones in West and Central Africa abound with species of timber that have for many years, been a major foreign exchange earner for the countries in those regions. Ghana, Nigeria, Cote d'Ivoire, Cameroon and Sierra Leone are among the countries in the tropical forest zones of West Africa. The timbers in the tropics are predominantly pored wood (hardwood) of many different species. A few of the common ones are covered here.

Abura (Subaha) is a West and Central African tree that grows to about 40 metres in height and 4 metres in girth. Its timber is light, pinkish brown and of medium weight. It has a distinct growth rings and moderately straight or interlocking grains with a fairly even and fine texture. It is resistant to acids, particularly sulphuric acid, but is easily attacked by pin-hole beetles. It is used for making plywood, doors, mouldings, turned products and furniture.

Agba is a yellowish brown timber mostly found in West African and Central Africa. It has straight, even and fine grain. It is easily seasoned and easy to work with. It is widely used for making cheap furniture and can be easily stained to resemble other timbers.

Danta is a West African tree that grows to about 27 metres in height and 3 metres in girth. Its timber is hard, tough and of medium weight, with a fine and even texture, and the grain is often interlocked. It seasons fairly well but has a tendency to warp. It is resistant to fungus and insect attack. It is easy to work with both hand and machine tools; takes screws and glues well; can easily be turned; gives a good finish and polishes well. It has numerous uses, including furniture construction, flooring, railway carriage and vehicle bodies (especially mammy trucks), railway boat sleepers, boat building, tool handles, gun stocks and general construction work.

Ebony is one of the West Africa trees that does not usually grow very high or huge, it is normally about 12 metres in height and 90 cm in girth. It looks pinkish when freshly sawn, but when dried, looks grey or grey-brown. It is heavy, tough and *resilient* (springy). It has a broad, light-coloured sapwood and small, dense heartwood. It is very durable but can be attacked by pin-hole borers. It works well but, owing to its hardness, it has a dulling effect on cutters. It finishes smoothly and polishes very well. It is used for making handles and hard-wearing tool bodies, pit props for mines and for carving and turnery.

Makore is a West and Central African timber with a pinkish brown to purplish colour and an interlocking grain that calls for care in seasoning. It is however a reliable timber when dry. It is predominantly used for the manufacture of veneers. It is also used for good class furniture and panelling.

Mansonia is a medium size tree that grows to about 30 metres in height and 3 metres in girth, and is common in West Africa. It has a highly coloured sapwood and light to dark-brown heartwood. It is strong and hard, with a fairly close-grained, fine and even texture. It is resistant to shock, loads, and decay, but the sapwood is attacked by pin-hole borers. It works easily with hand and machine tools and has less dulling effect on cutting tools. It takes nails, screws and glues well. When planing this wood, take care to prevent the sawdust from getting into your throat and nose, as it has an irritating effect. It is used widely for high-class furniture, interior decoration and piano cases. It is also used for dashboards of motor cars, shop fittings, carving and turnery.

Nyankom is a West African timber with a varying red, and sometimes purplish, colour. It has a medium to coarse grain and quarter-sawn boards are well figured. It is generally used for furniture and fittings.

Mahogany is a large tree that grows to about 58 metres in height and 6 metres in girth. It is found in West and Central Africa. When freshly converted (processed) it looks pinkish but it becomes reddish when seasoned. The sapwood is yellowish-brown. The grains are sometimes straight but generally interlocked to produce a striped figure on quarter-sawn boards (see Figure 9.6 on page 123 for different ways of sawing logs). It has a medium to coarse, but even texture. It seasons easily with less splitting or checking but is liable to warp. It is easily attacked by powder post beetles and pin-hole borers. It is easy to work using hand or machine tools, and has good holding power for nails, screws and glues. It is used for veneering and high-class plywood manufacture, furniture, panelling and interior fittings.

Obeche, also called **wawa** in Ghana, is a large West and Central African tree that grows up to 55 metres in height and 5 metres in girth. Its timber is soft and light and has a pale yellow or straw colour. It has no clear distinction between the sapwood and heartwood. It has typically straight-grained and medium, but even, texture. It is less resistant to decay and liable to blue stain. It is very susceptible to attack by borers. It works very easily with all cutting tools and finishes well. It is easy to nail and screw, but does not hold these fasteners strongly. It is used for kitchen and utility furniture, for making shelves, boxes and packing cases, It is also widely used for the corestock of plywood and for making formwork in building construction.

Odum, also called **iroko** is a tree that grows tall and large, reaching about 60 metres in height and 7 or 8 metres in girth. It is hard, strong, very durable, well figured and of medium weight. It is yellowish when freshly converted, but becomes fairly dark when exposed to sunlight. The texture is medium to coarse, sometimes interlocking. It seasons (dries) well under natural conditions or in the kiln. It sometimes has the tendency of developing hard calcareous deposits that can cause damage to the edge of cutting tools. It is widely used for constructional work and joinery such as door and window frames, cabinet work, interior fittings, staircases, flooring, boat building and agricultural implements.

Ofram (afara) is a large tree that grows in West Africa, up to about 48 metres in height and 4 metres in girth. Its sapwood and heartwood are indistinguishable and generally of yellowish-brown colour. It seasons well and is stable. However, it is not resistant to decay and is liable to attack by pin-hole borers and termites. The sapwood is particularly liable to attack from powder post beetle. It is easily worked using both hand and machine tools and has little tendency to blunt cutting tools. It is used for general interior joinery, light-duty flooring and the manufacture of veneers and plywoods.

Sapele is a large tree that grows up to about 60 metres in height and 6 metres in girth. It is predominant in West Africa and East Africa. It is moderately hard and of medium weight. It has a pale sapwood and a pinkish-red heartwood when freshly converted but becomes red-brown when seasoned. It has a double spiral grain that produces a very regular stripe or roe figure in quarter-sawn stock that shows off as long parallel light and dark streaks. The texture is usually fine and even. A quarter-sawn stock seasons fairly well and with less distortion than many woods (but plain sawn boards are liable to distortion and split). A quarter-sawn stock can reasonably resist insect attack but the sapwood is susceptible to attack by powder post beetles. It works easily with both hand and machine tools, takes nails, screws and glues well. It is used very largely in the form of veneer, in the production of furniture, cabinet works, interior decoration, panelling, shop fittings, boat building and fittings and floorings.

African Walnut, mostly found in West and Central Africa is a large, buttressed tree that grows up to about 45 metres in height and 4 metres in girth. It has a narrow light-brown sapwood clearly distinguishable from the golden-brown (often black-streaked) heartwood. It is moderately hard and of medium weight. The grain and texture are similar to those of mahogany (medium to coarse but even). It is susceptible to pin-hole borer attack and the sapwood can be infested easily by powder post beetles. It is easy to work using hand or machine tools and takes nails, screws, glue and polish well. It is used for furniture, panelling, decorative veneers, interior joinery, shop fittings and gun stocks.

The common timbers in South Africa include, Cape boxwood, Cape beech, bloodwood, muninga and stinkwood.

Most of the African countries export a great

number of their primary species of logs of timber to other countries, especially Europe, in order to earn foreign exchange to import other essential commodities. The export price of processed (converted) timber is higher than the price of unprocessed timber. In view of this, it is advantageous for timber-exporting countries to deal more in processed timber as this brings in more income to meet their import demands.

It must not be forgotten that trees play an important role in the environment and over-exploitation of wood for manufacture or export can cost society dearly, especially if the trees are not replaced or replanted immediately they are cut down. If this is not done, in the long run this kind of resource will be depleted to the detriment of mankind and society in general.

European timbers

The forest regions in Europe produce a variety of species of pored and non-pored trees, used as timber for export and local use. The most popular ones are discussed here. Note carefully their origins, characteristics and applications.

Pored wood (hardwood) timbers

Ash is a tough and durable timber obtained in Northern Europe. It has a cream to pale brown colour. It is noted for its good bending qualities and is used for making high quality furniture, hockey sticks and hammer handles, etc.

Beech is a close-grained timber, common throughout Europe. It has a pale pink colour and can bend easily. It is commonly used for making furniture, mallet shafts and rolling-pins, and is popular for wood turnery.

Boxwood is a very durable timber, with a yellowish-white colour. It is obtained in Northern Europe. It is hard, stiff and very tough, does not splinter, has good resistance to wear, and is fairly resistant to borers. It is commonly used for making chisel handles, mallets, plane stocks, flooring, weather boards and general construction.

Elm is obtained in Britain and is a tough and durable timber, having a pale brown colour. It is used for making coffins and for coach-building, boat-building and also for some types of furniture.

Hickory is a tough, elastic, and durable timber found in Northern Europe. Its qualities have made it a suitable timber for manufacturing sports goods such as tennis racket handles and skis.

Spanish mahogany is red-brownish in colour with excellent working qualities. It comes from Spain. It is extensively used for wood carving because of the excellent working properties. It is also used for furniture, veneers, wood turning and ship building.

Oak is a strong, durable timber obtained in Britain. It has a golden brown colour with a highly figured grain. It is mainly used for furniture, veneers, boat-building, wine casks, parquet and block flooring, decorative articles, panelling, mouldings and ornamental fittings.

Sycamore is a strong, close-grained, creamy-white timber found in Northern Europe. It is widely used in wood turnery, and also for wall panelling, parquet flooring and veneering.

Non-pored wood (softwood)

Among the non-pored wood obtained in Europe are: larch, pitch-pine, yellow pine, redwood, Sitka spruce, European spruce, and yew.

Larch is a very durable timber with reddish-brown colour. It is commonly used to make pit-props, gate-posts and fencing.

Pitch pine is a strong and durable timber with high resinous content and has a yellowish-brown colour, with conspicuous dark-growth rings. It is widely used for railway sleepers and heavy gate-posts. It is also used for making stairways and decorative panels.

Yellow pine is not as durable as the pitch-pine and has a light colour but if fully matured, it is yellowish. It works easily and does not warp or twist readily. It is widely used by engineer's pattern makers.

Redwood, sometimes called **Scots pine** or **red deal**, is reddish-brown is colour and a durable timber. It is used for flooring, rafters and for other building construction works.

Sitka spruce is a pale straw colour and reasonably durable. It is used in the aircraft industry, making ships' masts and in interior building work.

European Spruce also called **white deal**, is a fairly durable timber with white colour. It is used for coach building, making packing-cases and construction work in the building industry.

Yew is a durable, an expensive timber in limited supply. It is golden in colour and has varying grain. It is used for constructing door-frames, window sills, windows, cladding, and for interior joinery, such as skirting, staircases, doors, interior fittings, shop fittings and radio and TV cabinets.

Classification of timber

Timber is a natural product of solid wood from a tree which has been sawn to sizes suitable for building and construction purposes. Timbers are also used for making a variety of furniture in homes and offices and for tool bodies, such as for wooden planes and handles for chisels and saws.

A study of timber must begin with the **living tree** (Figure 9.1) from which it is obtained. A living tree basically has three main parts; the **roots**, **trunk** (stem or bole) and the **crown** that is composed of the branches and the leaves. The **roots**, which are embedded in the soil, support the entire tree. They also absorb water and mineral salts from the soil, which are transmitted through the sapwood of the trunk and branches into the leaves. The **trunk** supports the crown and it is from the trunk that the logs that are converted to commercial sizes of timber are obtained. The principal function of the leaves of the **crown** is to process the solution of water and mineral salts into food for the tree.

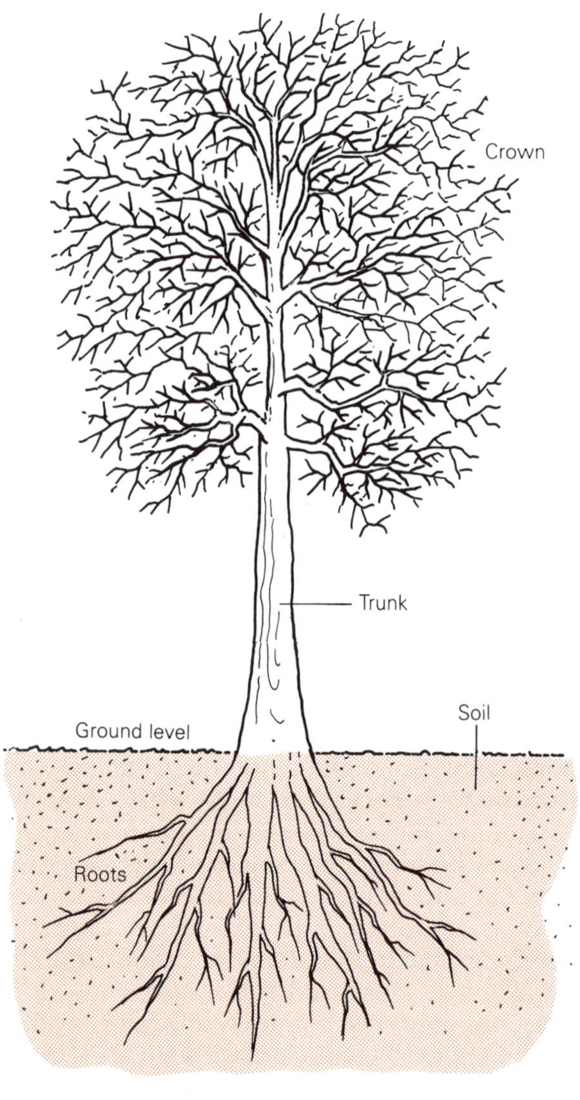

Figure 9.1 Parts of a living tree.

The **growth** of the tree takes place at the tips of the tree's branches, in its roots and around its circumference (girth). It becomes taller and taller as new buds produce growth at the tips of the branches, and thicker as layers of wood (**cambium layers**) are added to the perimeter of the tree. New wood tissue is formed on the inside and a new cover (**bark**) is formed on the outside. The process continues as long as the tree lives. This type of growth also accounts for expansion of the tree's root system. With the aid of the sun and by means of a process known as **photosynthesis**, the green pigment **chlorophyll** in the leaves uses carbon dioxide from the atmosphere while giving out oxygen. The carbon dioxide is used to change the sap chemically into a more refined sap which contains sugar and starches. The refined sap, that is the food for the tree, travels down the leaves and the branches through the cambium layer to feed all parts of the tree.

The tree, besides providing logs for timbers, is also a source of fuel; particularly in the developing countries where it is used for firewood and making charcoal for cooking. The fruits provide food for both human and animal consumption. The trees also adorn the environment and provide shade for both humans and animals.

Botanically, there are two classes of tree; conifers and deciduous (Figure 9.2). The **conifers**, that are botanically known as **gynosperms**, are distinguished by their tall slender trunks, narrow or needle-like leaves like those of the 'Christmas' tree, and naked seeds in cones. They are called **evergreen** trees because their leaves mostly remain green throughout the year without shedding. They grow mainly in cold climates such as are found in northern Europe and Great Britain but are not common in tropical countries where the climate is warm. Commercially, timbers produced from these trees are called **softwoods** though, physically, some are harder than the hardwoods. Other examples of softwoods are the pines, firs and spruces.

The **deciduous trees** can be as tall as the conifers but they have a bigger girth (diameter). The botanical name for this class of trees is the

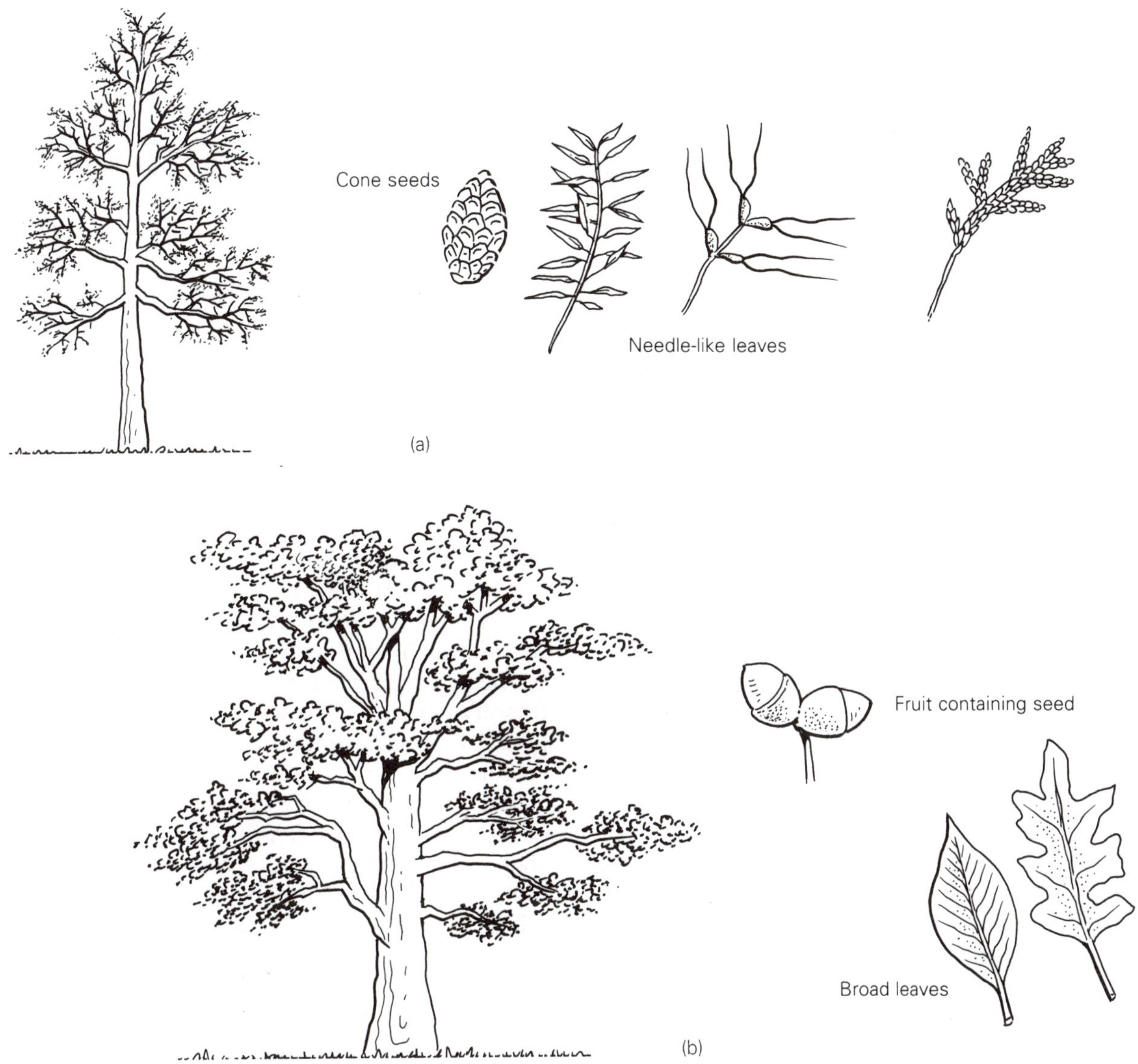

Figure 9.2 Types of tree: (a) conifers (softwood), (b) deciduous (hardwood).

angiosperms. They have broad leaves that dry up and fall during the dry season in tropical areas such as African countries. In temperate areas like Great Britain and other European countries the shedding of the leaves occurs in the autumn (from September to November). Some of them, however, do retain their green leaves throughout the year. Deciduous trees bear a great variety of flowers and fruits containing seeds, often called **dicotyledon** seeds. Timbers produced from this class of trees are known commercially as **hardwoods**. Some of them, such as wawa (obeche) and balsa are actually softer than some softwoods. Other examples of hardwoods are mahogany, oak, danta and odum.

Structure of a tree

The **wood** of a tree is made up of cells. The structure is of a complex nature as can be seen in cross, radial and tangential sections when magnified (Figure 9.3). The **pith (medulla)**, a cylindrical, brown, soft and spongy tissue is formed at the centre of the trunk and the branches. This is formed by the young sapling from which the tree grows. It is often dead and cannot be used as timber. The **heartwood (duramen)** is the mature part of the tree surrounding the pith. The cells making up the heartwood have hardened and contain no sap. For this

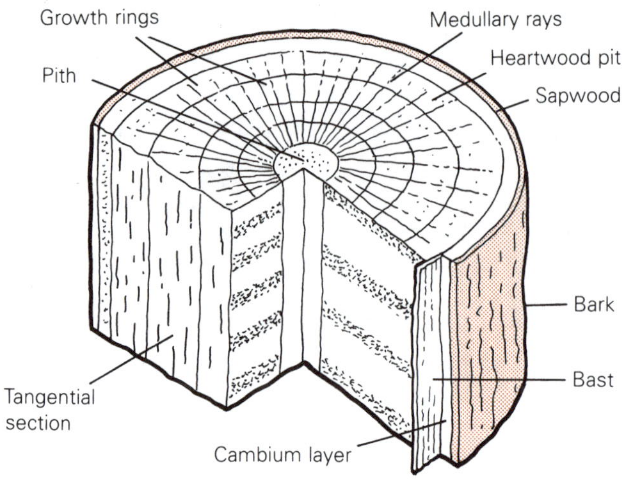

Figure 9.3 Structural parts of a tree.

reason, this part is less liable to attack by insects and other pests, or decay, and hence it is referred to as **truewood**. It is usually darker in colour and pleasant in appearance. It is mostly used for timber. The **sapwood** (**alburnum**) is the part that surrounds the heartwood and it is formed by the cambium layer. The cells of sapwood are alive and they carry water and mineral salts up to the branches and leaves to turn into food. As the tree grows, the sapwood matures to become heartwood. Until then, because of the starches contained in the cells, it is frequently attacked by insects. The **cambium layer** surrounds the sapwood, and it matures to become sapwood over the previous year's growth. It also forms new bark to replace the dead bark. This process of formation of a new layer of wood on the inside and new bark on the outside continues as long as the tree lives and brings about outward growth in the girth of the trunk and branches. **Bast** (**phloem**) is the part immediately under the bark of the tree and it surrounds the cambium layer. The main function of the bast is to convey food from the leaves to all the other parts of the tree. The **bark** (**cortex**) is made up of fibrous cells and is fed from the bast. It forms the outermost part and protects the growing tree, at the same time preventing the excessive evaporation of moisture. As the tree grows outwards, the bark splits and falls off the tree, giving way to a new one from the bast. **Rays** are bundles of thin-walled cells that convey food from the bast to the inner part of the tree and help to distribute the sap more evenly. They vary in size in different trees, providing in some of them a pleasing pattern in the cut timber. **Growth rings**, also called **annual rings** are cone-like layers of wood formed each growing season. In the temperate countries where there are four seasons with entirely different climates, there are two distinct densities of the cells (growth rings) in a year's growth (Figure 9.4a). The 'early wood', that is formed in spring and summer when growth is rapid and vigorous has wide, thin-walled cells and large cavities. The 'late wood', formed in the autumn and winter when the growth is slow and less vigorous, has narrow cells with thick walls and small cavities. Trees which grow in tropical areas, such as in West African countries, have even growth all the year round and the growth rings are less distinct.

The structural parts discussed so far are common in both hardwood and softwood. Apart from these, there are some structures that are peculiar to hardwoods and are not present in softwoods, and vice versa. The structure of **hardwoods** (pored woods) is more complex than that of softwoods. A hardwood is mainly composed of pores, fibres, parenchyma or soft tissue, and rays (Figure 9.4b).

Pores (vessels) are hollow tubular cells which extend vertically down the trunk of the tree. The size of the pores varies considerably. In some woods the pores can be seen with the naked eye, while in others they can only be seen with a magnifying glass. Sometimes they are difficult to distinguish even with the aid of a magnifying glass. On the end grain or a cross-section of wood, these tubular cells are seen as small holes (pores). They appear, however, on longitudinal surfaces as fine vessel lines and for this reason they are also known as **vessels**. The main function of the pores is to convey water and mineral salts from the roots to the leaves in the sapwood. Woods that have relatively large concentrated pores are called **ring-porous woods** while those with pores fairly uniformly scattered over the whole growth ring are called **diffuse-porous woods**.

Fibres are thick-walled cells of small diameter. They are elongated and tapered. They constitute the bulk of the wood tissue surrounding the pores. Their principal role is to provide strength and mechanical support to the growing tree. The fibres and the pores, when put together, determine the **texture** of the wood.

Parenchyma (soft tissue) is found in small, thin-walled cells which run vertically in the trunk of the tree. They are more or less rectangular in shape and are joined end to end. They form strands or covering around the pores and are seen as fine lines separating the growth rings. They serve as the food storage of the tree.

Medullary rays are thin-walled cells that run at right angles to the pores and fibres that form the growth rings. They are rectangular in shape and slightly elongated. They vary in size according to the type of wood. They also, like the parenchyma, serve as storage for food.

The structure of softwoods (**non-pored woods**) (Figure 9.4c) is simpler than that of hardwoods. It consists principally of two types of cells; tracheids and rays. Softwoods are referred to as non-pored woods because of the absence of pores in its structure. They do not contain fibres either.

Tracheids are narrow, tubular cells, approximately 3 mm long, which run parallel to the trunk and the branches of the tree (longitudinally). They are arranged in rows separated at

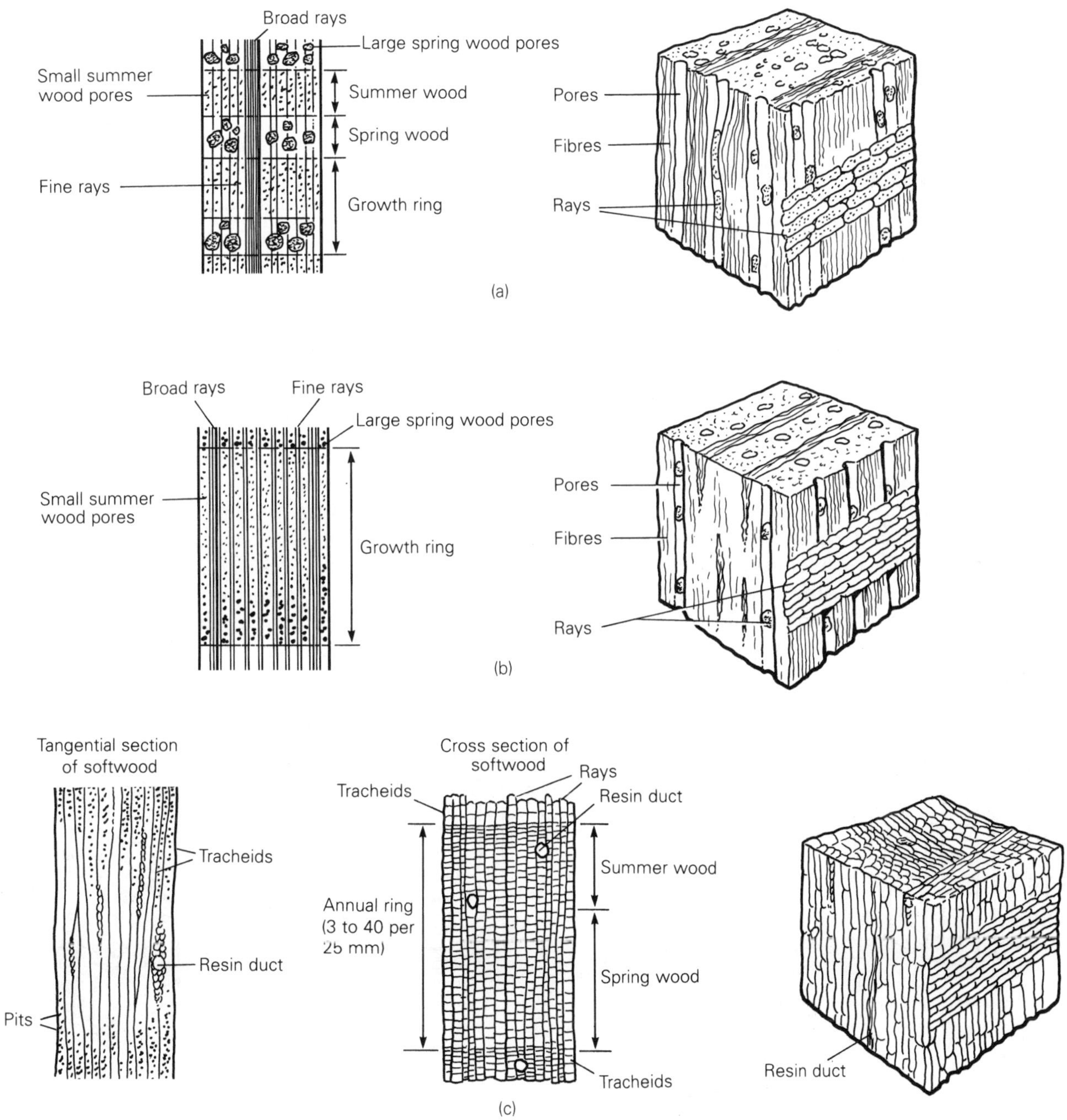

Figure 9.4 Structures of pored and non-pored wood: (a) ring-porous hardwood, (b) diffuse-porous hardwood, (c) softwood.

intervals by the rays. Tracheids perform two functions: they convey water and mineral solutions (salts) from the roots to the leaves in the sapwood and provide strength and mechanical support to the tree. The cells communicate with each other through small holes called **pits**. Water passes from one cell (tracheid) to the next through the pits.

Rays in non-pored woods are thin-walled cells. They form straight, narrow and radial bands. They also consist of the cellular tissue called parenchyma. Some of the rays in some softwoods have small holes called **resin ducts** or **canals**. These canals or ducts accommodate a waste product known as **resin**. The resin is secreted by cells immediately surrounding the canals.

Conversion of timber

Timber is obtained when a mature tree is felled in the forest and the trunk is cut into manageable sections or **logs**. The logs are then conveyed by heavy trucks or lorries to the sawmills, where the bark is removed or peeled off. The logs are then sawn by heavy circular saws or band saws into commercial sizes. This process is known as **timber conversion**. Logs that are prepared (barks peeled off) ready for conversion are termed **round timbers**. There are three methods of converting logs: live or through and through sawing, back or tangential sawing and quarter sawing (Figure 9.5). Conversion of timber should be carried out economically with minimum wastage. There are certain factors that influence the choice of method of conversion. These are: the class of timber required; the quality and size of the log to be converted; market requirements; and the effects of shrinkage and seasoning of some timbers.

Live sawing (**slab**, or **through and through** sawing) is a method of sawing a log into timbers by making a series of parallel saw cuts through the log without turning and re-turning it for re-cutting. It is the simplest and most economical method for light timbers and those which season easily. The disadvantage is that the boards tend to **cup** and **distort** badly during seasoning. This is because all the boards have sapwood along both edges which shrinks more than the heartwood because of its higher moisture content. It is also unsuitable for sawing logs which are knotty or otherwise defective.

Back sawing is a process of sawing a log such that the width of the boards is always tangential to the growth rings; hence it is also called **tangential sawing**. The log being sawn is turned and re-turned to new positions for the sawing to produce the tangential cut. Boards converted in this way produce pleasant figure markings, particularly suitable for cabinet work. Boards converted through this method have the advantage of not splitting easily when nailing and they usually season more rapidly. However, they shrink more in width while seasoning.

Quarter sawing is a method of sawing logs into boards in such a way that the boards have their width in the general direction of the rays. The logs are turned and re-turned to new positions frequently. Fully quarter-sawn boards have distinct rays showing decorative medullar ray

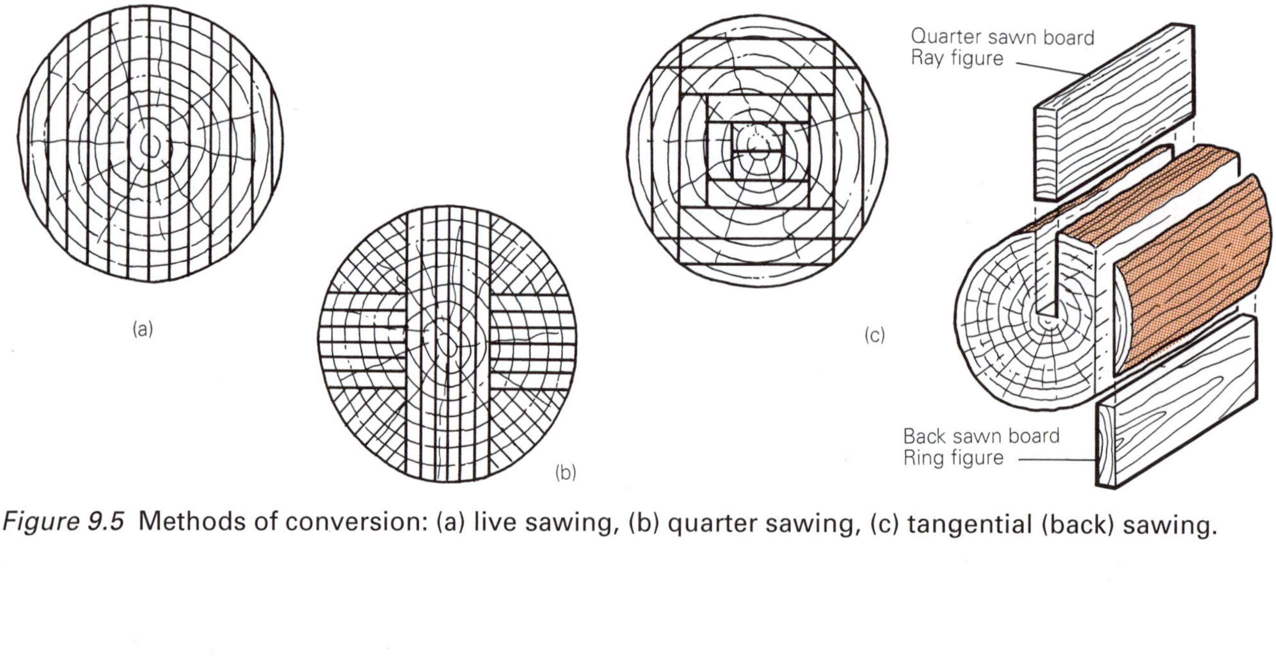

Figure 9.5 Methods of conversion: (a) live sawing, (b) quarter sawing, (c) tangential (back) sawing.

figures called **silver grain** or **flowers**. Quarter-sawn boards shrink less in width while seasoning. Because they have less shrinkage in their widths, they are often used for making the stocks of wooden planes.

Common, marketable sizes of timber

Round timbers, the unconverted timbers, are not just sawn into any shape or size. There are recommended commercial shapes and sizes to which they are converted or sawn (Figure 9.6). The following are the terms used to describe and identify the various marketable sizes.

A **baulk** of timber is a piece of log that has been roughly squared by sawing off part of the sapwood in readiness for full conversion. A baulk may have a cross-section of at least 115 mm by 100 mm. It can also be used in this form for heavy structural work.

A **half timber** is one-half of a baulk that has been sawn along its length (longitudinally). A half timber is mainly used for structural work, It may, however, be re-sawn into other marketable sizes.

A **flitch** is the term given to any baulk or half timber that has been suitably squared ready to be cut into veneers by slicing (knife-cutting) method. Flitches are not less than 100 mm by 300 mm.

A **plank** is a board of timber having a size of between 50 mm to 150 mm thickness and a width of at least 300 mm. Generally, it is used for supporting loads on the wide faces, as in scaffolding. It is also used on its edge for supporting deep joists and ceiling hangers.

Deal is the term used for softwoods of width between 250 mm and 280 mm, and thickness ranging from 50 mm to 100 mm.

A **batten** is a piece of timber cut to a thickness ranging from 50 mm to 100 mm and a width of between 100 mm and 200 mm. Battens are mostly cut from softwoods. They are used for holding roofing and tiling and ceiling linings.

A **board** is the term given to a piece of timber with a thickness less than 50 mm and a width of 100 mm or more. Boards are used in building construction, for painted woodwork and joinery, for flooring, for shelving and for furniture construction.

Scantling is a timber having a thickness ranging from 50 mm to 100 mm and a width varying from 50 mm to 115 mm. Scantlings are mainly used for smaller beams and frames structures, such as rafters, studs and joists.

Strips (laths or fillets) are pieces of timber less that 50 mm thick and less than 100 mm wide. Strips are usually dressed (planed) and used to cover joints in sheeting materials.

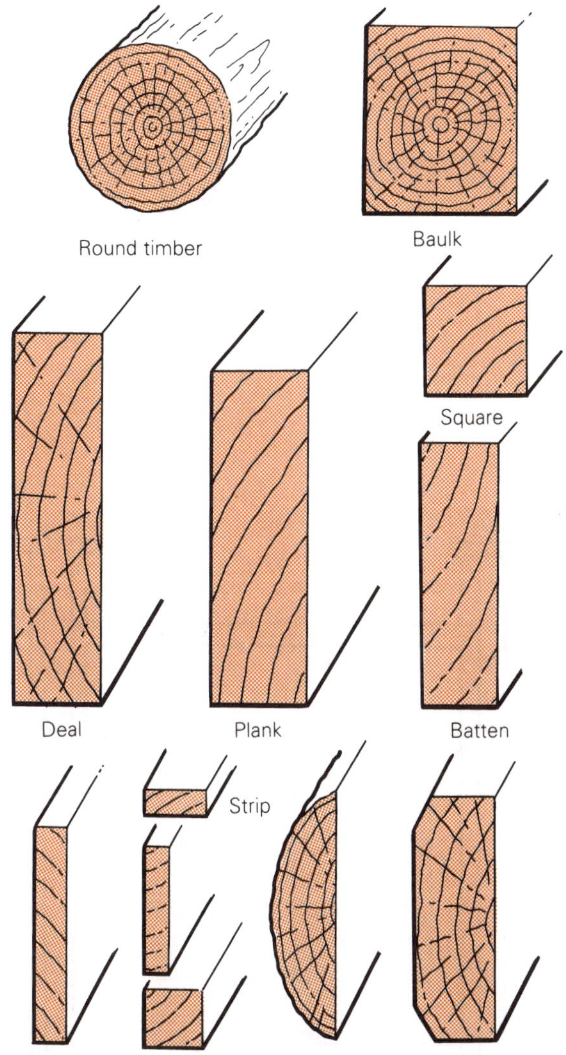

Figure 9.6 Marketable timber sizes: (a) round timber, (b) baulk, (c) deal, (e) plank, (f) batten, (g) board, (h) scantling, (i) half timber, (j) waney edge

Square (quartering) timber, as the name implies, is square in cross-section. The sizes range from 25 mm to 150 mm square. The large squares are often used for columns to support beams while the small sections are used for furniture work such as table and chair legs and turned pieces.

Seasoning of timber

Seasoning is the process of drying-out the excess moisture from a converted timber so that it can be

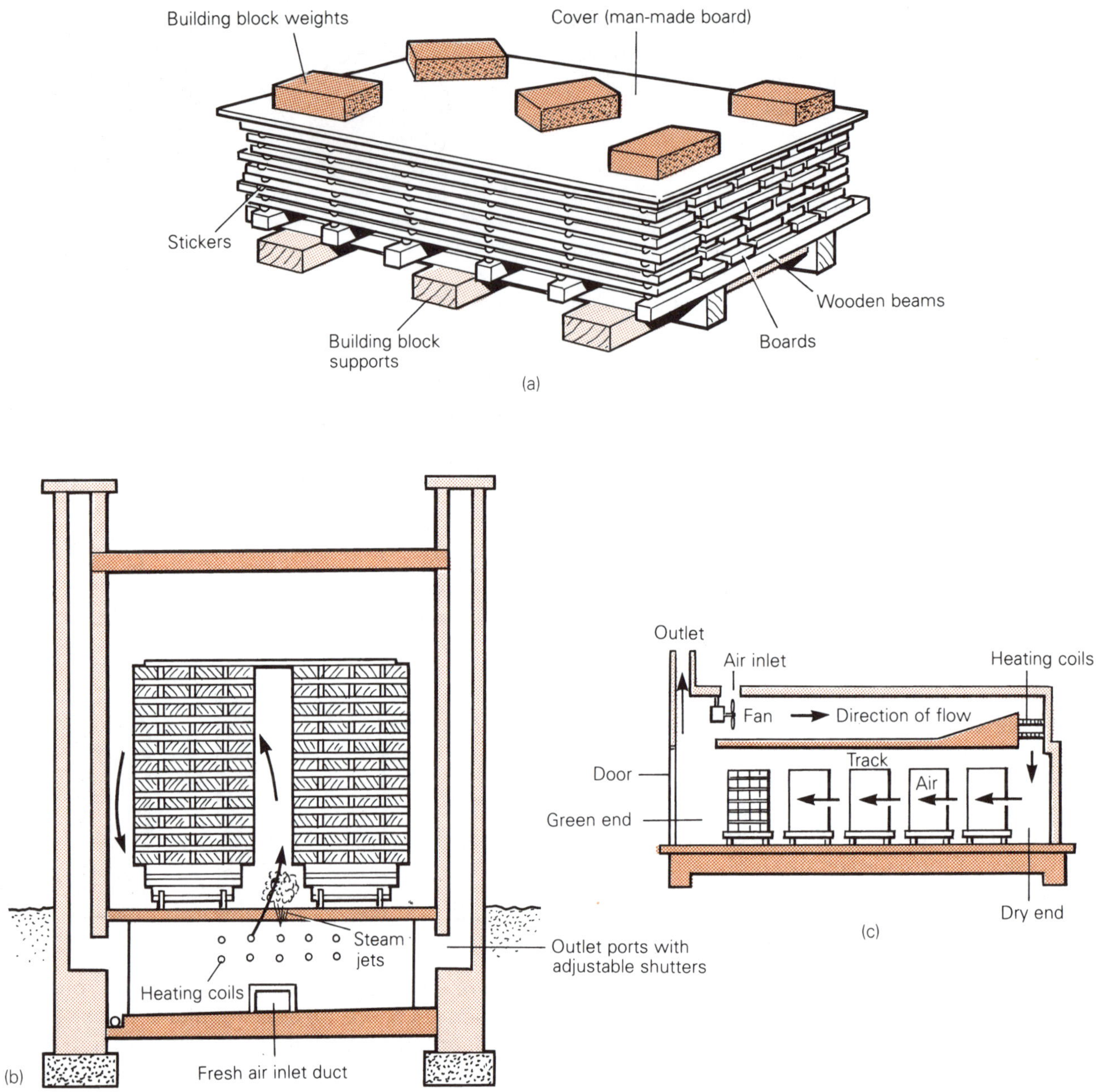

Figure 9.7 **Methods of seasoning: (a) natural air seasoning, (b) kiln seasoning, (c) progressive kiln seasoning.**

worked (Figure 9.7). The purpose is to reduce the moisture content (MC) of the timber to a stage at which it becomes stable in the environment where is to be used. The moisture content is quite critical when considering how and where the timber is to be used or the type of construction work to be done.

As **shrinkage** is likely to occur during seasoning, efforts must be made to control the rate and degree of seasoning to prevent undue distortion and splitting before the timber is actually put to use. A well-seasoned timber should reduce to minimum further shrinkage, checking and distortion. Conditions are less favourable for decay, mould stains and some kinds of insect attack when a timber is properly seasoned. Seasoned timber is lighter in weight and therefore more economical to transport and easier to handle than a **green timber.** Properly seasoned timber is strong and can take paints, varnishes and glues satisfactorily. There is a lower tendency for nails, screws or bolts in seasoned timber to corrode when it is used in construction and preservatives can penetrate more easily into the fibres.

The two main methods of seasoning are natural seasoning and artificial seasoning.

Air seasoning is one of the natural methods where the pieces of timber to be seasoned are carefully stacked with stickers (wood strips) between each layer and the air is allowed to circulate freely through the piles.

Kiln seasoning is the most widely used artificial (accelerated) method for seasoning. It is a faster method and the temperature and the humidity conditions of the kiln (oven) can be accurately controlled. The pieces of timber are stacked in a similar way as for air-seasoning in a controlled oven where the temperature and humidity are accurately controlled. When the pieces of green timber are first placed in the kiln, steam is used to keep the humidity high and after a period of time, the temperature is raised while the humidity is reduced. Fans are used to keep the air in constant circulation over the surfaces of the wood. The two types of kiln are the **compartment kiln**, where the conditions at any given time are the same throughout the kiln, and **progressive kilns**, where the conditions at one end of the kiln differ from those at the other end.

A kiln schedule is a carefully compiled set of temperatures, humidities and timings to be followed by the kiln operator. These will vary depending on the size and type of wood, and its initial moisture content. Improper seasoning, either by air or kiln, can result in such seasoning defects as splits, checks, warpage, loosened knots, honeycomb, and the internal stresses called case-hardening. Note that in the tropical regions, like Ghana, most timbers are air dried for all types of construction. However, furniture industries which export furniture parts to temperate countries have to kiln dry the timber because a lower moisture content is required, due to the fact that such parts will be used for assembling furniture that will be used in heated rooms.

Determination of moisture content

Moisture content (MC) is a measure of the amount of water contained in a timber and is expressed as a percentage of its oven-dry weight (Figure 9.8). A well-seasoned timber has a low percentage of moisture; that is, the lower the percentage the drier the timber. When a living tree is felled, more than half of its weight may be moisture and a large part of this must be removed before the timber can be put to effective use.

There are two main methods for calculating the moisture content: the oven dry method and the moisture meter method. To apply the **oven dry method**, obtain a sample piece of timber and weigh it; dry it in an oven to remove all the moisture and weigh it again; subtract the dried weight from the initial weight; divide the difference by the oven dried weight and multiply by 100. This will give you the percentage moisture content:

$$\%MC = \frac{\text{initial weight} - \text{oven dried weight}}{\text{oven dried weight}} \times 100$$

OR

$$\%MC = \frac{\text{amount of moisture}}{\text{oven dried weight}} \times 100$$

This method is more accurate than the moisture meter method, but involves cutting specimens and a delay for some hours before a result can be obtained. With the **moisture meter method a** portable instrument determines the moisture content by measuring the electrical resistance of

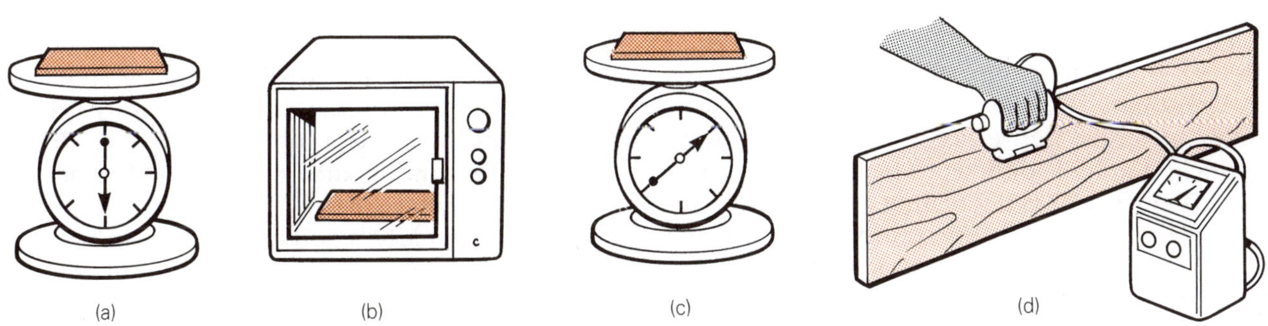

Figure 9.8 Determination of moisture content: (a) weigh before drying, (b) dry the sample, (c) weigh after drying, (d) determine MC with a meter.

the timber. The reading gives an immediate answer that is sufficiently accurate when in the range of 7–25%.

A piece of timber is apt to lose and gain moisture from the air around it until the moisture is balanced with or in equilibrium with the air around it. When a timber is in this situation, its moisture is said to be at equilibrium moisture content. Since timber is subject to daily and seasonal changes in the relative humidity of the air, it is continually making slight changes in its moisture content and this effects changes in its dimensions. This is evident during humid weather when doors and drawers often tend to stick or bind – and then work freely again during the dry part of the year.

Surface quality of timber

The surface quality of timber depends on the visible nature and arrangement of the wood elements such as fibres and vessels that are characteristic of each species of timber. Some of the natural grain features that are peculiar to certain species of timber are valued for their decorative figure markings, as for example those in sapele. In some cases the grain direction has a definite weakening effect.

Grain refers to the direction of the fibres and vessels (Figure 9.9) and varies according to the position the board occupies in the tree and the method of conversion used for cutting the log. For example, in quarter-sawn boards, rays are seen if they are distinct, and back-sawn boards show distinct ring markings. The grain may be straight, spiral, interlocked, diagonal or wavy. **Straight grain** occurs when the fibres of adjacent layers in the growth rings are parallel to the axis to the timber. This is the natural growth of the fibres and works smoothly in construction. **Interlocked grain** occurs when the fibres of adjacent layers in the growth rings are inclined at different angles to the axis of the timber. Boards cut from timber having interlocked grain show a ribbon or stripe figure, as in some mahogany. Interlocked grain is difficult to plane and can split easily. **Wavy grain** is undulating or wavy arrangement of the fibres; large waves produce a wavy figure and small, irregular waves produce a curly figure. Wavy grain is difficult to plane and work. **Diagonal grain** occurs when the fibres are arranged at a more or less constant angle to the axis of the piece, as a result of sawing at an angle to the back of the log. **Cross grain** occurs when the fibres have a varying inclination to the edges of the piece. Strength is reduced when a cross grain piece is used.

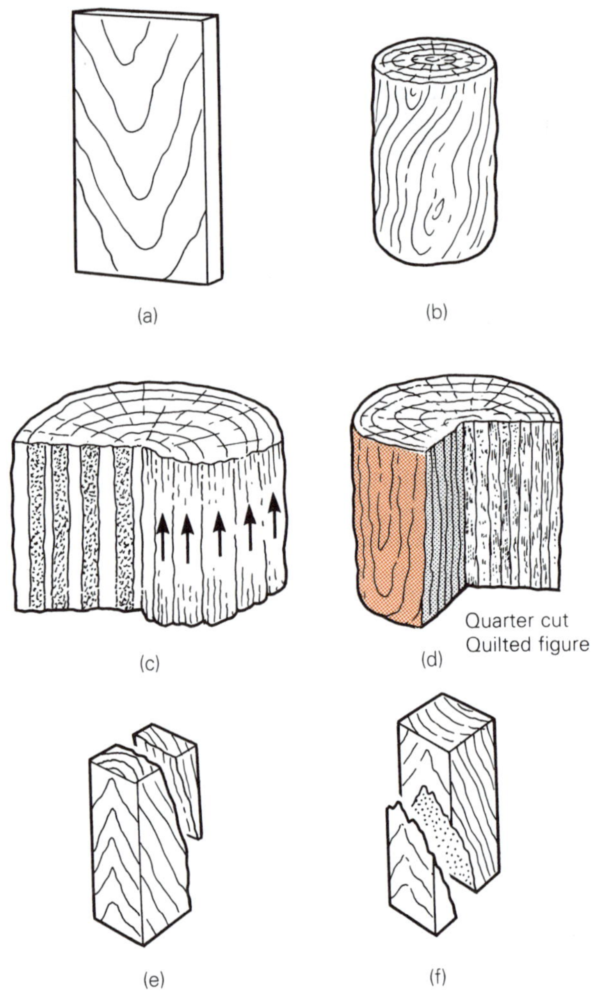

Figure 9.9 Types of grain: (a) straight, (b) spiral, (c) interlocked, (d) wavy, (e) diagonal, (f) cross grain.

Other surface qualities of timber are texture, figure and colour. A good knowledge of these can be an advantage to the woodworker. **Texture** refers to the relative size, quality and distribution of the wood cells. The texture may be coarse or fine and either rough, smooth or oily to the touch. Fine-textured timber has small closely spaced cells whilst coarse-textured timber has relatively large well spaced cells. When the difference between early wood and late wood is slight, the timber is even-textured whereas a timber with marked contrast in the growth rings has an uneven texture. **Figure** is a term usually applied to timbers that possess unusual or attractive designs resulting from: the difference in growth between the earlywood and latewood; variations in the colour, grain and texture, due to the natural arrangement of the cells; the method of conversion; or irregularities in the tree, such as knots and burls. **Colour** depends on the substances and chemicals in the cells and walls.

Defects in timber

A defect in timber is any irregularity appearing in or on the timber that may reduce its strength or durability if used for construction work. It may occur in the timber during growth, felling or seasoning. Defects may be classified as natural or artificial, depending upon their nature or form. Natural defects are features in the growing tree, such as grain irregularities, knots and shakes, while artificial defects are features caused by improper handling, conversion, seasoning or inadequate protection or preservation of the timber. Some defects are due to bush fires, parasites such as insects, fungi and strong winds. Defects, particularly the artificial ones, can be avoided by using correct methods for conversion and seasoning, and by careful selection of the timber in relation to its use.

Working (movement) of wood refers to an alternate swelling and shrinking that occurs in seasoned timber because of seasonal changes due to a humidity fluctuation. The working can cause defects such as checking and splitting when conditions become severe. Polishing or painting may minimise working of the piece. Proper construction methods should make allowance for the movement of the timber.

Natural defects

The following are some of the common natural defects. **Knots** (Figure 9.10) are the remains of tree branches that are cut through during the conversion process. They can weaken the strength of timber due to its short or cross grains surrounding them. It is difficult to plane knotty portions of the timber. Knots may cause warping and checking during seasoning. There are three forms. A **live knot** has its growth rings firmly intergrown with the surrounding growth rings, making it quite strong but difficult to work with. A **dead knot** (**loose knot**) may be decayed or loose and eventually falls out of place because its growth rings are not intergrown with the surrounding timber and greatly weakening the piece of timber. A

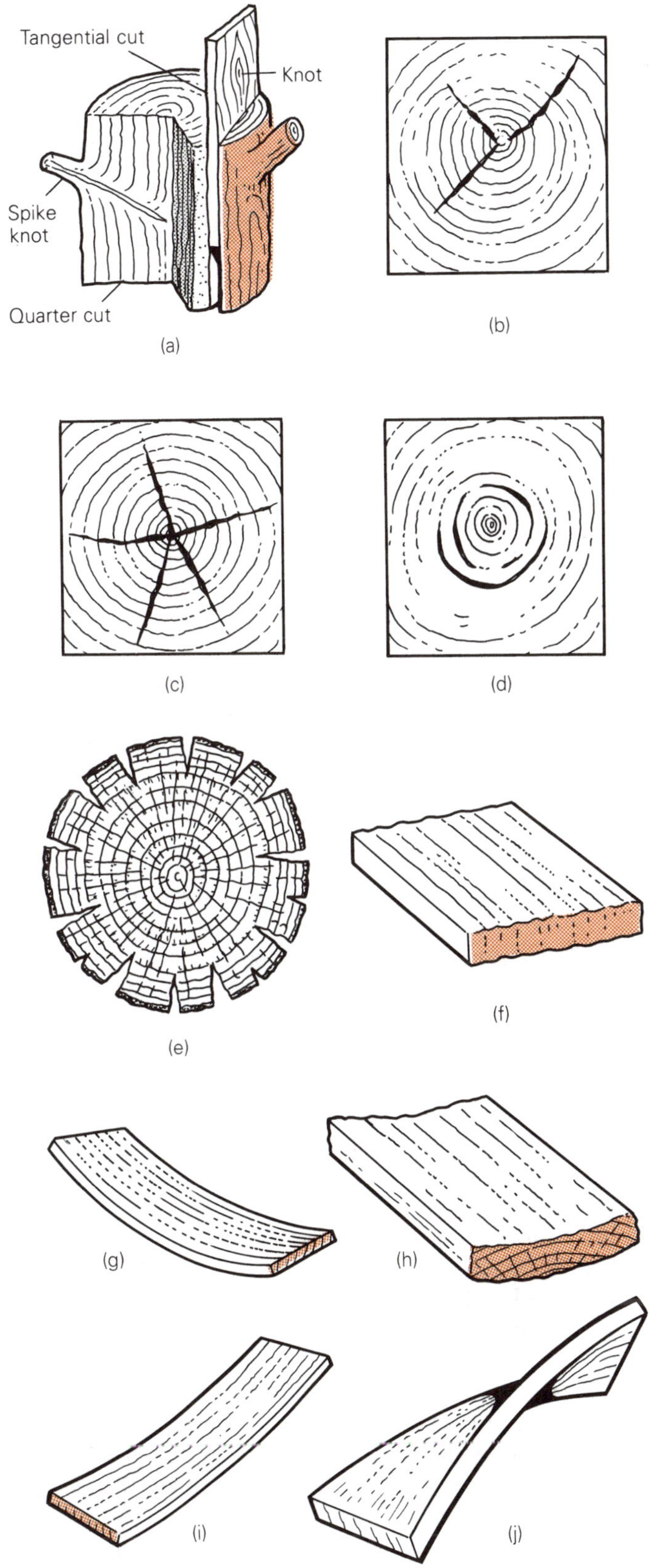

Figure 9.10 Types of defect in timber: (a) knots, (b) heart shake, (c) star shake, (d) ring or cup shake, (e) check, (f) collapse, (g) bowing, (h) cupping, (i) spring, (j) twist.

splay knot is obtained when a branch of the timber is sawn lengthwise and it is usually elongated in shape.

Shakes are partial or complete splits between adjoining layers of the timber. They may be caused by the relief of growing stresses in the tree, or may be a result of the tree falling heavily on the ground during felling. **Heart shakes** (Figure 9.10b) are splits occurring from the pith of a log along the path of the rays, and are caused by leaving the timber in log form for too long before conversion. A heart shake in the form of a star are often referred to as **star shake** (Figure 9.10c). **Radial shakes** are splits extending from the outside of the log inward along the path of the rays. **Ring** or **cup shakes** (Figure 9.10d) are the result of separation of the wood fibres around adjoining growth rings, where the cell density varies. They are caused by strong winds straining the tree while it is growing or by the shock sustained by the tree when it is felled. They can extend for some distance longitudinally through the timber. During conversion much of the log can be wasted because of cup shakes.

Artificial defects

Defects that occur during seasoning may be termed as artificial defects.

Case-hardening is a defect that occurs during kiln seasoning where the exterior surface (case) of the timber dries out and becomes hardened before the interior (inside) is dried. This phenomenon causes a condition of stress. The inner fibres become separated as they try to resist the shrinkage of the dried surface fibres. When eventually the inner fibres are allowed to dry, the shrinkage is resisted by the dry outer fibres. As a result of the timber shrinking in this manner, the tearing of the internal fibres forms short splits or **honeycomb checks**.

A **check** is a separation of the wood along the grain, usually in the direction of the rays. They are generally the result of shrinkage stresses set up by the rapid drying of the surface due to effect of the hot sun or wind on freshly sawn surfaces or during seasoning. Checks reduce the shear strength of the timber. The forms of check are: **end check**, a split at the end of a log; **heart check,** a split extending from the pith outward; **internal check**, a split occurring in the interior of a timber; **ray check**, a split that occurs along the rays; **surface check**, a split occurring mainly on the surface of the timber; and **through check**, a split extending from one surface to the another surface, passing through the timber (Figure 9.10e).

Collapse is a flattening of the cells during drying, showing uneven shrinkage on the surface of the timber (Figure 9.10f). The moisture removed from the cells has not been replaced by air.

Warping is any variation of a piece of timber from a flat surface that occurs as a result of uneven shrinkage, often caused by grain irregularities and incorrect stacking during seasoning. Forms of warping include bowing, cupping, spring, and twisting. **Bowing** (Figure 9.10g) is a curvature along the wide surface in the direction of the length of a board and it is caused by spacing the stickers too far apart during seasoning, causing the timber to sag under its own weight. **Cupping** (Figure 9.10h) is a concave curvature across the face of a board caused by shrinkage and bad stacking while seasoning. It is more prominent in back-sawn boards. The curvature is usually away from the heart. **Spring** (Figure 9.10i) is distortion or curvature along the edge of a board and usually, it does not affect the face. **Twisting** (Figure 9.10j) is another seasoning fault and is caused by spiral distortion of the board along its length. It may be a result of the board being weighed down, thereby allowing it to distort in length. Timber used in this condition-may cause the whole job to be distorted.

Deterioration of timber

Fungal decay

Defects may be caused by deterioration or destruction of the timber. The commonest defect of this nature is **decay**. It is the decomposition of a timber caused by fungi that feed on the cell walls and the cell content. The two main forms of decay are dry rot and wet rot. The **dry rot fungus** usually attacks timber that is badly ventilated, as is sometimes the case in the structure of a building or shed. It is the more damaging of the two fungi, able to penetrate porous materials such as brick and plaster walls to reach the timber. The characteristics of dry rot are that the affected timber becomes dry and light and eventually crumbles. To prevent dry rot, use only the best quality wood, keep it well ventilated and coat the timber with a suitable preservative. **Wet rot** is a decay that occurs in living trees and old dead logs. The fungus, yellowish in colour, needs water to thrive and cannot attack a timber of less than 20% moisture content. Wooden structures such as outdoor gates, doors and windows are susceptible to wet rot. To protect the timber, treat it with a suitable preservative.

Insect attacks

Some insects are pests that attack and destroy timber altogether or cause defects in it, reducing its value. The two groups of insects which cause the greatest amount of damage to timber are beetles and termites. **Beetles** infest the timber by laying eggs in openings in the timber. The eggs hatch into worms or larvae that bore into the timber or wood, creating tunnels behind them. A larva may exist in that form for a year or more depending upon the availability of starch for food. It then develops into a mature beetle and bores its way out of the timber, leaving holes in the timber or wood. Wood boring beetles (Figure 9.11a) which are destructive wood pests include the powder post (*Lyctus brunneus*), the furniture beetle (*Anobium punctatum*), the pin-hole borer (*Ambrosia beetle*), the longhorn beetle (*Hylotrupes bajulus*), and the death-watch beetle (*Xestobium rufovillosum*).

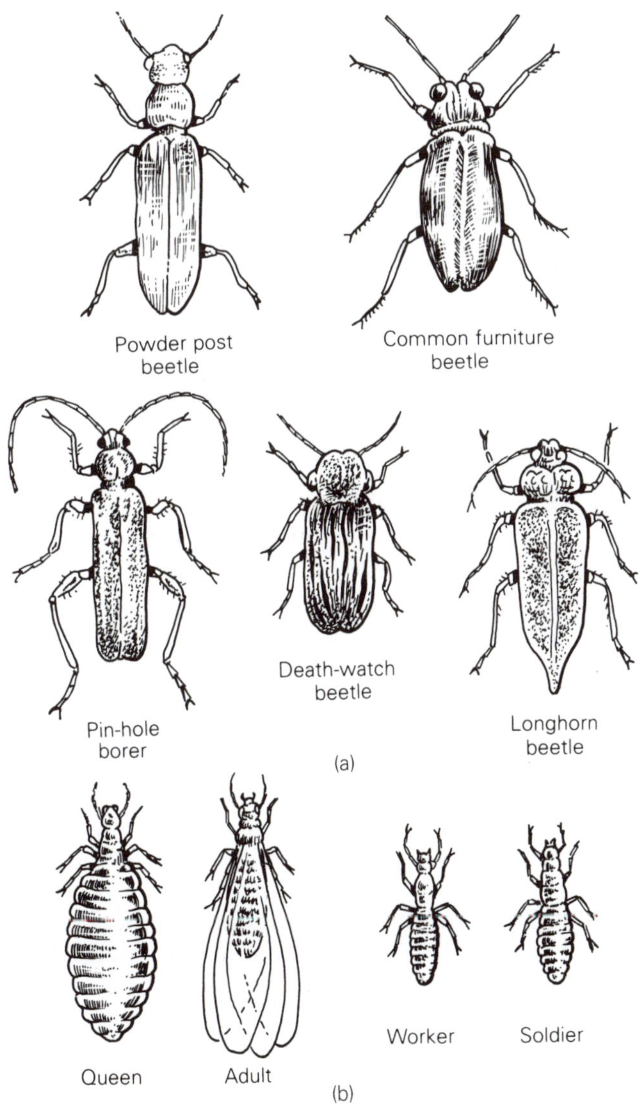

Figure 9.11 Insects pests (a) Wood borers, (b) termites.

The **powder post beetle** attacks seasoned hardwood. Its colour is between reddish-brown and black. It has a length of about 2 mm. The female lays whitish eggs in large pores in the wood where larvae can feed on the starch in the sapwood. Untreated sapwood can be attacked or re-infested at any time. It is the larvae that bore into the wood and do the damage. They produce fine powder, hence their name. Softwoods are not attacked by this beetle because its large eggs cannot not enter into the pores of the timber. Liquid insecticide can be forced into the flight holes with a syringe to penetrate the infested wood. Avoid using untreated sapwood in construction work.

The **furniture beetle**, dark brown to almost black in colour, attacks both the sapwood and truewood of seasoned timber, particularly old furniture, wall panelling etc. It is the larvae that destroy the timber. They excavate along the grain, honeycombing the wood and weakening its structure. The undersides of drawers and table tops that are not polished often or coated with preservatives are the most vulnerable to attack. The dust or frass from the bored holes feels gritty when rubbed between the fingers. Use a liquid insecticide and force it into the holes.

The **pin-hole borer** exists in all the tropical regions and only a few timbers are immune to its attack. Pin-hole borers attack the sapwood and truewood of freshly felled logs and unseasoned timbers, especially those of pored woods. The attack ceases when the timber starts to dry. Pin-hole borers make galleries less than 3 mm in diameter. The tunnels are usually fairly straight and across the grain. Efficient logging operations, immediate removal of the barks of the logs, quick conversion of the logs and effective seasoning may prevent pin-hole infestation. Chemical treatment may be used.

The **longhorn beetle** mainly attacks sickly trees and timbers that have been left lying in the forest. It is distinctive in appearance with two shiny spots on the top of its head and white hairs on the wing covers. The larvae live in the bark and sapwood. It bores large, oval-shaped tunnels, producing coarse fragments and small pellets of wood.

The **death-watch beetle** is dark greyish brown and about 6 mm to 8 mm long. It has a broad flat head. It infests old timbers and beams of softwood or hardwood. The larvae bore very large holes into the timber. The dust from the hole normally contains pellets.

Termites (Figure 9.11b) attack both the sapwood and truewood of most timbers. They attack converted timbers that are in contact with the ground. Even a timber that is separated from the ground (soil) may be attacked if the termites are able to build communication tubes from the soil to the wood. To prevent termite attack, use termite-resisting wood such as odum (iroko) for the structure; impregnate the timbers with suitable preservatives; and for treating infested timber, use arsenic trioxide powder.

> ▲ Arsenic trioxide is extremely ***poisonous***, so use it with great care. Wear a mask and avoid contact with the powder.

Other pests

Marine borers are not insects but marine organisms (molluscs and crustaceans) that attack timbers that are used in salt water, such as the sea water. The common marine borers are the ***Teredo navalis*** (**ship worm**) that is a **mollusc** and the ***Limoria lignorum*** that is of the **crustacean** type. The Teredo navalis causes rapid and considerable damage to underwater structures such as wharf and bridge piles, ships, boats, and other wooden water craft. To prevent attack, the part of the structure that will be submerged in the water must be covered with copper or muntz metal or else encased in concrete.

Wood preservation

Wood preservation means treating it with solutions that makes it poisonous to fungi and insects, such as beetles and termites, so as to prolong the serviceable life of the timber. In general, existing wood preservatives may be divided into two groups: oil-based and water-based. **Creosote oil** is a substance derived from the distillation of coal tar and penetrates well into the cells of the timber. It is a permanent preservative. However, it is also dirty and cannot be used where appearance is important. It tends to stain the timber and so, when applied to a surface, makes it difficult to paint or polish over that surface. It is flammable when freshly applied and has an unpleasant odour. Therefore it is more suitable for external than internal work. **Pentachlorphenol in mineral oil**, called **penta** for short is a chemical dissolved in light or heavy mineral oil. It is more permanent than the creosote. It is suitable for general application on external joinery and wooden structures. Other oil-based preservatives in use are **copper naphthenate** and **zinc napththenate**, both of which are either dissolved in light oil for brushing, spraying and dipping, or dissolved in heavy oil for pressure treatment. They have good penetration but are highly toxic, and they have a strong odour.

Water-based preservatives are chemicals dissolved in water. They are clean and odourless. Some are permanent but others may leach out of the wood when in contact with water. They are used on seasoned timbers and any piece treated with them must be re-dried after treatment.

Methods for applying preservatives

There are many methods by which preservatives can be applied, the commonest including the pressure method, hot and cold method, cold dipping, brushing and spraying.

The **pressure** method requires that the timber is placed in a metal cylinder and the preservative forced into the timber under pressure. This method permits control over the amount of preservative absorbed. It is applicable to most timbers and the sapwood of all timbers.

In the **hot and cold** treatment the timber is placed in a tank of heated preservative and left until it cools. This method is suitable for permeable timbers and the sapwood of all timbers.

With **cold dipping** the timber is immersed in cold preservative for a period of time. The longer the period of immersion the more effective the treatment, particularly with tar-oil. It is mostly used for permeable timbers and the sapwood of all timbers.

Brushing and **spraying** are normally used where deep penetration is not required.

CHECK YOUR UNDERSTANDING

- The timbers that are obtained from the tropical forests are mainly pored wood whilst the temperate zones contain non-pored as well,
- Pored woods, also known as hardwoods, are obtained from deciduous trees, botanically classified as angiosperm. These are identified by their broad leaves that are shed once every year.
- Non-pored wood (softwood) is obtained from a class of trees known as conifers (gymnosperms). They have narrow or needle-like leaves that remain green throughout the year without shedding.
- The main parts of a tree trunk are the bark (cortex), bast (phloem), cambium layer, sapwood and heartwood.
- A timber is obtained from a trunk of a fully matured tree that is cut into manageable sizes called logs and then conveyed to the sawmills where they are converted into commercial sizes such as planks, boards and scantling.
- After a log has been converted into commercial sizes, the pieces are seasoned to dry-out the excess moisture.
- Any irregularity appearing in or on a timber that may affect its strength and suitability when used for construction work, is regarded as a defect.
- Wood preservation involves treating wood with solutions to prevent fungi and insects attack and to prolong its serviceable life.

REVISION EXERCISES AND QUESTIONS

1. Explain why certain trees in the temperate forest areas are called evergreen trees. How are these trees different from those obtained in the tropical forest areas?
2. Name three essential parts of a living tree and state the function of each part.
3. By what methods can a log be converted into commercial sizes?
4. Name the methods by which excess moisture can be removed from a sawn pieces of timber and state reasons for removing the excess moisture.
5. What do you understand by the term ‘wood preservative’?
6. Name the commonly used preservatives and show how they can be applied.

Veneers and manufactured boards

10

Introduction

A solid board (timber) has many limitations and disadvantages in its use for constructional and cabinet work. It is limited in width and as such a number of pieces have to be joined together before it can be used for a job requiring a wide surface. It is also subject to twisting, warping, expansion and shrinking. For these reasons, alternative materials have been developed to overcome these deficiencies. There are now man-made or manufactured boards, such as plywood, laminboard and composite boards.

Veneers

A veneer is a thin sheet of wood obtained from a log of timber by rotary peeling, slicing or fine sawing. Traditionally, veneers were used for decorative purposes, but nowadays, they are widely used in the manufacture of plywood and for surfacing various types of composite boards. Before the veneer is cut, the log or flitch is steamed or dipped into hot water to reduce to a minimum any tearing or breaking of the fibres. In **rotary cutting** (Figure 10.1a) the log is rotated against a large, vertically held knife that peels off a continuous thin sheet of veneer from it. In the **slicing method**, the log or flitch does not rotate but instead it moves forward and backwards horizontally (or vertically) against a horizontally (vertically) fixed knife. By this, a thin slice of veneer is produced. (Figure 10.1b,c) It is also possible to **saw off** a thin layer of veneer.

The choice of timber for producing veneers is dependent upon the type of veneer required, which in turn depends upon the use to which the veneer will be put. There are **core veneers** used for making corestocks for multi-plywood. These are usually rotary cut from cheap, light, straight grained timber that do not possess decorative figure markings. Another type of veneer is the **cross-band (counter) veneer** that is also cut from straight grained, good gluing timbers. The

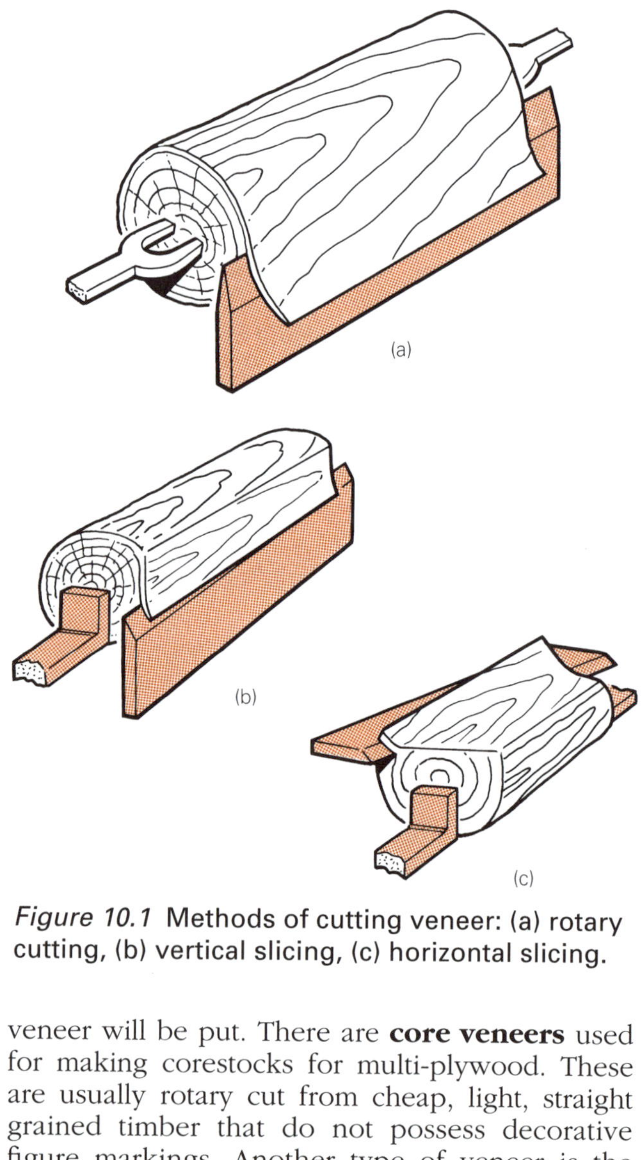

Figure 10.1 Methods of cutting veneer: (a) rotary cutting, (b) vertical slicing, (c) horizontal slicing.

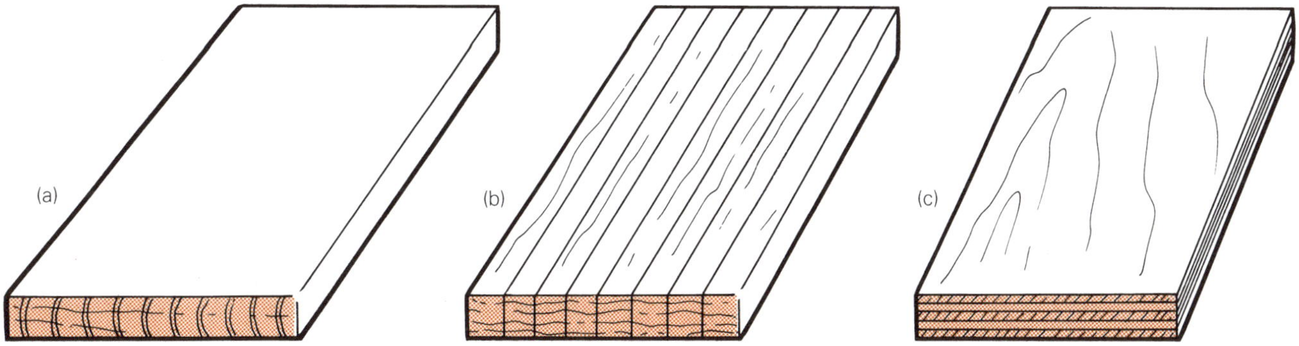

Figure 10.2 Grounds or cores for veneering: (a) solid wood (quarter sawn), (b) built-up core (solid core-stock), (c) veneers glued together (plywood).

cross-band veneers are used for gluing across the both faces of core of a plywood, laminboard, blockboard, and sometimes particle board to strengthen the core and reduce shrinking and warping. They are always laid with their grains running at 90° to the grains of the core. **Back veneers** are unfigured veneers glued to the back face of plywood, laminboards, or particle boards to balance the stresses that may be set up when the face veneers are applied. They may be cut by the rotary method or sliced. The **face veneers** are cut from purposely selected logs or flitches that possess interesting grain markings, texture and colour. These are the veneers used for the exposed (face) surfaces of manufactured boards.

Veneering is the art of beautifying the surface appearance of a piece of furniture by laying a thin sheet of handsomely figured veneer, over furniture made from unfigured or cheaper timber. The cheaper timber (wood) on which the veneer is laid is referred to as the **ground** (Figure 10.2). A ground may be a solid timber, plywood, laminated board or blockboard. Solid timber has a tendency to distort, warping or curling when glue is applied to it. This happens particularly when only one surface is glued. To eliminate this defect, it is advisable to veneer *both* surfaces when dealing with a solid ground. The other materials, plywood, blockboard and laminboard, are free from warping or curling and one surface only may be veneered.

The **tools** required to carry out veneering operations include most of the common woodwork hand tools and some special ones:

- The **straight edge** (Figure 10.3a) is a straight flat piece of wood used for marking out. It is also used as a cutting guide when cutting pieces of veneer.
- **Veneer knives** (Figure 10.3b) are penknives made from good quality steel or old hacksaw blades. They are usually ground and sharpened on one side only and used for cutting thin veneers.
- The **veneer saw** (Figure 10.3c) is a saw with a curved edged blade and small teeth without set (see page 0000 for details of saw setting). The blade is made from an old saw blade, screwed to a wooden handle with little projection. It is used mainly for cutting thick veneers. It should be guided by a straight edge to give a straight cut.
- The **veneer cramp** (Figure 10.3d) consists of two straight-edges with their insides slightly rounded. They are held together by two bolts and wing nuts, one at each end. It is used for holding a number of veneers together, first for cutting, and then for planing or shooting.
- The **bench plane** (general woodwork bench plane) is used for preparing the ground or the core joining the veneers. It is also used for cleaning up the edges of a finished veneered panel.
- The **toothing plane** (Figure 10.3e) has a wooden body similar in shape to the wooden smoothing plane. The cutting iron (blade), is inserted vertically, and has grooves on its face which produce a series of fine points on the surface of the ground. The plane is used diagonally, first in one direction and then in the other. This takes out irregularities and gives a key (that is, a good roughened surface) for the glue.
- The **veneer hammer** (Figure 10.3f) is made of wood and has a head and a handle. A rounded brass strip is inserted into the head, projecting slightly and is used to press down the veneer. The veneer hammer is used to flatten the veneer and force out air bubbles and surplus glue, particularly when the veneer has been laid with hot glue.

Figure 10.3 Veneering tools: (a) straight edge, (b) veneer knives, (c) veneer saw, (d) veneer cramp, (e) toothing plane, (f) veneering hammer, (g) flat iron, (h) cutting gauge, (i) taping stick, (j) gummed tape.

- The **flat iron** (Figure 10.3g) is a heavy piece of flat cast iron that is heated and used for melting the glue (particularly animal glue). It is also used to flatten the veneer just before the veneer hammer is applied to squeeze out the excess glue. In order to prevent the veneer from curling away from the ground, the surface is dampened (wet) with warm water before the iron is applied.
- The **cutting gauge** (Figure 10.3h) is used for cutting narrow strips of veneer, especially those used for veneering the edges of panels. Whenever decorative cross-bandings are required around panels, the cutting gauge is used for cutting the surplus veneer. The gauge should be set to the width of the cross-band and the cutting must be done before the glue sets. This will enable the surplus veneer to be removed easily.
- The **taping stick** (Figure 10.3i) is a flat stick with a rounded end, used to press the tape into place when joining veneers edge to edge.
- **Gummed tape** (Figure 10.3j) is a strip of self-adhesive paper that is stuck over edge joints in veneer to stop the M opening up while the glue dries.
- The **cabinet scraper** (see page 28) is a strip of metal used to level and smooth the veneer sur-

faces after the glue has dried or set. It is necessary to use the scraper when the veneer has wild grains such as crotch, butt, burl and ribbon figure.

- The **chisel** (normal woodwork chisel) is used to cut thin veneers and to remove waste strips of veneer.

Methods of veneering

The two common methods for veneering are the hammer method and the caul method.

The **hammer method** (Figure 10.4) is the more suitable when using animal glue. The glue should be freshly prepared and of a proper consistency. The veneering should take place in a warm atmosphere.

The procedure is as follows:

1. Cut the veneer to size, making room for an overlap of about 10 mm all round (Figure 10.4a).
2. Prepare the surface to be veneered. It must be perfectly smooth and have no large knots or blemishes. Use the toothing plane or a hacksaw blade to roughen the surface (Figure 10.4b). The plane should be used diagonally (to the grain) to take out all irregularities.
3. Wet the veneer slightly on both faces with clean, hot water. If the veneer is warm, it prevents the glue from cooling too quickly.
4. Apply the glue to the surface, making sure the entire area is properly covered.
5. Put the veneer in place. Press it down using your bare hands.
6. Use the veneer hammer to squeeze out the surplus glue (Figure 10.4c). Start from the centre and work towards the edge in a zigzag manner. Maintain a steady pressure throughout.
7. If the glue becomes cold, before it is time to squeeze out the surplus glue, wet the surface of the veneer with clean, hot water and pass a hot iron over the entire surface. This will melt the cold glue and allow it to be squeezed out (Figure 10.4d).
8. Use a cloth, wet with hot water, to remove the surplus glue from the edges. Take care not to allow the water to flow under the veneer as this will cause it to lift off the ground.
9. Trim off the overlap with a sharp penknife. This must be done when the glue has dried or set.

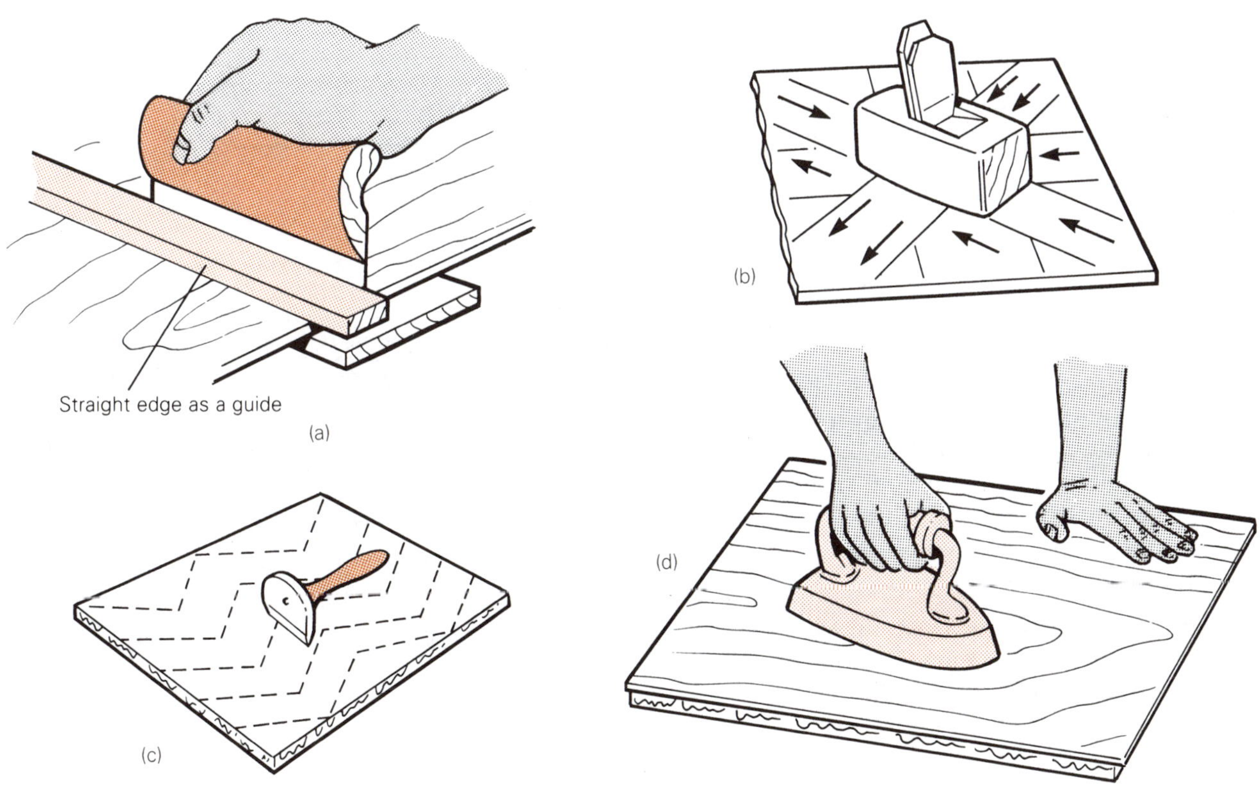

Figure 10.4 Hammer veneering: (a) cutting veneer, (b) preparing the ground, (c) squeezing out surplus glue, (d) Melting chilled glue.

10. If there is any blister on the surface due to trapped air below the veneer, use a penknife to slit it open. Insert a small amount of glue and then press the veneer down into place once more.

The **caul method** follows almost the same procedure as that of the hammer method. The only difference lies in the manner in which the veneer is held in place while the glue sets. A caul is a flat sheet of material such as blockboard, used to hold the layers of veneer while the glue sets. The method is mostly used when the surface is large or is to be covered with a pattern made up of several different piece of veneer. It is also very convenient for veneering curved surfaces. In this case, shaped cauls of the same radius as the surface being veneered are used.

The procedure for applying the caul is as follows (Figure 10.5):

1. Place the veneer in position on the ground, cover the surface veneer with a piece of paper to prevent the veneer from adhering to the caul.
2. Place heated cauls, one on the piece of paper on the surface of the veneer and another under the ground.

> **Note**, it is only when animal glue is used that the caul must be heated. Other kinds of glue, such as casein glue or any of the synthetic glues, that are used cold, do not chill and therefore do not need to be warmed up by the heated caul. Use cold cauls when these types of glue are used.

3. Place battens across the caul at short intervals and clamp them down (Figure 10.5a). The battens or bearers should be slightly convex in shape so as to exert pressure from the centre to force out the excess glue (Figure 10.5b). Note the order in which the battens are clamped (Figure 10.5c).

Matching panels are those in which surfaces are laid side by side or end to end with selected veneers which match each other in grain and figure to form attractive designs.

The procedure is as follows:

1. Prepare the surface of the ground to be veneered.

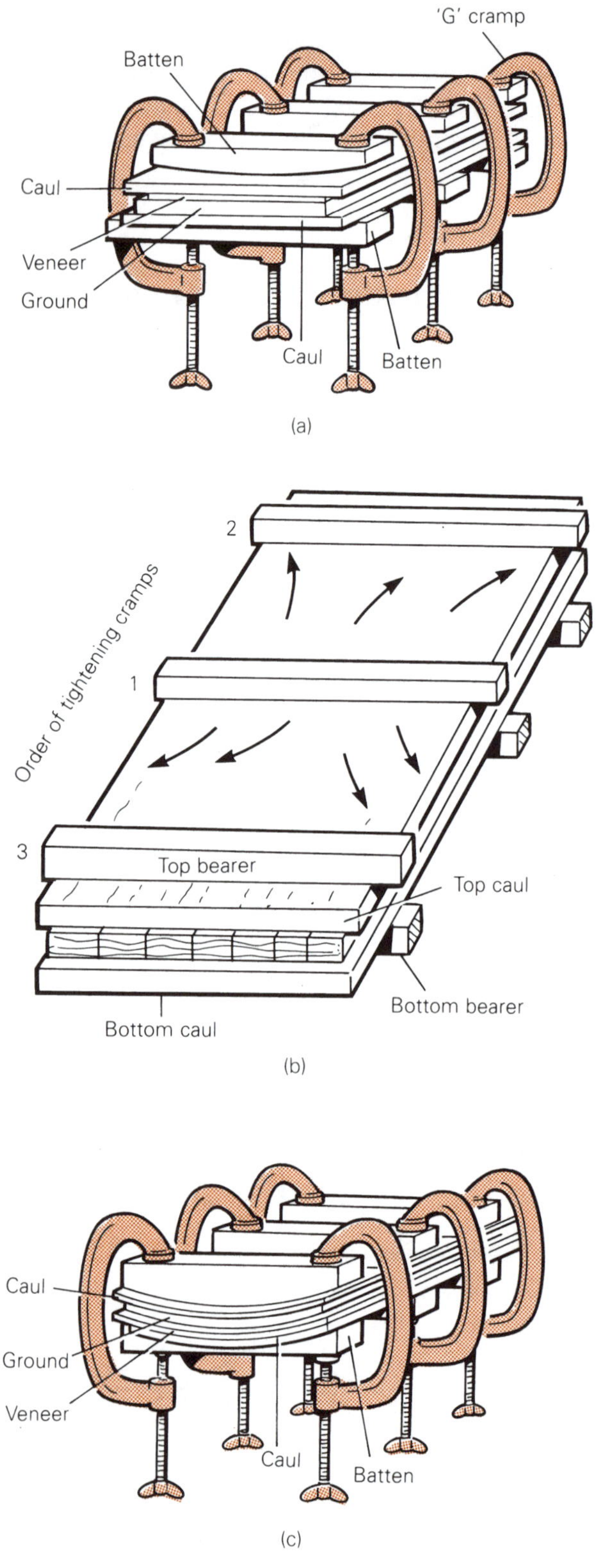

Figure 10.5 Caul veneering: (a) clamp the battens, (b) order of clamping battens, (c) force out excees glue.

2. Mark the position of the joint on the surface.
3. Cut the veneers, allowing for overlaps at the joint and at the edges of the ground.
4. Apply the glue to the surface.
5. Lay the first veneer to overlap the marked line.
6. Lay the second veneer to overlap the first one at the marked line (meeting line of the two pieces of veneer).
7. Place a straight edge over the veneer on the centre line, (i.e., on the joint) to serve as a cutting guide. Cut through both veneers using a sharp penknife.
8. Remove the top waste strip and then the bottom waste strip, while the edge of the second veneer is slightly raised.

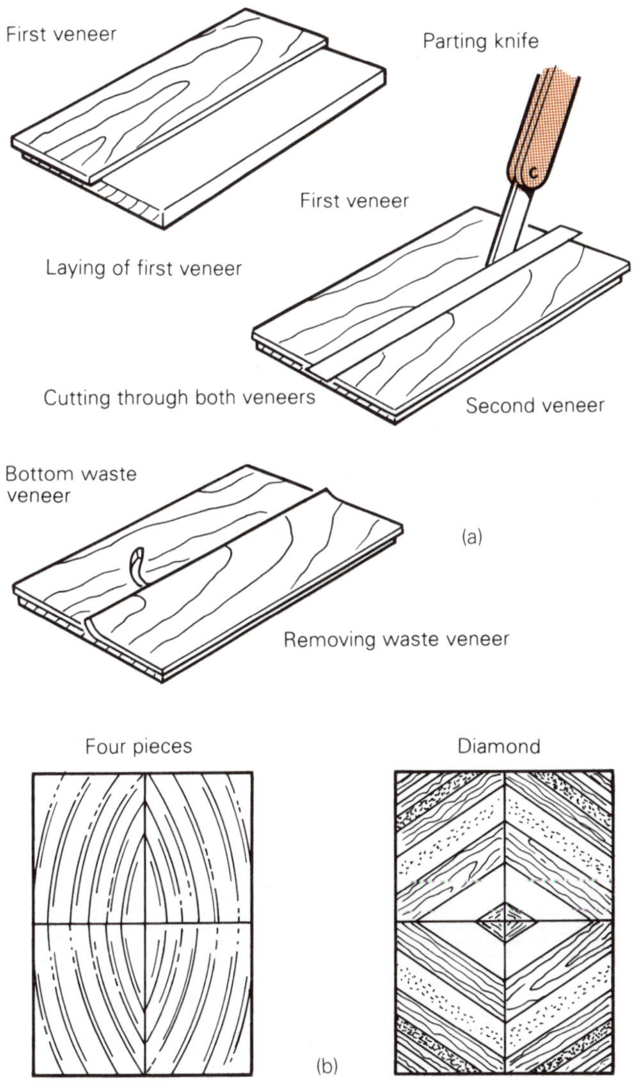

Figure 10.6 Matching and quarter panels: (a) preparing matching panels, (b) quarter panels.

9. Press down both edges of the veneer and hold them in position by strips of gummed tape until the glue sets. The caul method can be used to press down the veneers.
10. Trim off the edges and plane to size.

A **quarter panel** (Figure 10.6) has four pieces of selected veneers carefully laid to form an attractive pattern, such as a diamond. The procedure is similar to that for preparing matching panels.

Manufactured boards

Manufactured boards are purposely built up from wood products. They have several advantages over solid board: they can be obtained in large sheets; they possess high uniform strength; they are free from shrinkage, swelling and warping; they can be safely nailed or screwed near the edge without splitting; they are economical in the use of figured timber; they can be worked and cut to size easily; their smooth, machine-finished surfaces enable them to be easily veneered, polished and painted. The most common manufactured boards are: plywood, blockboard, laminboard, particleboard and fibreboard.

Plywood

Plywood (Figure 10.7) consists of an odd number of sheets of veneer glued together face-to-face with the grains of adjacent pieces intersecting at right angles. A plywood may be termed **three-ply** when it is built-up of three layers or veneers; namely the face ply, the core or inner ply and the back ply. Normally, the face and back plies are laid at right angles to the grain of the core. The thicknesses of the three layers can be equal or the core may be thicker than the outer layers. When the core is thicker the plywood is referred to as **stout heart three-ply**. The term **multi-ply** is applied to a plywood having a core of three or more plies. An odd number of plies is used for multi-ply plywood and the grain of each successive layer is laid at right angles to the last in order to provide a greater strength and stiffness in all directions and also for dimensional stability.

The manufacture of plywood basically follows these stages: cutting the veneers; drying the veneers (8%–10% MC); grading and preparing the veneers; gluing the layers of veneer; pressing the layers into close contact (by hydraulic

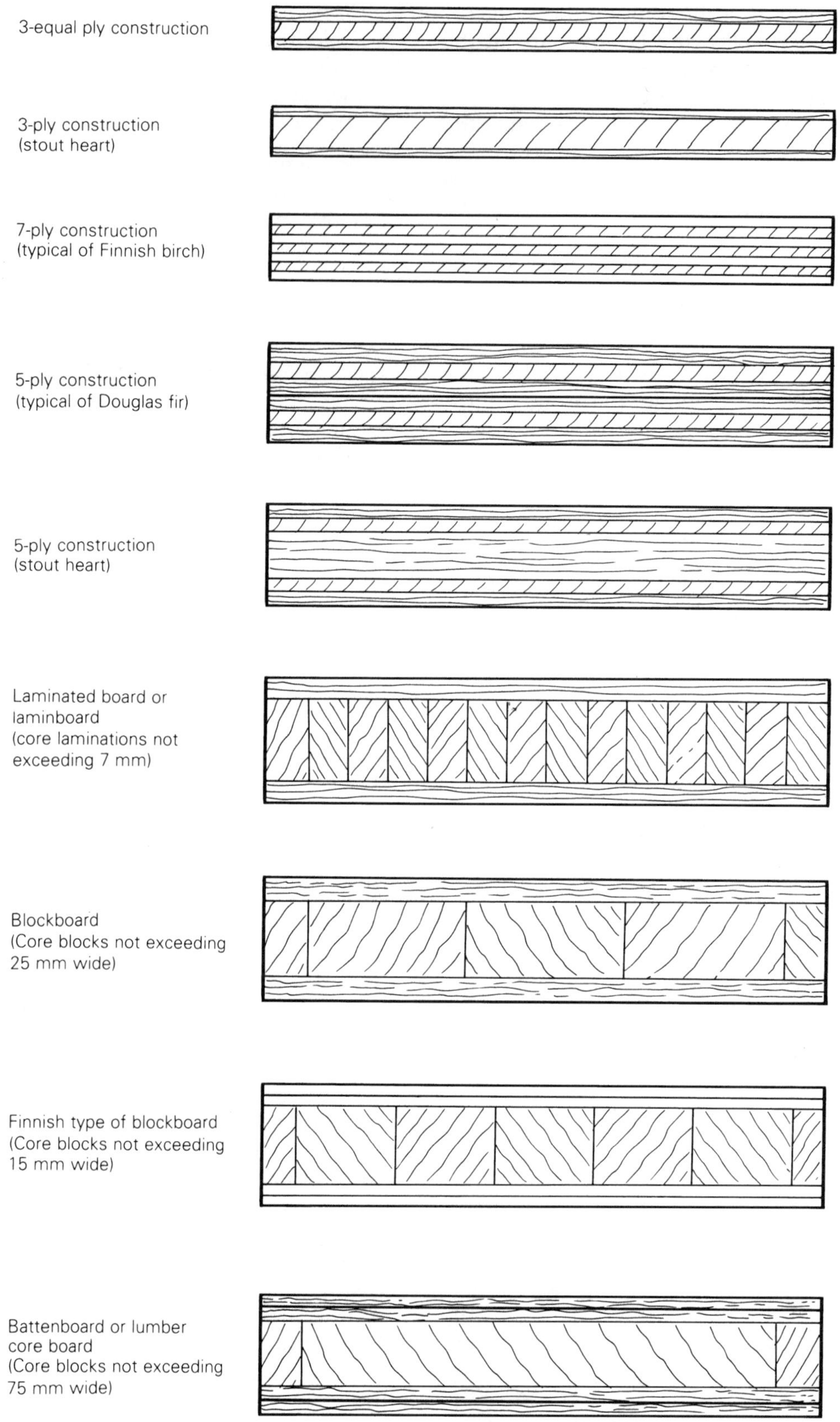

Figure 10.7 Sections of plywoods and core boards.

presses); reconditioning or re-drying to an equilibrium moisture content (8–15%); trimming and finally sanding to the desired surface finish.

Plywoods are classified and graded according to the of the adhesive used for bonding the veneers and the quality of the veneers used. The **strength and stiffness** depends upon the density of the species of timber used for the veneers and the quality of the adhesive used. The higher the density the greater the strength and stiffness. As a common saying has it, 'a plywood is only as good as its glueline, it is worthless if it de-laminates'. You must therefore, always select the correct grade of plywood to suit its intended use.

The three main grades of plywood are: marine grade, exterior grade and interior grade.

Marine grade is made from specially selected veneers of high class species of timber and has a high strength, uniform texture and surface hardness. It is bonded by either phenol-formaldelyde (P.F.) or resorcinol formaldelyde (R.F.) adhesives and highly resistant to weather, water, heat, most chemicals and micro-organisms. As the name suggests, it is suitable for boat building.

Exterior grade plywood comes in two grades, **'A'** and **'B'**. **Grade 'A'** is bonded by phenol-formaldehyde and resorcinol-formaldehyde adhesives. It is waterproof and suitable for permanent exterior applications. **Grade 'B'** is bonded by melamine-formaldehyde (M.F.) and urea-formaldehyde adhesives. It is moisture resistant, cold waterproof, and resistant to micro-organisms.

Interior grades can be moisture resistant or non-moisture resistant. The **moisture resistant** type is bonded by urea-formaldehyde, casein or albumen adhesives. It can withstand three hours in water at a room temperature, and can be a **selected** grade (selected face veneers) or **merchantable** grade (plain rotary-cut face veneers). The **non-moisture resistant type** may be bonded with animal glue, soya beans, casein and blood albumen glues, and can be of a **selected** or **merchantable** grade.

Plywoods have a range of applications, including construction of home and office furniture, toys, sports equipment, drawer bottoms, door panels, wall panelling and roof covering, floors, boats, aircraft, boxes, radio and television cabinets.

In order to prevent plywoods from buckling, they must be stacked lying flat on closely spaced stickers with a weight placed on top of them.

Built-up cored boards

Besides plywood, there are other kinds of man-made boards (Figure 10.8) whose cores are built up with a number of narrow solid wooden strips,

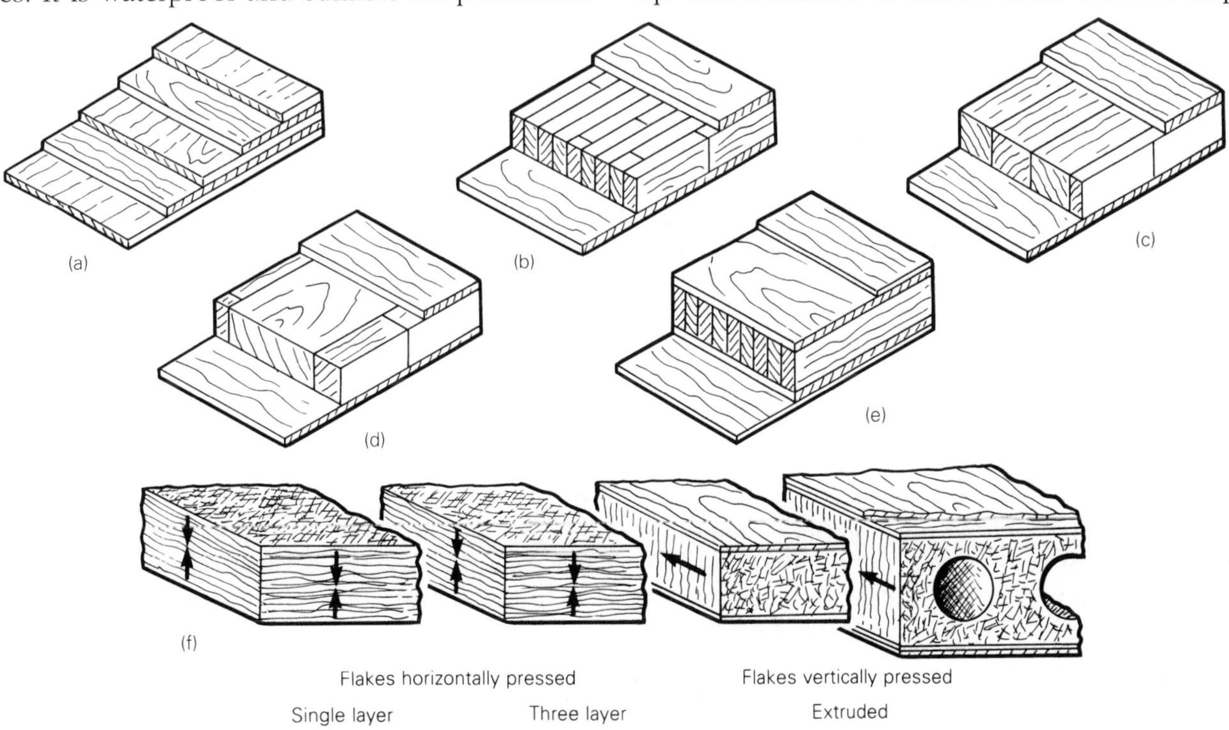

Figure 10.8 Manufactured boards: (a) multiply, (b) laminboard, (c) blockboard, (d) battenboard, (e) composite, (f) particle boards.

glued together to form the desired width. The strips are laid such that the heart side of adjacent strips face in opposite directions. The corestock is cross-banded (covered) with a sheet of plain veneer on both sides before the face veneers are applied. The grain of the cross-band is always at right angles to the grain of the corestock or strips. The three main types of solid core manufactured board are laminboard, blockboard and battenboard. The difference between them depends upon the widths of the strips constituting the corestock. **Laminboard** has a corestock of strips 9–10 mm wide. It is widely used for high quality furniture and joinery. The corestock of **blockboard** consists of strips 18–25 mm wide, while the corestock of a **battenboard** is made up of battens about 75 mm wide.

Particle boards

Particle boards are manufactured boards made from a number of wood waste products such as chips, splinters, shavings, flakes, etc. They are machine bonded together with a synthetic resin adhesive under heat and pressure. Some of the common names of particle board are **chipboard, shaving board** and **flakeboard**. Particle boards may be classified according to the layer of the structure. A **single layer board** is a type of particle board made from chips and resin bonded together, pressed flat under heat and pressure to form one homogeneous (same consistency throughout) layer structure. It is generally used for building construction. The **three-layer particleboard** consists of three layers of small flakes of wood, each resin bonded together and pressed flat. The core layer is thicker and is sandwiched by two finer layers, one on each side. **Extruded board** has a single homogeneous layer that has been forced through an opening formed by an oil-heated metal plate corresponding in size to the thickness and width of the board.

Fibreboards

Fibreboards are, as the name implies, manufactured from fibres of wood. The fibres are rearranged to form grainless sheets and are bonded together by their own natural adhesive properties or by the addition of other adhesives. The two types of fibreboard are **hardboard** and **softboard** (also called **insulated board**). The hardboard has its fibres compressed whilst with the insulated board (softboard) the fibre are not compressed.

CHECK YOUR UNDERSTANDING

- A veneer is a thin layer of timber produced by rotary peeling, slicing or sawing.
- Veneers are grouped into four types: core veneers, cross-band veneers, back veneers and face veneers.
- The veneering process involves laying a thin sheet of figured veneer over a piece of furniture constructed from unfigured and cheap timber, giving it a more decorative appearance.
- The two widely used methods for veneering are by the use of a hammer and a caul.
- Manufactured boards are man-made boards including plywood, laminboard, blockboard, particle board and fibreboard.
- Plywoods are classified and graded according to the nature of the adhesive used for bonding the veneers and the quality of the veneers used for the manufacture.
- The materials used for edging include, solid wood, flexible extruded plastics, and metal.

REVISION EXERCISES AND QUESTIONS

1. State the advantages of manufactured boards over solid boards for constructional and cabinet work.
2. Give reasons why cross-band veneers are laid with their grains at right angles to the grains of the corestock.
3. Explain why blockboards and laminboards are free from warping or curling. How can you differentiate a blockboard from a laminboard?
4. Name five veneering tools and state the application of each.
5. How are particle boards classified? Name the classes.
6. State the two types of fibreboard and explain the main difference between them.

Shaping and surface decoration

11

Shaping processes

Most parts of furniture and many other wooden items are shaped to improve upon their aesthetic appearance. There are many methods for shaping wood (Figure 11.1), and those discussed here include: shaping by wastage, (i.e., cutting shapes from solid wood), shaping by lamination, shaping by kerfing, shaping by bending and shaping by moulding. Before you decide which method to use, you must consider the type of timber to be shaped, the strength required, the degree of curvature and the type of finish required.

Shaping by wastage (Figure 11.1a) is the removal of material from a given piece of wood or timber to provide a a form or shape better suited to the required design. Flat curves, such as those used for shaped chair arms or table legs, can be sawn out of solid, straight-grained wood. Some of the tools used for shaping or removing the waste are saws, such a bow or keyhole saw; chisels; planes such as spoke-shaves, surform planes; rasps; and files. Machines such as routers, circular saws, band saws, jigsaws and planers, may be used to speed up the production process as in industry, where they are used for mass production.

This method of shaping is by far the quickest but it has two major disadvantages: much material is wasted and it often results in **cross-grain weakness**, owing to the fact that the grain direction will not always follow the direction of the curve. It is therefore only suitable for components in low stress situations or which will be supported by other structural members.

Shaping by lamination involves gluing a number of thin strips of timber, their thicknesses ranging from 0.5 mm to 5 mm in school (but up to 50 mm in industry), into a desired shape. The process involves spreading glue over each piece and bending the pieces simultaneously over or around a bending former.

The following points should be taken into account when designing work for lamination:

- all curves should be regular and smooth;
- avoid sudden changes in direction;
- use thin laminae for small radii and thicker ones for larger radii;
- any number of laminae may be used but remember, if thick sections are required and thin laminae are used, much more adhesive is required and it becomes expensive.

Examples of items that are made by this method include: hockey sticks, archery bows and tennis racquets. In the boat-building industry a variety of parts are made by lamination and in the furniture industry the technique is widely used to make parts of chairs and other furniture where curvature is required.

Lamination may take three forms:

- **Laminated blocking** (Figure 11.1b) where the required shape is broken down into smaller segments so that the cross-grain weakness is reduced to a minimum. Waste is considerably reduced and the strength of the shaped wood is further enhanced.
- **Laminating thin layers** (Figure 11.1c), a process that makes use of relatively thick laminates sawn from solid wood, spread with glue and put together by cramping them against a prepared former.
- **Laminating veneers** (Figure 11.1d), the method often used when forming shapes with thin veneers. The veneer layers are spread with glue and are pressed together by means of hand cramps, a platen press using matched male and

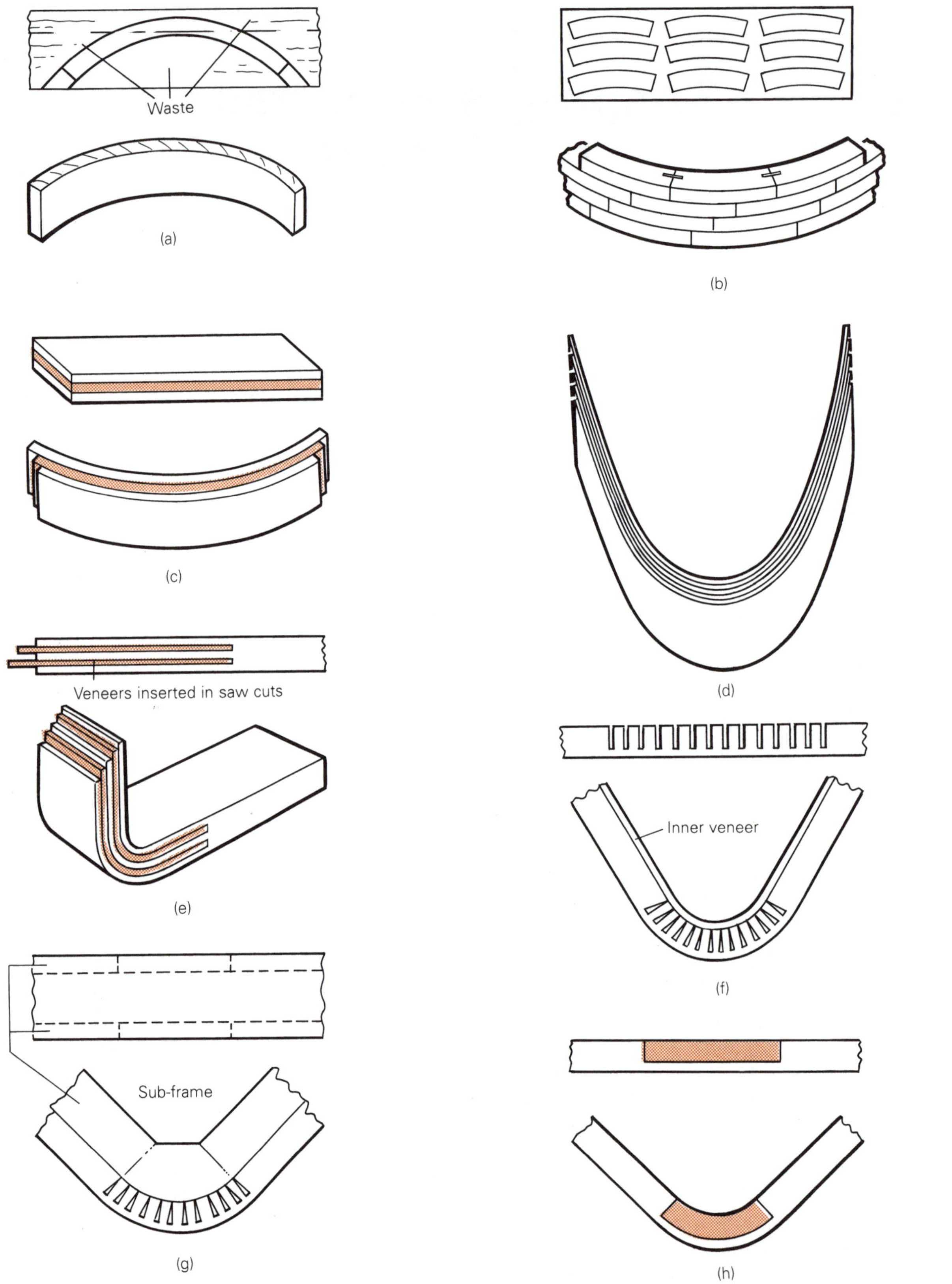

Figure 11.1 Methods of shaping timber: (a) shaping by wastage, (b) laminated blocking, (c) laminating thin layers, (d) laminating veneers, (e) infilling saw kerfs with laminates, (f) reinforced kerfing, (g) kerfed bend and sub-frame, (h) trenched and infilled.

female moulds, or by use of a vacuum bag, vacuum press or dome press. With this process, the grain of the laminates all lie in the direction of the curve thereby providing a greater strength. See shaping by moulding (pages 0000) for the detailed descriptions of these processes.

Shaping by kerf bending (kerfing) consists of making a number of parallel saw kerfs (cuts) part-way through a piece of timber to relieve the normal bending stresses. It is then bent over or around a former without steaming. Though this method reduces the strength of the job, it may be used quite satisfactorily for jobs that do not require much strength. For example, curved drawer fronts, curved small doors, rounded corners of cabinet carcases, curved plinths and edging for curved table tops may all be made by kerf bending.

Sometimes, when a bend is required at the end or ends of a length of a piece of work, saw kerfs are made from the ends to just beyond the start of the curve and then infilled with glued laminates of the same timber and pressed to a shaped former. This process may be referred to as **saw kerfing with infilling laminates**. Other kinds of kerf bending are: **reinforced kerfing**, **kerfed bend and sub-frame**, and **trenched and infilled**. See the illustrations in Figure 11.1e–h.

Shaping by bending is based on the fact that wood in its natural state has elastic properties enabling it to be bent. Many timbers, if subjected to heat and moisture, become semi-plastic and the shape of the cells will *permanently* change to the form or shape of the bent. When a piece of wood is bent, the cells on the inside of the curve are compressed, while the cells on the outside are stretched.

The usual method for bending solid wood is as follows:

1. prepare a jig to hold the shape;
2. heat the timber in steam under pressure for a period of time (approximately 45 minutes for a piece of timber 25 mm thick);
3. remove from the steam and quickly cramp it in the pattern required;
4. leave it to cool completely.

Bends made at a very high moisture content will need drying out and there is a risk of drying defects. Though a low moisture content would eliminate drying defects, there is the risk of bending stress defects. Moisture content around 20% is recommended.

Another method of bending wood is by the use of a **tension strap**. The strap effectively becomes part of the piece being bent and will carry the stress on the tension side of the bend thereby shifting the neutral axis to or very near to the strap and making most of the bending stresses become compression strain.

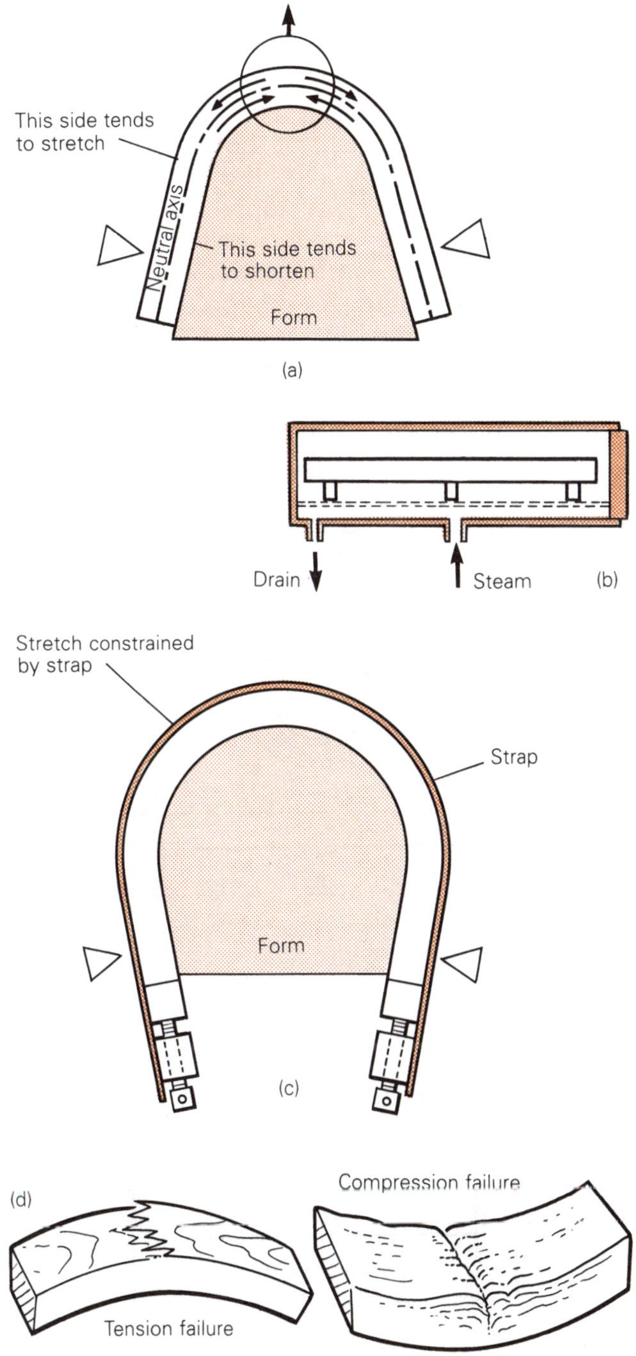

Figure 11.2 Method of bending solid wood: (a) applying steam under pressure, (b) bending on a former, (c) using a strap,(d) types of failure.

The two distinct merits of bending solid wood to a curved shape are that less wood is wasted, and it reorientates the longitudinal grain to the axis of the curved thereby avoiding cross-grain weakness.

The major problem associated with bending solid wood is how to bend it to the required curvature without failure. The illustration in Figure 11.2 shows this fundamental problem. When a simple beam is centrally loaded the resultant bending tends to shorten the upper (concave) surface and stretch the lower (convex) surface. Under this situation, half of the wood fibres will be in compression and the other half in tension. The stresses developed are axial: compression along the upper surface and tension along the lower surface. Both stresses are maximum at the surface, diminishing to zero at the horizontal central axis – the neutral axis – of the beam. Axial stresses along both the top and bottom surfaces are maximum at mid-span, decreasing to zero at the end of the beam. Therefore, in a 'free' beam under load the compression and tension stresses are in equilibrium. The stresses are called bending or fibre stresses. When a beam is bent beyond its elastic limit, considerable deformation can occur in compression before eventually resulting in compression failure, but in tension it takes only a 1% elongation to cause tension failure.

Successful bending of wood therefore depends upon overcoming its comparatively low tensile strength and taking advantage of, and if possible enhancing, its superior compression strength. This can be achieved in two ways as described: by steaming the wood to improve its plasticity and by the use of the tension strap to constrain the elongation of the wood.

Shaping by moulding may be accomplished in three ways: the use of matched male and female moulds, vacuum forming and dome press.

When **matched male and female moulds** are used (Figure 11.3a) pre-glued veneers are laid in

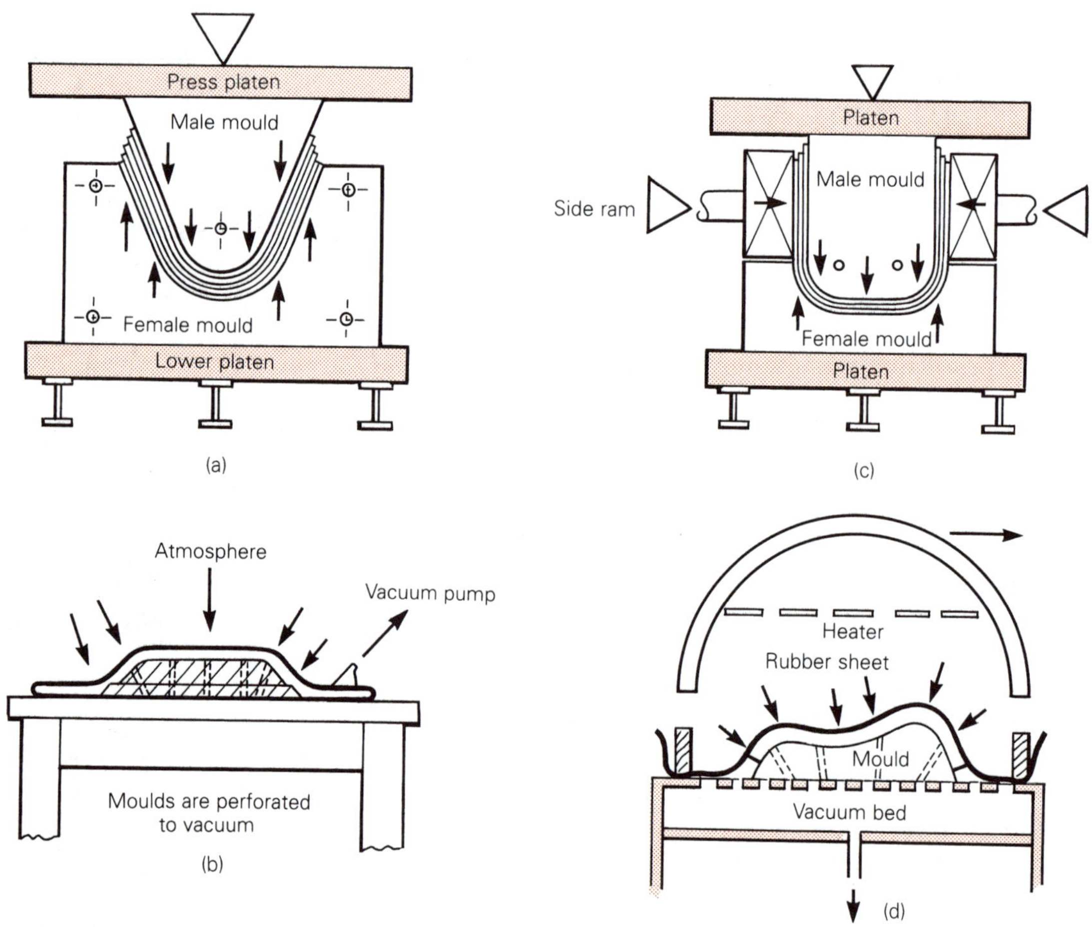

Figure 11.3 Methods of moulding: (a) matched moulds, (b) shell press, (c) vacuum bag, (d) vacuum press.

the female mould and the matching male mould is then pressed home to form the shape. It must remain under pressure until the glue is cured (dried). The pressure may be applied by using hand cramps for small simple mouldings or a screw or hydraulic platen press for larger work.

Vacuum forming requires only one mould, usually male. The methods used are the vacuum bag or vacuum press (Figure 11.3b). In the **vacuum bag** (Figure 11.3c), the veneers are positioned on the mould and enclosed in a sealed. The air is then extracted by a vacuum pump to subject the contents to the prevailing atmospheric pressure. The vacuum is maintained until the glue is cured. For the **vacuum press**, the single mould rests upon the perforated bed of the press and a rubber sheet attached to a frame is lowered and sealed to the bed. The air is extracted and the atmospheric pressure shapes the veneers to the mould. It is necessary to maintain the vacuum until the glue is cured. The curing time can be accelerated by heaters contained in a sliding dome (Figure 11.3d).

The dome press method incorporates the vacuum process within a pressurised and heated chamber called an autoclave. The first part of the process is the same as for vacuum forming, air being extracted from a rubber bag to shape the veneers to the mould. The dome of the autoclave is then lowered and sealed to the bed of the press. It is pressurised with compressed air and steam, subjecting the assembled veneers to heat and very high pressures. The heating process causes the wood fibres to become more plastic and, together with the increased pressure, enables sharper mouldings to be produced.

Decorative processes

Surface decoration, as regards to furniture, refers to the nature and the form of the outer appearance that enhances its beauty to attract and appeal to the eye. There are many ways of decorating the surface and these include: using timber with an attractive figure and grain to construct the artefact; shaping the profiles; chamfering the arisses; carving or incising the surface; veneering the surface; and inlaying the surface.

Inlaying is one of the oldest practices of decorating surfaces of cabinet or furniture by means of planting thin strips of coloured material in the surface. The strips are generally sold in lengths of about 900 mm and about 1 mm to 5 mm in width. Those with widths within the range of 5 mm to 18 mm are called **bandings**. The strips are often in two or three colours in chequered or sandwich arrangements (Figure 11.4). They are very thin and the recesses made for them are

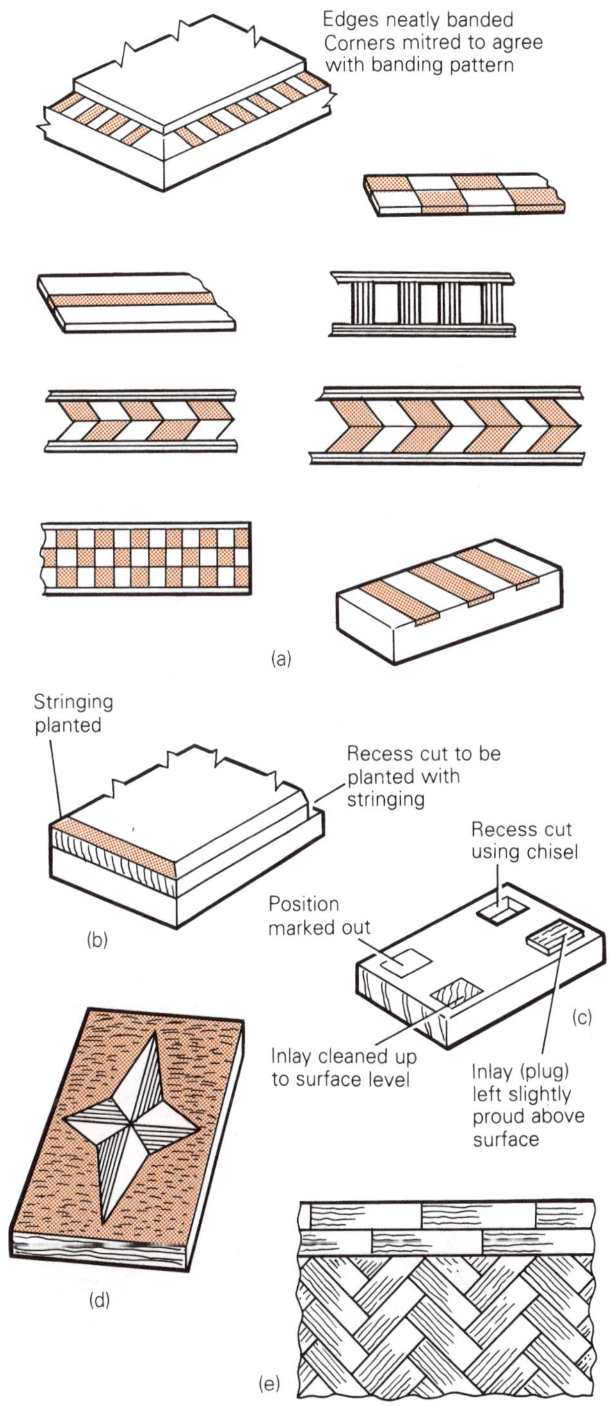

Figure 11.4 Forms of surface decoration: (a) kinds of banding, (b) stringing, (c) inlaying, (d) marquetry inset, (e) parquetry floor.

usually shallow. These strips can be inlaid either in solid or veneered surfaces. If used in a veneered surface they can be included during the veneering process but on a solid work a recess must be cut specially for them.

Stringings are plain strips of wood of simple colour which are square in section, the smallest being 4.5 mm wide. The timbers commonly used include **ebony** (black), **boxwood** (creamy white) and **rosewood** (red brown). They are glued into rebates cut along the edges of veneered work to protect the corners or edges from breaking off. The **inlay** (thin strips of coloured material) may be made of plastic, pearl, metal, ivory, bone, or wood.

Among the tools required to carry out the inlaying process:

- are the **cutting gauge**, used for cutting the boundaries of a groove and a trench;
- the **scratch stock**, used to finish off or to clear out the groove or trench; the scratch stock is in the form of a try square and it consists of two pieces of wood fastened together with screws. The cutter is held in between the two pieces, as shown in Figure 11.5. The vertical part serves as a fence when being used. It can also be used for finishing off mouldings when the appropriate cutter is inserted.

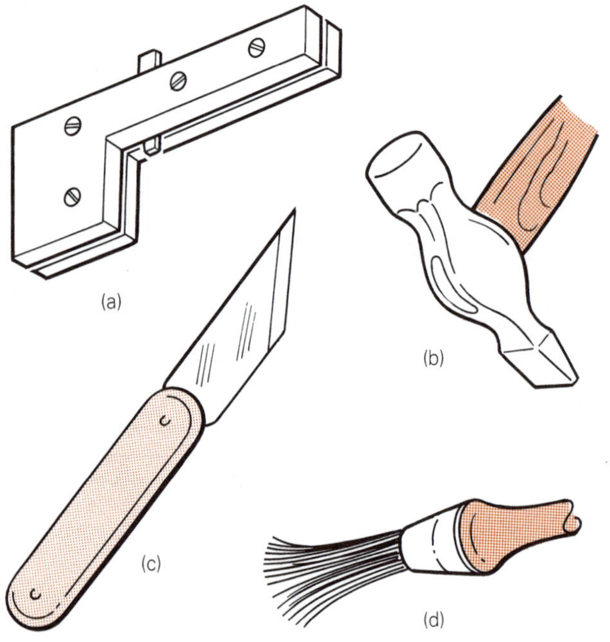

Figure 11.5 Tools used in surface decoration: (a) scratch stock, (b) cross-pein hammer, (c) knife, (d) brush.

- the **penknife**, used for cutting off the strips and banding;
- **brushes**, used to apply glue in the groove or trench and on the back of the strip or banding;
- the **cross-pein hammer**, used for pressing the strip or banding down into the groove or trench;
- the **smoothing plane**, used for cleaning off;
- **scrapers**, used for cleaning off.

The procedure for inlaying includes the following steps:

1. Prepare the surface to be inlaid.
2. Mark out the positions on the surface.
3. Set the cutting gauge at the correct position and cut the boundaries of the groove.
4. Set the cutter to the correct position in the scratch stock and finish off the groove.
5. Apply glue to the strip or draw it through a glue brush loaded with glue.
6. Press one end of the strip into the groove.
7. Press the remainder into place by drawing a hammer along it.
8. Allow sufficient setting time, say 24 hours, for the glue to dry.
9. Clean off with a finely set plane and scraper. When cleaning off inlay of intersecting grooves, plane the one across the grain first to avoid any splitting at the junction of the two grooves.

Marquetry is a form of decoration whereby the entire surface required to be decorated is covered with coloured veneers arranged to form an attractive design. The design may be in the form of a geometrical figure or may be based on pictorial motifs, such as fruit, buildings, figures, flowers or scenes. The procedure is similar to that of veneering.

An **inset** consists of flat pictorial patterns made from veneer of a variety of colourings, that are planted into a recess or recesses made on the surface of the work.

Overlays are made up of shapes or designs cut out of wood, metal or plastic and glued to the surface of the work. They may be in the form of carved, fretted (pierced) or split turning.

Wood carving is the oldest form of decoration and has been practised since ancient times for surface decoration. Carving as a craft requires a great deal of skill in the use of cutting tools. The two main form of carving used for the decoration of woodwork are relief carving and incised carving.

Relief carving consists of cutting away the back ground of the pattern in order to make the design itself appear raised above the surface. This is of course a difficult form of carving for sharp corners easily break off.

Incised carving consists of cutting a pattern or design into the surface of the work-piece by means of gauges and vee-tools so that the pattern lies below the surface. Designs for this type of carving are generally more simple than the type of design found in relief carving, and for this reason, it is the best type of carving to be introduced in normal school work.

CHECK YOUR UNDERSTANDING

- Parts of furniture and other wooden items are shaped in order to improve their aesthetic value.
- Shaping wooden items may be accomplished by wastage, bending, and kerf bending.
- Surface decoration is a means of enhancing the appearance of a piece of furniture or cabinet in order to make it more attractive and appealing to the eye.
- Surfaces can be decorated by using well-figured and attractive board for the construction, chamfering the edges, or arisses, carving, veneering and inlaying.

REVISION EXERCISES AND QUESTIONS

1. What factors are worth considering when you want to shape a piece of work?
2. Explain why a piece of timber shaped by sawing is weaker than one shaped by bending.
3. Briefly describe the process of inlaying.
4. Differentiate between the terms 'relief carving' and 'incised carving'.
5. Distinguish between stringings and bandings.

12 Non-wood materials

Introduction

In modern design, the use of one type of material for manufacturing artefacts is giving way to combination of several materials. This concept is being practised in woodwork design as well. Many non-wood materials such as ferrous and non-ferrous metals, plastics, plastic laminates, leather and glass are now prominently used in the woodwork industry. It is therefore essential that you know something about these materials.

Metals

Metals can be grouped into two categories: ferrous metals and non-ferrous metals.

Ferrous metals are those containing iron and the most important examples are **steel** and **cast iron**. Both of them are **alloys** (combinations) of iron and carbon but cast iron contains a greater proportion of carbon than steel. The addition of certain elements such as chromium and tungsten to steel improves its properties considerably. For instance, when chromium is added, the steel becomes resistant to corrosion and is called stainless steel, while tungsten increases its hardness. Both steel and cast iron are used in woodwork for the manufacture of machine components, tools and stands for various furniture items.

Non-ferrous metals include aluminium, copper, zinc and brass.

Aluminium is a non-ferrous metal (contains no iron). It is silvery white in colour, and is light, soft, malleable, ductile and a good electrical conductor. It can be rolled into thin foil, drawn into wire, spun into containers, bent into boxes, cast easily and polished to a high finish. It is widely used for making cooking utensils, electrical work, and small fittings for woodwork.

Copper is a non-ferrous metal of salmon pink colour when polished. It is very malleable and ductile, and a good conductor of heat and electricity. It can be forged, rolled, drawn into tubes, and wires, cast or beaten into bowls. It is mainly used for making soldering bits, bowls, boxes, electrical and wireless parts, wires, rivets and woodwork fittings.

Zinc is a bluish-white non-ferrous metal that is widely used for coating or galvanising steel for making roofing sheets and woodwork fittings that need protection from the weather, such as hinges, locks, stays and nails.

Brass is an alloy of copper and zinc. Mixed in different proportions they poroduce different kinds of brass. It is yellow in colour and shines when polished. It is hard, resists corrosion and easy to work on. It is used for plaques, ornaments, water taps and woodwork fittings such as door and drawer locks, hinges and stays.

Ceramics

Ceramics are very hard substances that are normally used when materials that can withstand high temperature are required – for example, machine cutters.

Plastics

Plastics are non-metal materials that are fast superseding metal and wood in the manufacture of many household and other utility articles or items. The two main classes of plastics are thermoplastics and thermosets.

The **thermoplastic materials** are distinguished by their hardness and rigidity after being shaped or formed into articles and their ability to become soft again when heat is applied. They can thus be re-shaped into another form. Among the kinds of thermoplastics are:

- **cellulose nitrate celluloid**, used for making articles such as mallet heads, knife handles, pumps, spectacle frames and set-squares;
- **cellulose acetate**, used for transparent wrappings, lampshades, containers, safety-glass interlayers, welding goggles, and machine guards;
- **polyvinyl-chloride (P.V.C.)**, used for a variety of articles such as garden hoses, the insulation of electric cables, factory ducting, radio and television components, protective clothing and car battery cases;
- **acrylonitrite butadiene styrene (A.B.S.)**, used for making tool handles, instrument covers, pipes and pipe fittings;
- **polythene**, a tough, flexible, resistant to acid and alkalis and suitable for good insulators.

Thermosetting materials are those that undergo a permanent change during the process of being shaped or formed. Once shaped, they cannot be softened for re-shaping into another form. There are several kinds:

- **Silicone** is water repellent, withstands extremes of hot and cold, is inert to most chemicals and maintains its strength at high temperatures. It is used for making high-temperature electrical insulators, additives to oils and waxes, rubbers, bushes, gaskets, and valve diaphragms.
- **Epoxy resins** are excellent adhesives used for many purposes, have high strength, good resistance to moisture and chemicals, good electrical conductor. They are used as a base for adhesives, as sealing of electrical parts and as glass fibre laminates.
- **Bakelite** is a strong and rigid material, a good insulator, and has good heat and chemical resistance. It is used widely for electrical plugs, switches and radio cabinets.
- **Ureaformalde** is a hard, rigid material that absorbs water and is resistant to petrol, oil and grease. It is often used for electrical plugs, sockets, switches and lamp holders.
- **Polyester resin** is strong, hard and tough, a good electrical insulator, and resistant to dilute acids and alkalis. It is used for making glass fibres, aircraft parts and building partitions.

Plastic laminates

Plastic laminates, also called **formicas** are made from base materials such as cellulose, cotton, nylon, or fibreglass. This base material is covered with a resin binder and subjected to heat and pressure to form a hard, strong panel. There are different brands in a variety of patterns and colours. They are tough, lightweight, and strong but rather brittle. They resist most abrasive wear, denting, cracking, heat, alcohol, water and fruit acids. They can be sawn, drilled, routed and planed. Plastic laminates are widely used for overlaying tops of restaurant and kitchen counters and tables, coffee-table tops and bar tops. Due to their insulating properties, they are often used in electrical work.

Care should be taken when you are bonding plastic laminates. The ground (work) should be flat and strong. It is also necessary that the surface of the ground is free from warping or moisture. The preparation of the surface of the ground is similar to that when veneering and the steps follow a similar order. The work should be clamped or have a weight put on it while the glue sets. In most cases, contact or impact adhesives are used when surfacing with plastic laminates. Spread an even layer on both the ground or base and the back of the laminate. Allow ten to twenty minutes for both surfaces to become dry and tack-free. Press the laminate into position. If you are not using a contact glue then the work must be clamped or pressed down by putting a weight on it while the glue sets.

Fibreglass

Fibreglass is a short name for a material that is more properly called **glass fibre reinforced polyester resin**, or **GRP**. Polyester resin is a syrupy liquid that turns solid when a catalyst (hardener) is added. To speed up the process of hardening (curing), an accelerator is also added. With careful measuring of the quantities of hardener and accelerator, the curing time can be controlled accurately. Once the curing time has been decided, it is important to remember not to mix more resin than can be used in the time available. The glass fibre is introduced into the material to provide strength and toughness. Fibreglass can be in a number of forms:

- **chopped strand**, a short glass fibre that are randomly fixed with a resin-soluble binder into a thin sheet. It is the most commonly used form of fibre.
- **cloth type**, a woven form of fibres that possesses greater strength than the chopped strand mat. It needs very thorough laminating as the layers will de-laminate if great care is not taken.
- **ribbon type**, used for edging and local reinforcing.
- **tissue type**, a finer form of chopped strand mat used as a finishing layer to cover the irregular appearance of the mat.

The most common way to use fibreglass is by laminating into a mould. The mould is normally made of wood and is a female form of the finished piece. It should be prepared to a high degree of finish as the resin will show up any imperfections.

The procedure is as follows:

1. Prepare the mould thoroughly and apply a parting agent to stop the material from sticking to the mould.
2. Apply the first layer of resin (this is called the **gel coat** and may include a pigment if a coloured moulding is required).
3. Allow the gel coat to harden and then apply the next layer of resin. It is in this that the glass fibres are laid.
4. Stipple the fibres with a brush or roll with a laminating roller to ensure that they are thoroughly wetted with the resin and all the air is excluded. This process of resin and fibre application is repeated until the required thickness of moulding is obtained.
5. When the final layer is cured remove the completed moulding from the mould.

Glass

Glass is produced by mixing together soda ash, limestone, sand and a quantity of cullet (recycled glass). Other materials, such as metal oxides, can also be added to colour the glass. The mixture is then melted in a furnace at a temperature of 1500°C. The molten glass is then drained into a drawing tower, where it is drawn up into sheets between rollers and gradually cooled. It can also be moulded into containers. The moulding of glass is normally done using a blow moulding process. If a patterned glass is required then the molten glass is rolled between patterned rollers. Clear sheets of glass can also be made by floating the molten glass on a bath of molten tin. This process produces a much flatter sheet than the drawing process.

Glass can be cut using a diamond cutter to score the surface, thereby producing a weak line through which it snaps leaving a very clean edge. Where the sharp edge of the glass will be exposed and could cause injury to the user then it must ground smooth. Glass is used for glazed doors and windows, louvre blades, picture frames, sliding doors of cabinets and showcases.

Leather

Leather is generally obtained from animal skins. To produce leather, the hair or wool is removed. The skin is then treated. Different treatments and different animals produce different types of leather. Goat skin makes a very supple leather. Leather from pig skin is very hard-wearing. A horse-skin leather is the thickest and strongest.

Leather can be cut using a craft knife. The direction of cut is not subject to grain direction as leather has no grain and so it is easily cut in all directions. Pieces of leather are joined together by stitching, either by machine or by hand. When hand stitching, it is advisable to prepare the material with small, evenly spaced holes, the holes being produced by an awl. When larger holes are needed for joining by thonging, a **revolving hole punch** should be used. This punch has a series of different size punches suitable for the various sizes of the thong being used. If it becomes necessary to strengthen the punched holes in the leather then brass eyelets can be placed into the holes before applying the thongs. The surface of a leather can be decorated by using punches to produce a regular pattern, or special leather tools may be use to emboss the surface. Another method is to use a hot tool to slightly burn the surface of the leather, producing darker lines.

In woodwork, leather is mostly used as covering material for upholstery work such as lounge seats, chair seats/tops and backs, padded table tops, etc. It is also used for making other items such as bags, belts and shoes.

Fabrics

Fabrics are materials made from yarns either by weaving, knitting or bonding. The yarns are often made from natural and man-made fibres spun together into threads. The natural fibres are of two types: those made from parts of plants and those that are made from hairs of animals. The common fabrics, made from natural plant fibres are cotton, and linen.

Most **cotton fabrics**, made from the boll of cotton plants are strong, hard-wearing, smooth, easy to wash, resistant to alkalis and therefore can withstand bleaching and strong detergents in laundering. They take dyes and finishes well. They are widely used for household articles, clothing, dresses, cushion covers and handkerchiefs.

Linen fabrics, made from the stalks of flax plants are very smooth. They are stronger and have better appearance than cotton fabrics. They are good conductors of heat and are very absorbent. They are used for the same purposes as cotton.

Wool and **silk** are the most widely used fabrics made from natural animal fibres. **Wool** is made from the wool shorn from the coats of sheep, goats, and rabbits. Woollen fabrics are more expensive than those of cotton and linen. They give comfort and warmth and therefore suitable for winter underclothing and warm outer wear for people of all ages. They are also used for household articles and furnishings, including upholstery.

Silk is made from the secretion of silk worms that hardens into a fine thread winding round the worm to form a case or cocoon. Silk fabrics have a beautiful soft, glossy surface and drape into folds. They are expensive, very strong and elastic, excellent conductors of heat, wash and wear well and absorb perspiration readily. Silk is an ideal fabric for dresses for special occasions due to its lovely appearance.

The **man-made fibres** are of two kinds: **regenerated cellulose fibres**, where the liquid used for spinning is obtained from plant sources, and **synthetic fibres** where the substances are made entirely from mineral sources.

The regenerated cellulose fibres are used for making viscose fabrics, acetate fabrics and triacetate fabrics.

Viscose fabrics are made from viscose fibre yarns chiefly obtained from wood pulp. The fibres are produced with a *lustrous* or a *matt* surface and have fair absorbency. The attractive appearance and texture combined with their low cost, make them popular fabrics for all types of clothing, linings, braids, ribbons, house linen and furnishings.

Acetate fabrics are made from acetate yarns to obtained from cotton linters or wood pulp treated with chemicals to form cellulose acetate. They are absorbent and fairly non-static but melt if they come into contact with hot objects. They are used for making satins, jersey, taffeta, linings, and soft furnishings.

Triacetate fibres are strong and softens at a high temperature than acetate fibres. They have a low absorbency and dry quickly. Triacetate fabrics are used for making items such as dresses curtains, trousers and skirts

Synthetic fibres are of various type: **polymide fibres**, are used for making nylon, celon, Brinylon, Enkalon and Ultron fabrics; **polyester fibres**, used for making such fabrics such as terylene, Dacron, Trevira, Crimplene and Lirelle; **acrylic fabrics**, used for such fabrics as orlon, acrilan, Courtelle and Dralon;

Elastofibres, made from polymers based on polyurethane and polyisoprene. Elastomer yarns are used for Lycra, spanzelle and vyrene fabrics.

CHECK YOUR UNDERSTANDING

- The two types of metal are ferrous metals, that contain iron, and non-ferrous metals, that do not.
- Ferrous metals are basically composed of iron and carbon and the most widely used ferrous metals are steel and cast iron.
- When steel is alloyed with chromium a stainless steel is obtained, whereas an alloy of steel and tungsten produces a hard and heat-resistant steel called tungsten steel.
- Examples of non-ferrous metals are aluminium, copper, zinc and brass.
- Plastics are broadly classified into thermoplastics, those that can be shaped and reshaped when subjected to heat and thermosets, those that are permanently shaped when first made.
- The kinds of thermoplastics are cellulose nitrate celluloid, cellulose acetate, polyvinylchloride, acrylonitrite buladiene styrene (A.B.S.) and polythene.

- Thermosets include silicone, epoxy resins, bakelite, ureaformalde and polyester.
- The base materials for manufacturing plastic laminates (formica) may be cellulose, cotton, nylon or fibreglass.
- Glass is produced by mixing together soda ash, limestone, sand and cullet.
- Leather is generally obtained from skins of animals such as goats, pigs and horses.
- Fabrics are materials obtained from yarns either by weaving, knitting or bonding.
- Yarns are obtainable from either natural fibres or artificial (man-made) fibres.
- The sources of natural fibres are plants or animals.
- Man-made fibres are of two kinds: regenerated cellulose fibres and synthetic fibres.

REVISION EXERCISES AND QUESTIONS

1. Name two classes of metal and give two examples of each.
2. Name the two types of plastic materials, giving two examples of each type and stating the uses of each example.
3. Name three different types of glass fibre.
4. Explain briefly how glass is made.
5. From what is leather obtained? Describe how the face of leather can be decorated.
6. Name two fabrics that are made from natural plants and list the characteristics of each one.
7. State two fabrics that are obtained from animal sources and describe the characteristics of each one.

13 Wood finishing

Introduction

The term **wood finishing** applies to the application of selected materials such as stains, fillers, varnishes, lacquers, paints, enamels and waxes to a wooden surface. The application of these materials prevents the tendency of the wood to absorb moisture, fumes and oils that may cause shrinkage, swelling, checking, warping and discolouring. It also prevents fungal and insect attacks. In effect, finishes provide a protective coating that helps to stabilise the condition of timber at the time of its finishing and also serve as preservatives. Some finishes are fire and chemical resistant. The ultimate reason for applying finishes by an artist or a woodworker is to create desirable effects, enhancing the natural beauty of the timber or hiding its blemishes.

The surface to be polished requires a thorough preparation including levelling, smoothing and cleaning.

Prepare the surface as follows:

1. Smooth the surface using a well sharpened smoothing plane.
2. Where possible raise the grain by applying water to the surface.
3. Where there are blemishes that are not too deep, scrape the surface with a hand scraper until all such defects are removed.
4. Remove any grease and oil by sponging with a lacquer thinner or benzene.
5. Remove any traces of glue from the surface by either scraping or sanding.
6. Stains that are not removed by sanding may be bleached with oxalic acid or commercial bleaches.

> ▲ Remember that oxalic acid and bleaches are ***toxic***. All such chemicals must be handled and applied with great care. Wear goggles, rubber gloves and an apron at all times when bleaching.

7. Fill nail holes, checks or open joints with patching material such as plastic wood, glue mixed with sawdust, stick shellac or putty. The colour of the patching material must match the finished colour of the wood to be polished.
8. Glass-paper the surface thoroughly using rough, medium and smooth grades respectively. Use a glass-paper block for flat surfaces and large curves where it may be possible. Glass-paper along the grain of the wood to avoid scratches.
9. Remove any dust from the surface and from the pores of the wood by brushing thoroughly.

Applying bleaches

Bleaches are toxic oxidising agents that break down the binding substance between cell walls (lignin) on the surface of wood to make it become white. They may be mixed in different proportions to lighten the colour tone of the wood to be treated. They may be used to reduce the contrast between sapwood and heartwood and also to remove dark stains or blemishes. A bleach may be applied by brushing, swabbing, or dipping in on the affected portion.

> ▲ Wear goggles, rubber gloves and an apron whenever you use bleaches.

Applying paints

Paints are basically composed of pigments, vehicles, thinners and dryers. The pigments provide the opaque and colour characteristics of the paint. The **vehicles** provide the suspension fluid for the pigment and also act to bind the pigment to the surface being painted. The **thinners** regulate the viscosity (flow) and penetration of the paint. The **dryers** increase the drying speed of the paint. In general, paints are used for the purpose of decorating, preserving and protecting the work (wood or metal). They may be applied by brush or spray.

The two main classes of paint are the oil-based paints (with oil as the vehicle) and the latex paints (a mixture of water and resin acting as vehicle).

Oil paints provide an opaque, coloured finish ranging from flat to glossy. They are water resistant and provide greater protection against moisture entering the wood.

Latex paints provide tough, durable finish and have very good colour retention. They are resistant to water and alkali, odourless, fast drying, mildew resistant and poor adhesion to smooth surfaces. Latex paints are of two types: **polyvinyl acetate** and **acrylic**. They are both pigment-binding and film-forming.

Applying stains

The two types of stains are oil stain and water stain. **Oil stain** is a transparent colouring agent, made with colour ground in raw linseed oil, a solvent or a japan drier. The solvents used include linseed oil, benzene, mineral spirits and turpentine. Oil stain is the easier to apply. It is slow drying, preserves the wood and brings out the beauty of the wood without raising the grain, producing uniformity of tone or colour.

Follow the following procedure when you want to apply oil stain:

1. Clean the surface to be stained, keep it smooth, dry and free of oil, wax and glue.
2. Select a brush with moderately soft bristles.
3. Dip the brush into the stain and brush it onto the surface. If it is a vertical surface, start from the bottom and work upward to prevent the stain from running down.
4. Allow the stain to set for about 20 minutes and then wipe of the excess stain, using a soft, lint-free, absorbent material. Wipe in the direction of the grain.
5. Allow the stain sufficient time (about 18 hours) to dry out thoroughly before you apply any other finishes.
6. Oil stains may also be applied by spraying or dipping.

> Wear safety glasses and gloves when applying stain.

Water stain is a transparent dye consisting mainly of aniline or coal-at powder dissolved in water. It is a *fast-drying* colour agent with very good penetrating qualities. It may raise the grain of a surface that has not been sponged, sized and sanded before application. It too is used to bring out the beauty of wood grain, to secure more uniform tone or colour. It is also used in order to harmonise with other finishes for the purpose of matching contrast or imitation. The procedure for applying water stain is similar to that for oil stain except that extra care should be taken to prepare the surface in order to avoid the grains from raising when the stain is applied.

Applying paste wood filler

The **oil-based paste wood fillers** consist of linseed oil, silex, japan drier and turpentine. Colour may be added to give different shades. **Synthetic-resin based filler** is also available and is both quick-drying and hard to rub off the surface. (The oil-based filler is easy to rub off due to its lubricating qualities.) Both types of filler are used to fill the pores of open-grained woods, producing a smooth surface for the application of the finishes such as shellac, varnish or lacquer.

Follow this procedure when applying the paste wood filler:

1. Prepare a smooth surface.
2. Clean the surface with a stiff brush, a tack rag or a vacuum cleaner.
3. Thin the paste filler with turpentine if it is too thick.
4. Apply the paste filler by brushing evenly along or across the grain.

5. When the filler begins to set (it becomes dull-looking), rub it with a piece of soft cloth across the grain to prevent the filler from being lifted out of the wood pores.
6. Rub again with a clean, soft cloth to remove all excess oil and filler.
7. Clean out all corners with a pointed hard wood stick.

Applying shellac

Shellac is made from a natural yellow-brown resin that is secreted by the lac insect mostly found in parts of India and Thailand. The resin is purified and then made into flakes. To make the shellac, the flakes are dissolved in methylated spirits to obtain a clear and transparent polish. Colours may be added if required. It provides a good preparation for polished surfaces as it forms a thin seal on the surface, so preventing subsequent applications from soaking in. It adheres well to the wood and it is quickly and easily applied. It can be applied with a good quality brush (squirrel or camel hair), but is more commonly applied using a 'rubber' made from a piece of lint free rag. A piece of cotton wool is placed in the centre of a 300 mm square piece of the rag and the four corners of the rag are brought into the centre. The new corners are now brought up to form the rubber. This is dipped into the shellac and applied to the work. Care should be taken to ensure that the shellac is applied in a thin, even coat, for if applied in excess, the grain of the wood will be raised. The first coat should be carefully rubbed down before subsequent coats are applied.

Applying varnish

Varnish is a transparent gum finishing material. The four types of varnish are oil varnish, polyurethane varnish, acrylic varnish and spirit varnish.

Oil varnishes are obtained by dissolving fossil gums or resins in oil, turpentine or mineral spirits, and dryers. An oil varnish that has high percentage of oil, classified as **long-oil varnish**, is highly resistant to water and alcohol and forms a fairly elastic film. **Short-oil varnish** (with a low oil content) rubs well but is quite brittle. **Medium-oil varnish** (medium oil content), such as floor varnish, is an all-purpose varnish. It is reasonably tough and elastic. It is used as a finish on furniture, floors, woodwork, boats and musical instruments, and as an insulator for electrical conductors.

Polyurethane varnish is a synthetic varnish obtained from polymers and resins thinned with mineral spirits. It is moisture curing and is oil free. It is very tough, hard, flexible, resistant to chipping, abrasion, solvents, detergents, alcohol, acids and water. It is used on furniture, woodwork, boats and floors.

Acrylic varnish is colourless and is composed of acrylic-resin glycols, mineral spirits and water. It is clear, dries in two hours, has little odour and good resistance to abrasion. It is used as a primer and a finish (except on floors).

All the types of varnish may be applied by brushing. Use a well-made, clean, chisel-type brush with fine bristles. The surface to be varnish must be made clean and dry. The varnish must be first applied in the direction of the grain, then across the grain and finally along the grain once more.

Applying lacquer

Lacquer is a synthetic transparent or opaque gum finishing material and is composed of **nitro-cellulose** (derived from cotton treated with nitric and sulphuric acids); a **solvent** such as acetone that dissolves nitro-cellulose, **varnish resin**; a **diluent** (a diluting agent) that dissolves the varnish resins (toluene or benzene), **plasticizers** or **softeners** (derivatives of phosphoric and plittralic acids) that produce a more elastic finish. Lacquer provides a hard, celluloid-like finish that dries very quickly and is durable, moisture and, alcohol resistant.

It is used as a finish on furniture, hardware, jewellery, automobiles, sports equipment, toys etc. It is best applied by spraying, but it may also be applied by brushing, dipping or padding. The surface to be lacquered should be clean, smooth, dry and devoid of oil and wax. The surface may first be undercoated with shellac or lacquer sealer.

> Never apply lacquer over a surface that is painted, enamelled, paste filled, oil stained or oil finished as it may peel off or lift off.

Follow this procedure when you use the spray gun method:

1. Thin the lacquer to be used with a recommended thinner (lacquer thinner), use a proportion of two-parts lacquer to three parts thinner.
2. Adjust the nozzle of the spray gun until the desired spray is obtained.
3. Press and hold the trigger of the spray gun until the nozzle is 12–15 mm above the surface and applied the spray along the whole length of the surface with a slow , sweeping motion. Each stroke should overlap the last sufficiently to blend in with the previous coat.

> ▲ For safety, wear a mask and gloves. It is essential to work in a well-ventilated room or spray booth.

Follow this procedure when you use a brush:

1. Select a good-quality brush with soft bristles.
2. Thin the lacquer in the correct proportion: i.e. one part lacquer to two parts thinner.
3. Apply the lacquer gently along the grain, working quickly as it dries fast.
4. Use a minimal amount to prevent lapping or lifting previous coats.
5. Clean the brush soon after use. Use thinner first and then water and soap.

CHECK YOUR UNDERSTANDING

- The term 'wood finishing' refers to the applications of various materials such glass paper, fillers, stains, varnishes and lacquer to a wooden surface in order to enhance its natural beauty and to preserve or protect the surface.
- A bleach is a toxic oxidising agent used for breaking down the binding substance between cell walls on a surface of wood in order to reduce the contrast between dark and light spots (heartwood and softwood).
- Paint is composed of a pigment, a vehicle, a thinner and a drier and is of two types; oil paint and latex paint.
- A stain is a colouring agent applied to the surface of a piece to harmonise with other finishes for the purpose of matching contrast or imitation.
- A varnish is a transparent gum finishing material used for coating the surface of a piece of furniture or a cabinet to beautify its aesthetic features and to protect the surface.
- A lacquer is a synthetic transparent or opaque gum finishing material that is composed of nitrocellulose, a solvent, a varnish resin and a softener (plasticizer).

REVISION EXERCISES AND QUESTIONS

1 Briefly describe how you may prepare a wooden surface for application of finishes.

2 What are the constituents of paints? State the two main types of paint and show the difference between them.

3 Explain why you would wish to apply wood filler to a piece of furniture you are making.

4 What are the qualities of polyurethane varnish?

5 Explain why lacquer should not be applied over a surface that has been stained or painted.

14 Basic upholstery

Introduction

Upholstery is the process of stuffing and covering chairs and other forms of furniture with various materials to make them comfortable to sit on and pleasing to the eye. There is much more to upholstery than can be covered in this chapter and so only the basics are considered

In upholstery large numbers of expensive tools are not required: you only need: a hammer, needles and pins, a ripping chisel, a webbing stretcher, scissors, a tape measure and a staple gun (Figure 14.1).

The introduction of new and improved materials for upholstery work has drastically simplified the process of upholstering and reduced the time it takes. Some of the new materials are latex and plastic foams. They are resilient and durable when covered and protected from light. Synthetic foams, another improvement, are cheaper but less resilient and slower to regain their shape when depressed. Heavy density foams are generally used for chair seats while the light ones are suitable for the backs of chairs.

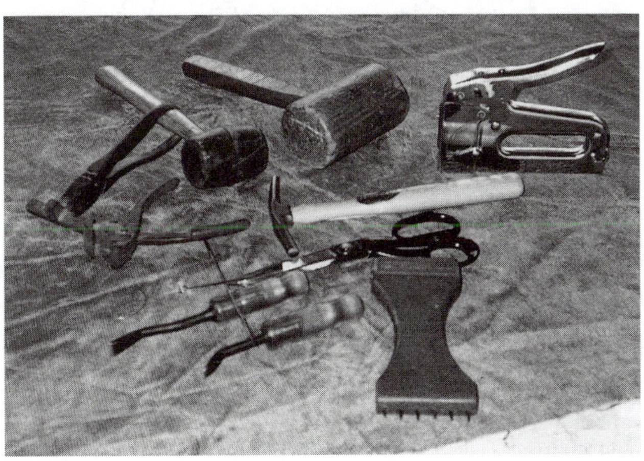

Figure 14.1 Upholstery tools.

> **Note well**: It is absolutely essential that synthetic foams used in upholstery are *fire retardant*. Fumes given off by burning foams are ***lethal***.

Upholstery materials may be grouped as follows: webbing materials, padding materials, covering materials and tacking materials.

Webbing materials

Webbing materials are those used for forming the base of the upholstery.

- **Jute webbing** is a strong, closely woven strip of jute about 75 mm–100 mm wide. It is used to form a base for coil springs and stuffing materials.
- **Steel webbing** is a steel strip, 20 mm wide, that is perforated for nails or corrugated with small loops by which coil springs may be attached.
- **Plastic webbing** is a plastic strip used to support cushion or to form finished seating surface, particularly in modern furniture.
- **Rubber webbing** is a rubber strip used for the same purpose.
- **Burlap** is a coarse canvas strip made from jute or hemp and used for covering springs, for webbing or for forming rolled edges.
- **Denim** is a twilled cotton fabric, strong and wear resistant that is used for making seat surfaces to support cushions.

- **Coil springs** made from steel rod, are used for seats, backs and cushions. Those in continuous rolls or pre-cut lengths, known as **sinuous springs**, do not require webbing or other bases. They are fixed with special clips.
- **Canes** and **raffia** are common, easily obtained and cheap in most African countries such as Ghana and Nigeria where they are widely used as webbing materials. In some cases, platforms that are woven with raffia or cane need no padding or covering.

Padding

Padding or **stuffing materials** (Figure 14.2) are those used for covering springs to form soft, smooth contours. **Rubberised hair** and **Spanish moss** are the most widely used padding stuffing materials. **Latex foam** (foam rubber) and **plastic foam** are also padding or stuffing materials. They have the advantage of being waterproof, durable and available in different thicknesses and degrees of softness. A layer of cotton batting (cloth sheet) is used between the stuffing material and the final cover.

Coverings

Covering materials are those used for covering the padded or stuffed materials to provide pleasant and attractive finish. The covering materials include, fabrics of various kinds, porous or non-porous woven leathers or plastics. When using a non-porous cover for a solid platform, air holes should be made in the platform. **Muslin**, made of cotton is used for covering stuffing materials. **Cambric** is a light, dust-proof cotton cloth, used for enclosing the underside of upholstered furniture.

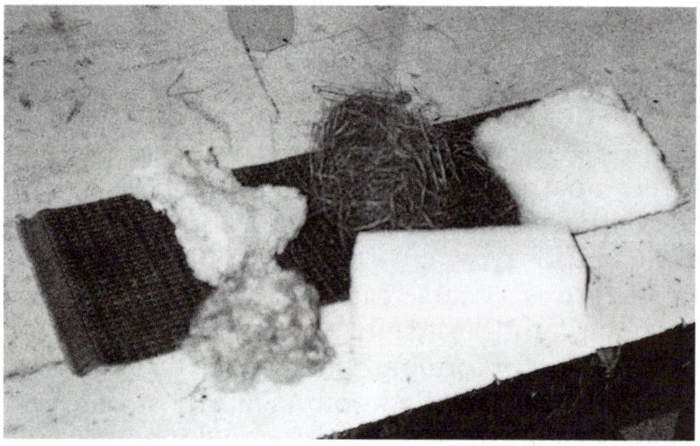

Figure 14.2 Padding materials.

Tacking materials

Tacking materials are mainly nails available in a range of types, and designs. There are the **blue cut tacks**, also called **upholster tacks**, **webbing tacks**, with barbs on the shank to provide greater holding power, and **decorative** or **fancy tacks**, used where their heads can form a decorative feature, the **improved tacks** having a bigger head.

Platforms

A **platform** (Figure 14.3) is the structure that supports the base, padding or stuffing materials and the covering materials. It may be fixed to a wooden or steel frame. The timber or wood for constructing the platform should have the ability to be easily tacked and to hold the tack nails firmly. A **hardwood** of soft texture that can provide enough strength to hold the tack nails is recommended. The joints must be sturdy and should be strengthened with screws, dowels, glue blocks, etc. All edges and corners that are likely to come in contact with the upholstery material must be chamfered or rounded.

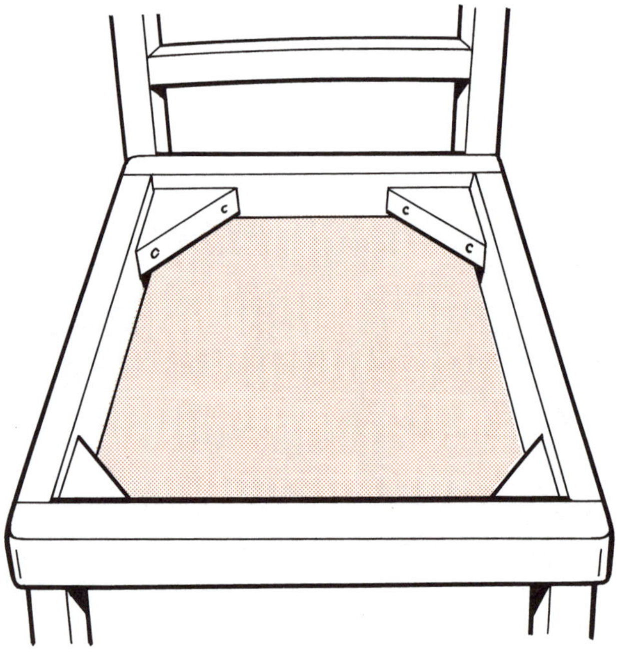

Figure 14.3 Platforms.

Loose seats and backs, are made separately and do not form an integral part of the furniture frame or chair frame. They can be removed for cleaning or re-upholstering. They are commonly used for dining table chairs, dressing chairs or stools etc.

Procedure for upholstering a loose seat using foam padding (Figure 14.4):

1. Prepare the plywood base. Bevel the edges to fit the rebate in the chair frame. To ensure that the base fits as required, fit the seat frame with the covering fabric in position.
2. Cut the foam 6 mm wider than the plywood base on all sides (Figure 14.4a).
3. Prepare the foam before you fix it to the base. Cut the sides of the trenches with a hand-held hacksaw blade and chisel out the waste to equal depth.
4. Treat the foam edges to form a durable edge.
5. Fix the foam to the plywood base (Figure 14.4b).
6. Fix the fabric cover onto the prepared base. The front edge should be tacked on first (Figure 14.4c). Work on a sheet or clean newspaper to prevent damage to the fabric. (Check that the newsprint will not rub off onto the fabric.)
7. Tack along the back edge. The fabric must be tensioned evenly over the foam to the reduce the thickness of the seat to about 25 mm at the centre.
8. Tack the sides down, maintaining an even tension throughout.
9. When working around a corner, pull the fabric gently and just tack it lightly, but do not drive the tacks all the way home (Figure 14.4d).
10. Remove excess material at joinery by pulling and cutting with scissors (Figure 14.4e). Cut the material depending on the fold.

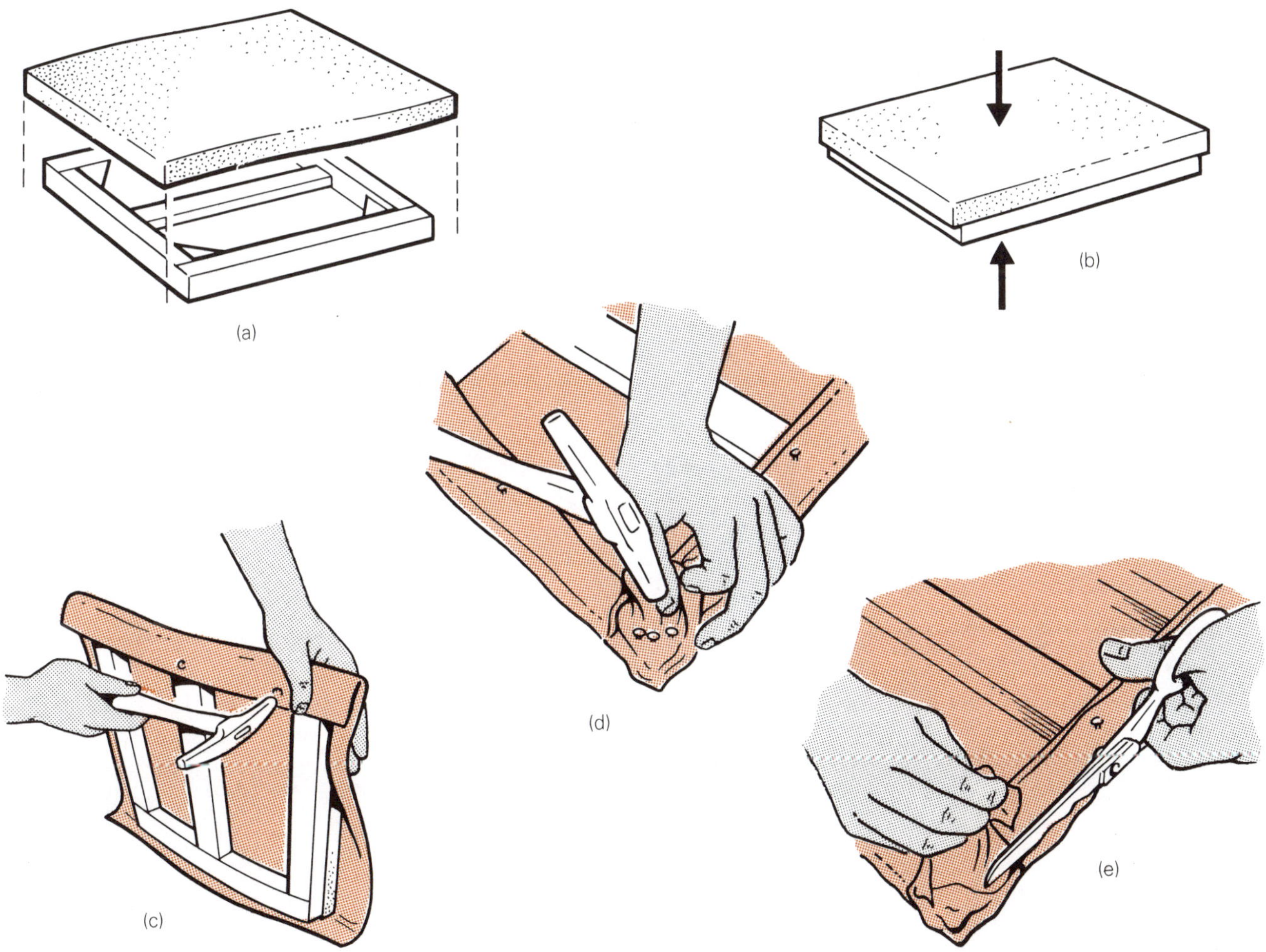

Figure 14.4 Upholstering a loose seat: (a) cut the foam, (b) fix the foam to the base, (c) tack front fabric first, (d) work round a corner, (e) trim excess material.

11. Fold down and tack, pull over the first tack as shown.
12. Fold down the other side and tack.
13. Finish off the underside of the chair seat carefully. A linen cover should be fixed only on fine work.

CHECK YOUR UNDERSTANDING

- Upholstery is the process of stuffing and covering pieces of furniture to make them comfortable and pleasing to the eye.
- Webbing is the process of forming a framework to support the springs and stuffing materials. The materials used for webbing include jute, steel, plastic, and rubber.
- Padding is the means of covering or stuffing a framework to provide soft and smooth contours. The materials used include rubberised hair, Spanish moss, latex and plastic foams.
- Platforms for upholstery should be constructed with materials that have the strength to take and hold tack nails firmly.

REVISION EXERCISES AND QUESTIONS

1 What are the characteristics of latex foam, plastic foam, and synthetic foam?
2 Name five materials that are often used for webbing.
3 Explain the term 'padding' and state any three materials that can be used for padding.
4 State the types of tack nails used for upholstery.

Design, drawing and construction

Introduction

The principal activities of designing and making items draw on your knowledge and skills and develop responsible attitudes of both a personal and co-operative nature. You may be required to make individual responses to problems which can be specific or general. When finding solutions to any problem you will have to make decisions whilst bearing in mind the constraints of time, resources, properties of materials, your own craft skills and other pertinent factors. Aesthetic, technological and functional principles form the basis of the thinking. The activities through which the basic knowledge, skills, and aptitudes are acquired may be grouped into three broad sections: designing, making and evaluating. **Designing** involves activities such as: investigation, clarification and specification of the task; developing and communicating the design ideas and modelling. **Making** consists of activities such as: planning and organising resources; making products using various materials. **Evaluating** involves testing and modifying products.

The **design and making** process follows the stages illustrated in Figure 15.1.

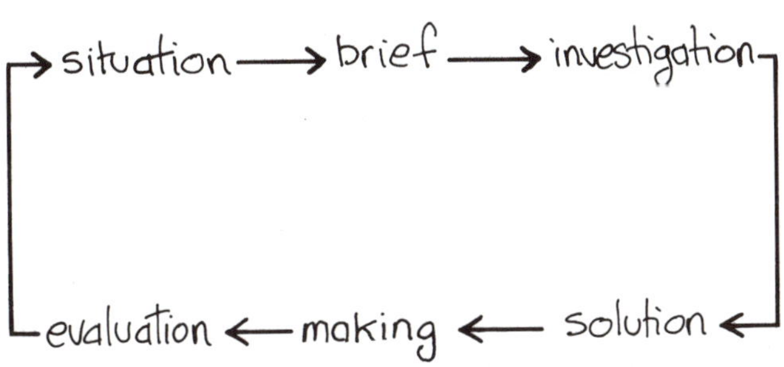

Figure 15.1 The design and making process.

In order to help the student to get a better understanding of what is entailed in this process, a brief explanation is given as to the nature of the activities at each stage.

Stage 1: Investigation, clarification and specification of the task

This stage of the design process consists of the following activities:

Situation

The first consideration at this stage is the situation in which the problem to be solved exists. This situation may be our home, our environment, such as schools, hospitals, public parks, the rural area, industry and others. It could be about how we spend our leisure, such as games and sports. Yet again it may be about other people such as elderly people, children, the sick or the handicapped.

Design brief

The student is required to develop the right approach to exploring and investigating these situations in order to identify existing problems that need solving. He/she then has to make a written statement of the problem which has been identified, describing concisely the nature of the problem. This written statement is known as a **design brief**. For example, it might be: design and make a container that can be used for disposal of rubbish in the home.

Analysis

Having identified and stated the problem, further work needs to be done in order to throw more light on the precise nature of problem to enable a best solution to be found. The process of doing this is known as analysis. The analysis must be made against factors such as:

1. The functions of the artefact or system – for example, you will have to know who will use it, where and how it will be used.
2. The appropriate and available materials to be used for the construction – is it to be made from wood, metal, plastic, concrete or some other material?
3. The appropriate constructional methods likely to be employed, taking into account the available facilities.
4. The appearance of the finished product – for example, is it to be polished, painted or plated?
5. The cost of construction – how will it affect the selling price? Will it be cheap enough to meet the average pocket?
6. The total time to be spent on the work – the more time it takes, the higher the cost of labour.

The analysis may take the following form:

(A) Functions

1. What will be the main functions of the container?
2. What is the size of the container?
3. How heavy/stable should it be?
4. Does the container need to be strong or rigid?
5. How can it be made safely?
6. Does it need to be portable?
7. Will it need a handle?
8. Will it need a cover?

(B) Ergonomics

1. Who will use the container – adults or children?
2. What will be the average height of the persons at sitting and standing positions?
3. What will be the appropriate position of the handle in relation with the user?

(C) Economics

1. How long will it take to produce the container?
2. How much will it cost to produce it?
3. How many will be sold?

(D) Material

1. What material(s) will be suitable/available for realisation?
2. What properties should the material(s) used possess?
3. Will the material(s) need any special treatment?
4. What type of finish will be required?

(E) Construction

1. Will the construction require any special skill/equipment?
2. How can it be made sufficiently stable?
3. How can the stand be fixed to the body?
4. What method will be used for fixing the handle?
5. What shape should the handle be?
6. How can the cover be fixed onto the body?

SPECIFICATION

Having considered all these factors and having decided what to do, a statement is then made embodying the major conclusions regarding functions, controls, limitations, etc. This statement is referred to as a **design specification**. For example, the container must be:

- able to hold reasonable amount of rubbish safely;
- very stable and rigid;
- easy to move about;
- easy to operate by foot (i.e., the lid);
- easy to construct with available materials;
- attractive and inexpensive.

Stage 2: Developing and communicating the design ideas

This is the second stage of the design process and it takes the following forms:

Formulation of ideas

Various ideas conceived in the mind as possible solutions of the problem are presented graphically in three-dimensional sketches. It is important at this stage to generate more than one solution.

Each must be an alternative solution of the problem.

A critical examination is made of each of the alternative solutions with regard to the specifications and objectives. The one emerging as the *simplest but the best* to meet the stated requirements of the problem is then chosen for further development into a final design or solution. Sometimes, it will be necessary to make models or mock-ups to assist in the selection of the best solution.

Development of the chosen solution

Further consideration is then given to the selected solution which is developed into a final design. The development includes making value judgements on the shapes and sizes of the components, the appropriate materials to be used in constructing the parts, the appropriate methods of joining the parts together, the available facilities to be used for the construction, and the expertise required for making the whole product or system. With these factors in mind, appropriate modifications are made to the various parts of the chosen solution. A full set of working drawings (assembly diagrams, exploded diagrams and detailed drawings) embodying all the modifications is made and presented as a final design of the solution to the problem.

Stage 3: Making the final design

The solution emerging as the final design must be constructed or manufactured as a prototype, preferably, full size and using the materials and the constructional/manufacturing methods specified on the working drawing of the design. The making process consists of three stages: planning and organising resources; making the product; testing, modifying and evaluation.

Planning and organising resources

PLANNING

It is often helpful if the manufacturing activities are planned carefully ahead of time. This will ensure early completion and avoid unnecessary delays. The planning may be facilitated if questions such as these are asked and acted upon.

1. What will be the appropriate order in which to make the various parts?
2. What are the main steps (procedures) involved in the making of each component part?
3. Will any special template, jig, tools or equipment be required to carry out any of the operations?
4. What kinds of skills and knowledge will be needed to perform each particular operation successfully?
5. Do I have the competence and confidence to work with a particular machine or equipment all by myself?
6. For which aspect of the work will help or demonstration be needed from the teacher?

ORGANISING RESOURCES

The next step is to organise the resources and the following procedure may be used:

1. Obtain all the required materials and prepare them in readiness. Bring the appropriate ones to the workshop only when they will be needed.
2. Get all tools and equipment ready and in good condition and bring them to the workshop when they are required.
3. Make arrangements ahead of time to secure any special tools and equipment that need to be borrowed.
4. Arrange the tools and the materials properly on the working bench or area to ensure safe working conditions.
5. All safety devices and guards on machines must be put in place before work begins.

Making the product (manufacturing)

With these advanced preparations and organisation, you can carry out the various operations with confidence. Remember to refer to your operational sequence and as far as possible go by it; call for help where you are in doubt about the proper way of performing an operation or using a tool, equipment or a machine; and observe all safety regulations.

Testing, Modifying and Evaluating the Product

The artefact or system which has been made must be tested and evaluated to ascertain whether or not it is able to function satisfactorily as stated in the design brief and the specification. The testing and the evaluation exercise must be carried out by the student (designer) himself or herself. The student should be encouraged to be fair in criticising his or her own work. A written report has to be prepared on the outcome of the testing and evaluation. It must include proposals for modification or improvement. Where possible, the student should be encouraged to carry out the proposed modification or alteration by himself/herself.

Example

As an example we give a design folio prepared by a student. The **situation** is that of patients at a teaching hospital. They generally have to leave their personal effects, food and medicines lying littered around their beds. This makes for untidy and congested wards. The **design brief** is to design and realise a suitable unit which can hold their personal effects, food and medicines and be close to their beds.

Several points are raised in the **analysis**. These include:

1. How many sections should the unit have?
2. How easily can it be moved?
3. What materials can it be made of?
4. Are the materials readily available?
5. What will be the size of the unit?
6. How long will it take to produce?
7. What type of joints will be used?
8. Will construction require any special skills?
9. Will construction require any special tools and equipment?

The design specification is shown in Figure 15.2. Initial ideas are shown in Figure 15.3 and the final idea is in Figure 15.4.

SPECIFICATION

1. THE UNIT SHOULD BE PROPORTIONAL TO HEIGHT OF THE BED.
2. IT SHOULD BE EASY TO USE.
3. IT SHOULD BE ABLE TO HOLD THE PATIENTS' PERSONAL ITEMS AND MEDICINES.
4. IT SHOULD HAVE PERMANENT JOINTS.
5. IT SHOULD BE MADE OF WOOD.
6. IT SHOULD BE AFFORDABLE.
7. IT SHOULD HAVE AN EASY LOCKING SYSTEM.
8. THE UNIT SHOULD BE SUITABLE FOR USE BY EXPECTANT AND NURSING MOTHERS.

Figure 15.2 Design specification for hospital bedside unit.

Figure 15.3 Initial ideas for hospital bedside unit.

Figure 15.4 The final idea for a hospital bedside unit.

Pictorial drawing

Pictorial drawing is a method of drawing a solid object in such a way that three of its faces are seen at the same time. This method produces a 'picture' view of the object. The common forms of pictorial views are called isometric, oblique and perspective. These forms are often used when you are presenting initial design ideas or possible solutions of design problems. The final assembly of a design solution is also presented pictorially, preferably in perspective.

Isometric drawing is a type of pictorial drawing in which the sides of the object lie on three axes or lines, one of the lines is vertical and the two others are inclined at 30° to the horizontal. An advantage of this method is that is much easier to visualise an object from this drawing than it is to understand the orthographic projections of the same object. See the illustration in Figure 15.5a.

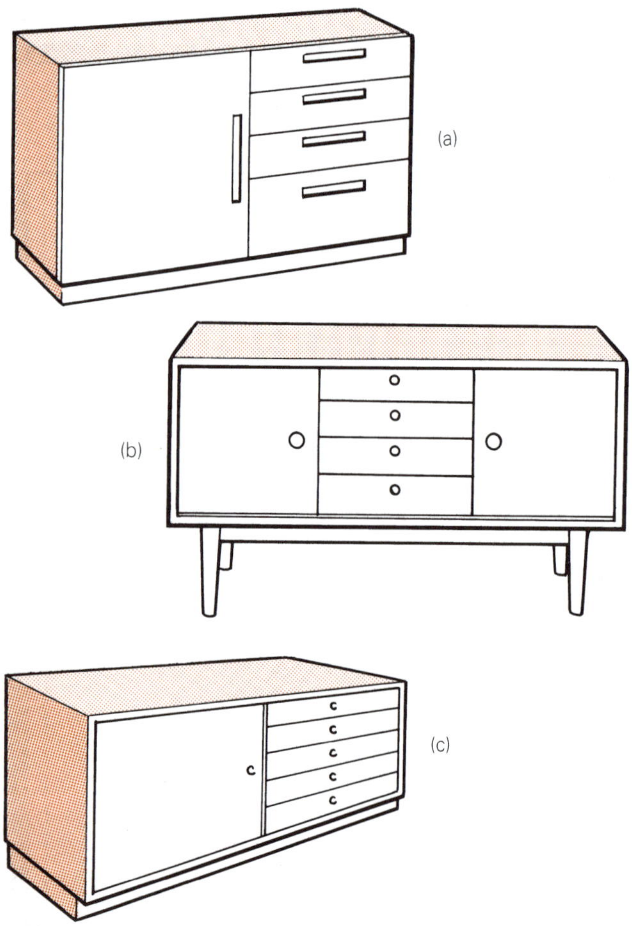

Figure 15.5 Pictorial views: (a) isometric, (b) oblique, (c) perspective.

Oblique projection is another form of pictorial drawing in which the object or a figure is drawn such that one of its sides (front view) is seen at right angle to your eyes and the adjacent side (side/end view) is inclined at 45° to the horizontal line. In this situation, the front view, the side view and the top view can be seen at the same time. The two kinds of oblique projection are the **cavalier** (inclined side given full dimensions) and **cabinet** (inclined side given half dimensions). The cavalier projection looks distorted in appearance. See the illustration in Figure 15.5b.

Perspective drawing is a type of pictorial drawing that has two of the adjacent sides of the object presumed to be converging at opposite points on a horizontal axis. The lines representing the sides vanish to a chosen point or pole, while the vertical lines representing the height, all run parallel to one another. Perspective views can be drawn in three positions: on the horizontal axis; below the horizontal axis and above the horizontal axis. See the illustration in Figure 15.5c.

Freehand sketching

A sketch is a drawing in which the proportions of size are judged by the eyes and all lines are drawn freehand; that is, without using any drawing instruments. The result is not as good as a scale drawing where you can use instruments. An H.B. pencil is recommended for sketching. In freehand sketching, you can use either pictorial or orthographic projection. Even though the use of instruments is not allowed, you are still required to apply all the rules in drawing.

A freehand sketch can be produced far quicker than scale drawing, and you can sketch quickly on the spot, the only materials you need are a pencil and paper. It also enables you to clarify an explanation, convey an instruction or express an idea quickly: in designing an artefact for example, the initial ideas are expressed in the form of sketches. Most objects have outlines composed of straight lines and circular arcs; if you can sketch these accurately, you are halfway towards producing good sketches.

Orthographic projection

The word 'graphic' means using visual symbols, letter, lines and diagrams to enable someone to have a clear picture in the mind of an object being described. Orthographic projection means using symbols and diagrams to communicate ideas visually in the form of two-dimensional diagrams. It is the medium to convey the shape, size, material, finish, and so on of engineering or building components. The two systems of orthographic projection are the **1st angle** and **3rd angle** projections. They are used to present views of engineering and building components. The views are represented in two-dimensional form (i.e., the length and breadth/height of the object is shown). To explain the principles of the 1st and 3rd angle projections, we use the idea of projection planes (Figure 15.6a).

The projection planes used for orthographic projection are the **horizontal** and **vertical** planes**.** They intersect at right angles (i.e., at 90°) forming four quadrants or angles known as 1st, 2nd, 3rd and 4th angles.

It is possible to project views of an object in the 1st and 3rd quadrants (angles) because it is always possible to see the views projected on to the vertical plane and the horizontal plane when the latter is folded down to provide a completely flat surface (such as a sheet of drawing paper). The principle of projecting in the 1st quadrant (angle) is known as **1st angle projection** and the projection in the 3rd quadrant (angle) is referred to as **3rd angle projection**.

The 2nd and 4th quadrants are not used for projection because they close up when the horizontal plane is folded down or up. In this case it is not possible to see distinctly the views

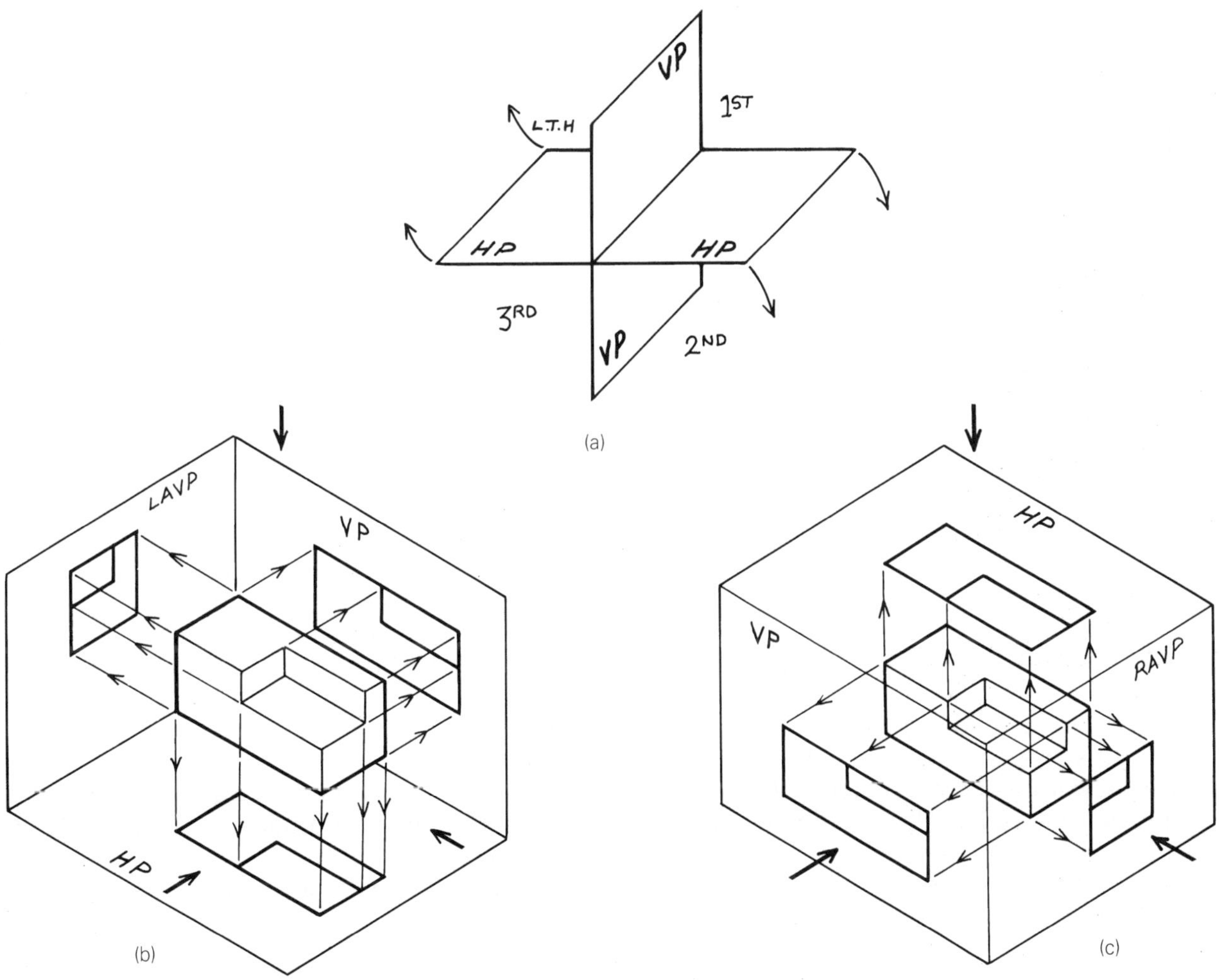

Figure 15.6 Projection planes and boxes: (a) principal projection planes, (b) first angle projection box, (c) third angle projection box.

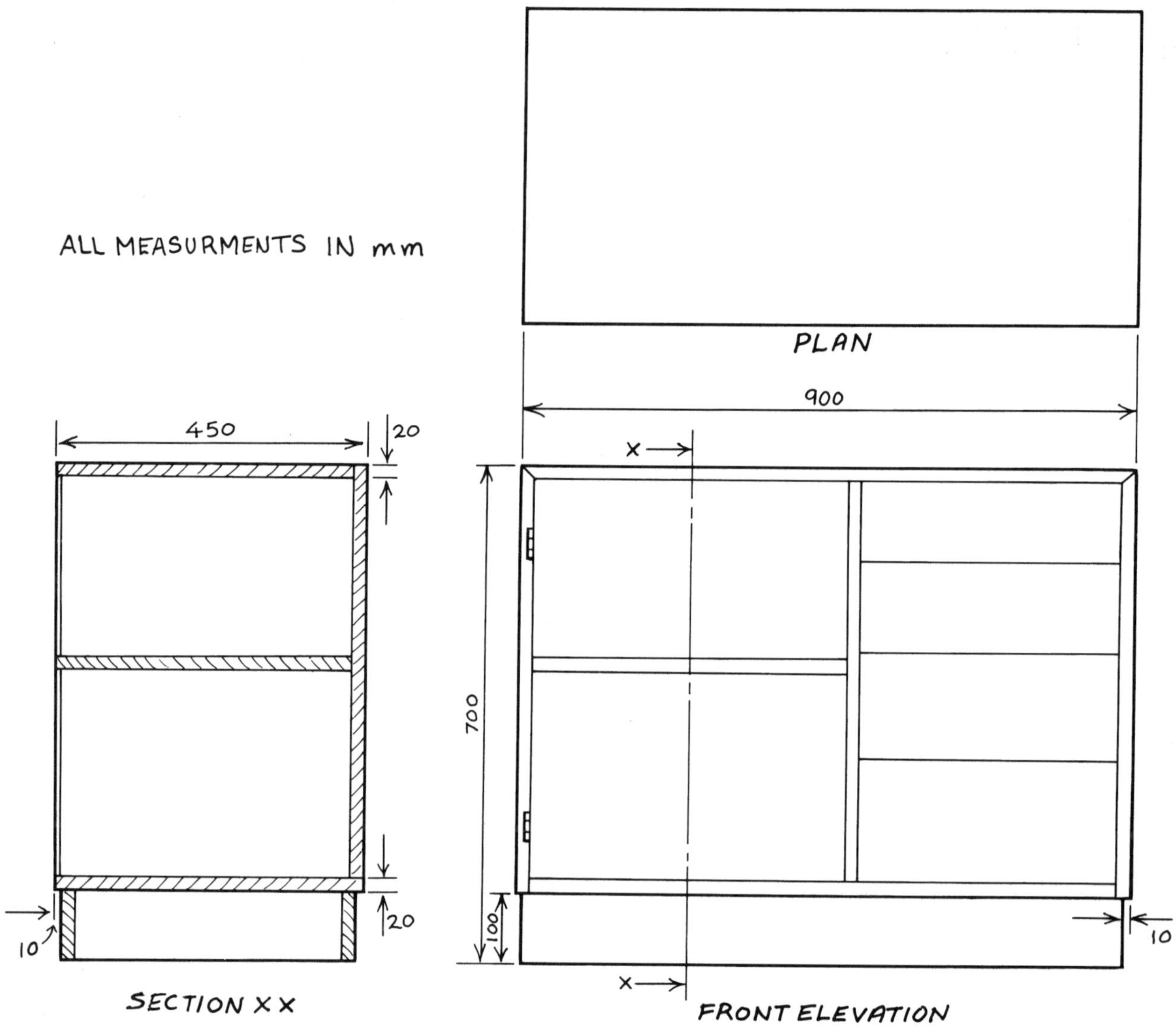

Figure 15.7 Sectional view.

projected on vertical plane and the horizontal plane. They coincide with one another.

Principles of first angle projection

With the first angle projection, the object is imagined to be suspended inside an opaque box. (Imagine this box sitting in the first angle, in figure 15.6a, with its base parallel to the horizontal plane and one of its sides parallel to the vertical plane.) The front and the top of the box are removed so that you can clearly see the front view and the top view (plan) of the object without obstruction. You can see in your mind (imagine) the vertical plane behind the object and the horizontal plane below it.

To obtain the front elevation (view), you must stand directly in front of the box and view the face of the object. It looks as if it has been drawn on the vertical plane behind it.

The plan (top view) can be obtained when you stand over the box and look down on the object. It looks as if it has been drawn on the horizontal plane below it.

To obtain the left end elevation, you must stand on the left side of the box and view the left face of the object at right angles on the right auxiliary (additional) vertical plane behind it. Similarly, to obtain the right end elevation, you must stand on the right side of the box and view the right face of the object at right angles on the left auxiliary vertical plane behind it. The practical way of *drawing* these views is illustrated in the diagram in Figure 15.7.

Third angle projection

To obtain the 3rd angle projection, imagine the object suspended in a transparent projection box in the 3rd angle of Figure 15.6a. The planes now come between your eyes and the object, that is, the object is always behind transparent screens or planes and is viewed through them. The relative positions of the plan and the elevations of the third angle are different from those of the first angle.

To obtain the front elevation, stand in front of the projection box and look at the front face of the object at right angle through the front screen. Since the view is taken through the screen, the image is presented (drawn) on the same screen.

To obtain the plan, stand over the box and view the top face of the object at right angle through the top screen and draw on that screen.

To get the left end elevation or view, you must view the left face of the object through the left auxiliary vertical screen to obtain the left end elevation. In the same way, the right end elevation is viewed through the right auxiliary vertical screen and presented on that screen.

What you see from the front you draw on the front; what you see from the left, you draw on the left side of the front elevation; what you see from the right you draw on the right side of the front elevation and what you see from above you draw on the top of the front elevation. See the illustrations of the 3rd angle projection in Figure 15.6c.

See Figures 15.8 and 15.9 for an exploded view and an orthographic projection of the unit designed for the hospital patients as part of the student project described on page 164.

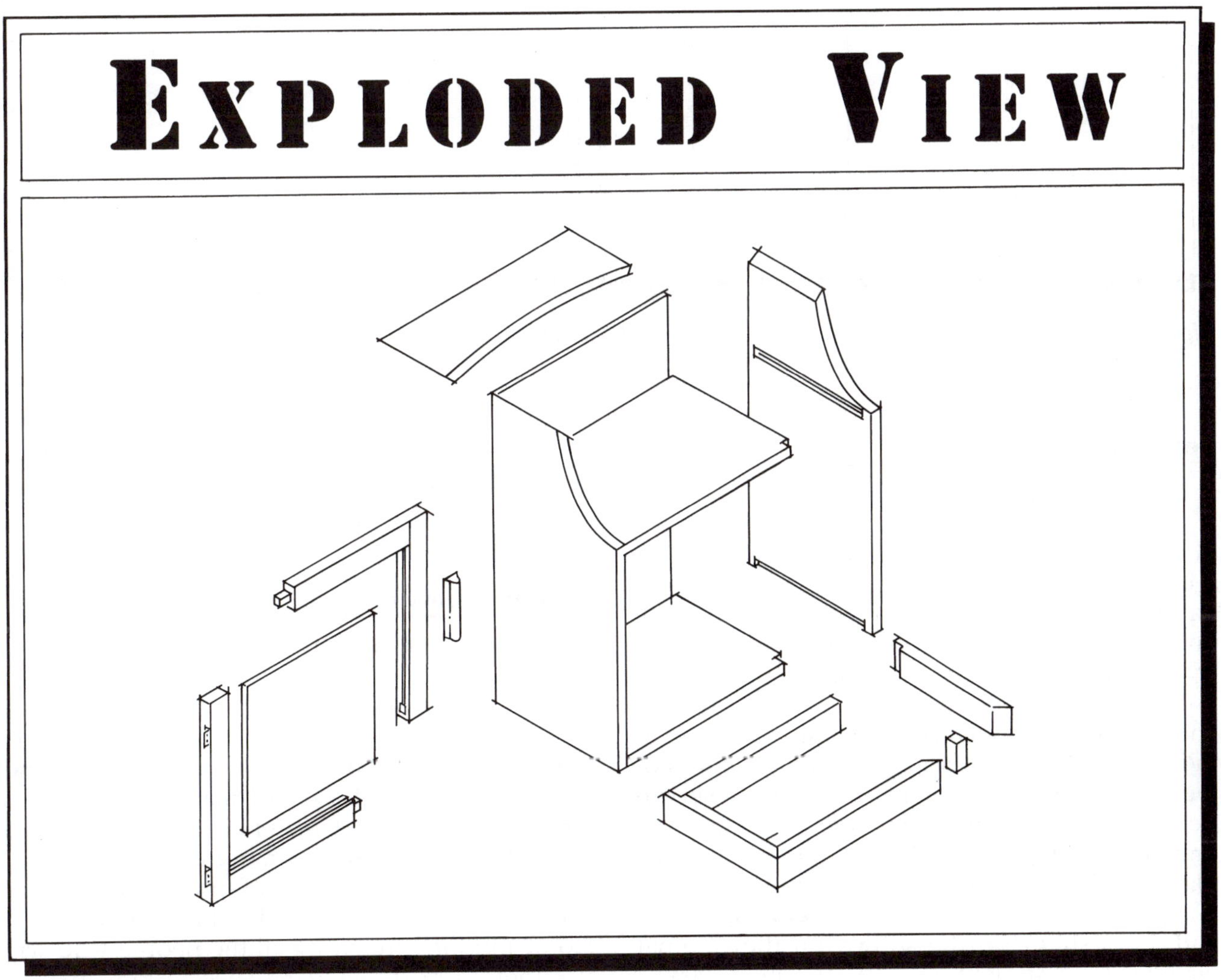

Figure 15.8 Exploded view of a hospital bedside cabinet.

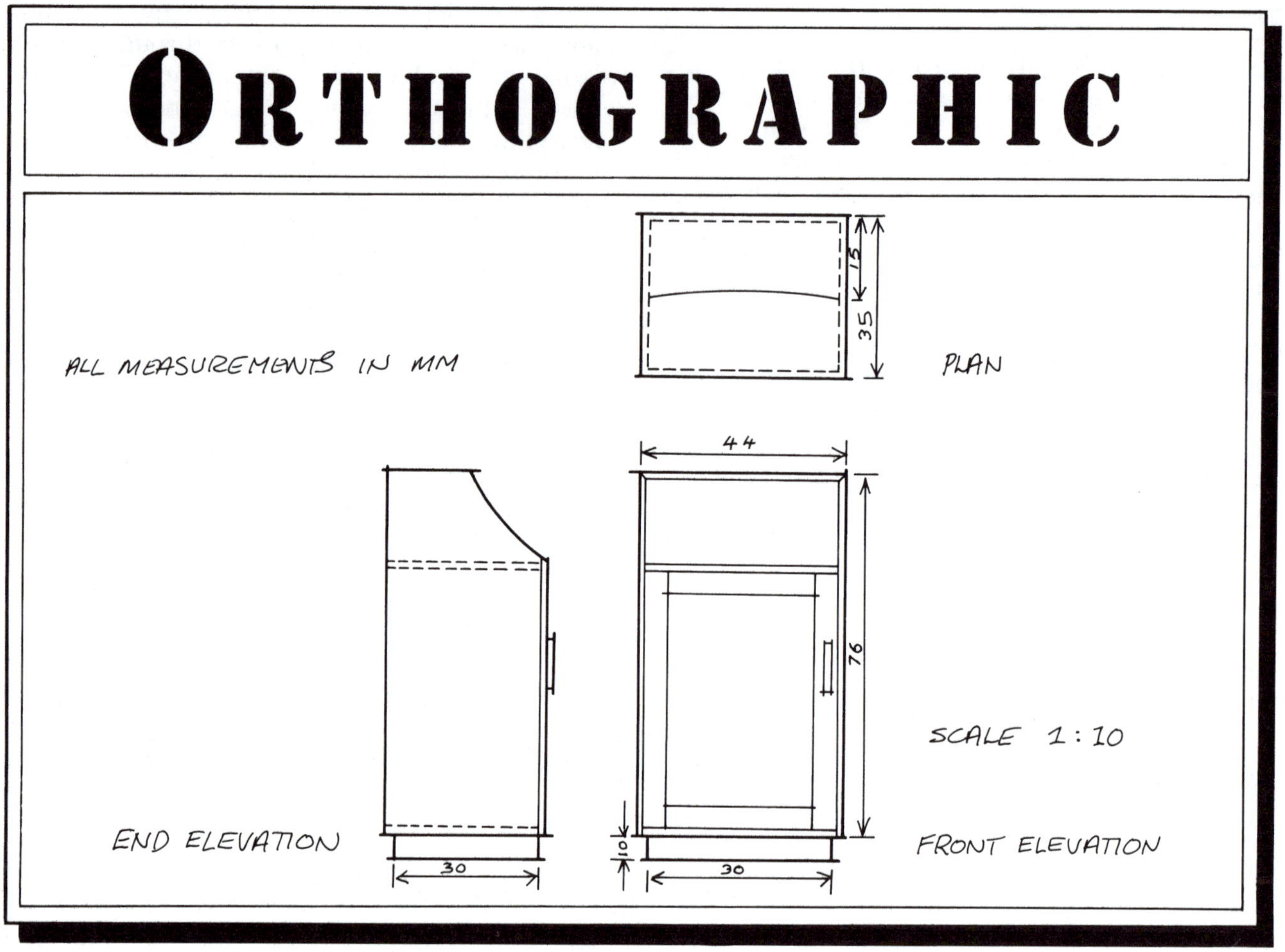

Figure 15.9 Orthographic projection for a hospital bedside cabinet.

Sectioning

In technical drawing, sometimes there is a need to show how an object will appear when it is cut in a way that exposes some hidden features. For a example, both the front and end elevations of an upright hollow cylindrical object (pipe), will only give a rectangular shape and the thickness of its walls will not be shown. On the other hand, if the pipe is sawn through using a hacksaw, the end elevation will change. The hollow portion in the pipe will now be visible and the thickness of the walls where it has been cut through is shown 'sectioned', using **hatching lines** (parallel lines drawn at an angle of 45° to the horizontal). The plane used to indicate the part or parts where it is imagined to be cut is known as a **cutting plane** and its position can be shown on the original diagram by a straight line. See the line marked XX in the illustrations in Figure 15.7. The section line is normally drawn thicker than the lines depicting the object itself in the drawing (the object lines).

Working drawing

A working drawing may consist of various drawings prepared by a designer to illustrate and explain details of the component parts of an artefact or a system he or she has designed. The details may include: types of materials and fittings; shapes and sizes of materials; estimates for quantity and cost materials and fittings; and instructions for construction or manufacture.

The drawings must conform to universally accepted standards and codes that can be understood and interpreted by all the people who will use them. British Standard Specification, BS 308 should be used. Generally, a working drawing

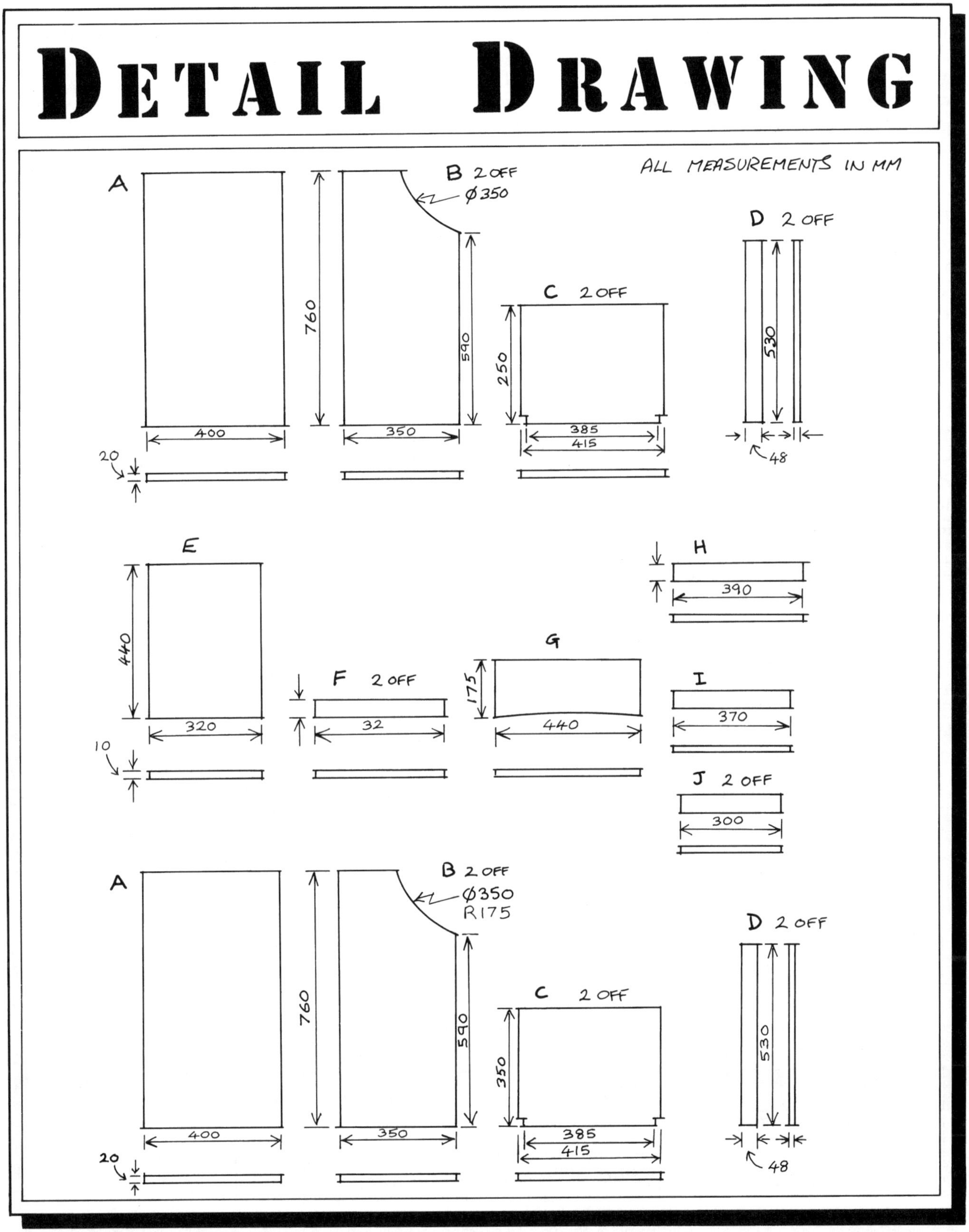

Figure 15.10 Assembly drawing for a hospital bedside cabinet.

consists of two parts: a detailed working drawing and assembly working drawing.

The **detailed working drawing** as the name implies, is a drawing that shows the details of every single component part of the artefact/system. The primary purposes of the detailed drawing are: to outline the shape of the object; to provide a size description (it indicates the dimensions of the object); to describe the type of material(s) to be used for its manufacture; and to provide any additional information required for construction or manufacture.

An **assembly drawing** illustrates how the individual components fit together to make a complete unit. It is drawn orthographically or pictorially (including an 'exploded isometric' or perspective view). The main functions are: to provide description of the shape of the assembled unit; to identify each component part of the unit; to indicate the *relative* position of each component part of the unit; to supply a list of the parts; and to act as a source of reference information. See the illustrations in Figure 15.10.

Preparation of a cutting list

A cutting list is a summary of the materials required to make an artefact or a system. It is compiled by taking the sizes of individual members from the working drawing. Generally, the individual members of the various units for example, carcase, doors, drawers, etc. are grouped together and listed in tabular form under the following headings:

members: the name of the member and/or the corresponding letter on the working drawing;
quantity: the number of each member required;
length: the length of each member required, stated in millimetres (allow an extra 3 mm for horns on stiles);
width: the width of each member required, state finished sizes in millimetres;
thickness: the thickness of each member, stated in millimetres, based on planed pieces;
material: type of material of each member and remarks regarding quality, figure, etc.

CUTTING LIST

PART No.	DESCRIPTION	No. OFF	DIMENSION	MATERIAL	FINISH	REMARKS
A	BACK PANEL	1	760 × 400 × 20	PLYWOOD	PAINT (OK)	
B	SIDES	2	760 × 590 × 350 × 20	PLYWOOD	✓	
C	SHELF/BOTTOM	2	385 × 350 × 20	PLYWOOD	✓	
D	STILE/DOOR	2	350 × 48 × 20	MAHOGANY	✓	
E	DOOR PANEL	1	440 × 320 × 10	PLYWOOD	✓	
F	RAIL (DOOR)	2	320 × 48 × 20	MAHOGANY	✓	
G	TOP	1	440 × 175 × 20	PLYWOOD	✓	
H	FRONT RAIL (PLINTH)	1	390 × 50 × 20	MAHOGANY	✓	
I	BACK RAIL (PLINTH)	1	370 × 50 × 20	MAHOGANY	✓	
J	SIDE RAIL (PLINTH)	2	300 × 50 × 20	MAHOGANY	✓	

Figure 15.11 Example of a cutting list.

Figure 15.11 shows the cutting list for the student project described on page 164.

Costing

Costing is a very important factor in any kind of design work. An idea may be acceptable in every other way but if it is too expensive to make or for someone to buy, then the designer has failed in his or her task. It is advisable to make sure that the cost of materials will be acceptable before making the final choice of solution. It may be helpful to draw up a chart so that the cost of each part of the artefact can be seen and a total worked out. The main cost elements associated with manufacture of products are: cost of materials; labour costs; production costs; packing and carriage; overhead expenses and profit.

The cutting list, as well as the materials list, should be used in calculating the cost (making allowance for waste in timber, plywood, etc.), using the current prices of the items. **Labour cost** will depend on the approved rate for wages. **Production cost** should take into account the preparation of templates and jigs. **Overhead expenses**, such as light and power, depreciation of tools and machines, and rent and rates (apportioned), should be considered.

It should be noted that costing and estimating form a very important aspect in any manufacturing enterprise. In a medium-scale furniture or wood working industry, the control of costs through expenditure budgets falls under three main sub-headings: **fixed costs**, those such as rent and rates, that are not normally affected during the year by management decisions; **managed costs**, those such as salaries, expenses and product changes, that can be adjusted by management decisions during the year; **variable costs**, those such as materials, factored components and packing, that vary with the level of manufacturing activity.

In large-scale woodworking industry, the cost of manufactured products are calculated under the following headings: materials costs; labour costs; overhead expenses; packing; carriage; sales expenses; advertising, and net profit.

In a school or college situation, where the items being produced are mainly class-work or projects intended to teach skills, it may not be necessary to consider all these elements when costing a project work. The only essential element to take into account is the material cost. If someone wants to buy an item produced by a student, the labour cost (time spent on the job) and the net profit should be added to the material cost to obtain the selling price. In a school or college where there is a production unit for making items for the public, then all the elements considered in the industry may be included.

Planning the construction of a design

Having developed the design and completed the working drawing, it is necessary to plan for its construction. This implies giving careful thought to such items as the amount and type of materials and fittings required, their cost, tools required for its construction, assembling and finishing. In industry, further planning is necessary to reduce labour costs and co-ordinate hand and machine techniques for construction, assembling, finishing, packing and transport. If the design is to be repeated several times, for example, in the production of chairs, tables and wardrobes, it may be necessary to prepare a full size working drawing on a sheet of plywood. This sheet is referred to as the **setting-out** or **working rod** and can be stored for future use.

Selection of materials

Selection of materials for the realisation of the design can greatly affect its utility, appearance and cost. The materials selected should suit the purpose of the artefact, taking into consideration size, strength, durability, acceptable in every other way to meet the aspiration of customers. It is therefore advisable to make sure that the cost of materials will be acceptable before making the final choice of solution. It may be helpful to draw up a chart so that the cost of each part can be seen and a total cost worked out.

Construction processes

A general procedure that can be followed for most jobs is as follows.

Preparation of pieces

Preparation consists of sawing and planing the pieces ready for setting out. A study of the design (working drawing) will determine the amount of preparation required for each member. Check sizes with the cutting list and drawing. The direction of grain and figure should be noted for strength and appearance.

Setting out

1. Place identification marks on members and joints, for example, letters or numbers.
2. Set out positions and shapes of joints. Mark waste with crosses. Check sizes from scale drawings or mark directly from the cutting rod if one was prepared.
3. Mark the positions of grooves, rebates, screw holes, etc.
4. Set out any shaped pieces and decorative features.
5. Mark all squaring and gauging from the face sides and face edges.
6. Place similar members, such as rails, stiles, legs and drawer sides, together to mark lengths and positions of joints.
7. Make sure of the grain direction for shaped pieces, keeping short grain to a minimum. Trace curved shapes from a template.
8. Arrange for wide boards, that are liable to cup, to be set out so that tops and shelves are always heart side up.
9. Doors and drawers are prepared and set out to fit the carcase, therefore finish the carcase assembling before doors and drawers are made.

Cutting out and shaping

1. Always cut on the waste side of the lines.
2. Cut waste from joints. Cut along the grain first and saw the shoulders last.
3. Plane grooves and rebates and bore holes for screws.
4. Cut shapes and decorative features, such as chamfers and mouldings.
5. Curved shapes, tapered legs and turned members are often difficult to hold, hence it is usual to cut joints, bore holes for dowels, etc., before shaping.
6. Make sure identification marks are not cut off.

Assembling

The assembling of large jobs has to be carefully planned and often consists of four stages.

1. **Trial assembly** of the members and joints without glue to see that they fit together properly and that the whole job is square. Any adjustment necessary is made at this stage.
2. **Clean up all internal surfaces and edges** which cannot be planed after assembling. Replace locating marks so that they can be removed later.
3. **Sub-assembly** is required for such items as chairs, tables, framed carcases, where it is necessary to glue up one or two units of the job at a time. For example, glue together the front legs and back legs of a chair separately, allowing them to dry and then complete the assembly by gluing in the side rails. Clean up the external faces and edges of the glued-up sub-units.
4. **Final assembly** consists of assembling the various units to complete the job. Many small articles can be completely assembled after the trial fitting and cleaning up. When assembling, make sure all materials and tools are ready so that the assembly can proceed without delay. Cramps opened to the right size, mallet, hammer, nails, screws, dowels, glue, wet and dry rags, etc., should all be ready at hand. Follow this procedure:
   - apply glue to both surfaces of joints;
   - check carcases and frames for squareness with a diagonal measuring stick;
   - sight in all directions for winding (if there is any, it may be necessary to correct it by moving cramps);
   - flush off joints and edges and glue on veneer edging or lipping if required;
   - wash off surplus glue immediately after assembling.

Fitting doors and drawers

Doors and drawers are constructed and fitted after the final assembly of the carcase. Fit any hardware, such as door locks and drawer pulls.

Finishing

Prepare the surfaces and edges by fine sanding and damping. When they are dry, sand them again ready for polishing. The polishing can be done either by spraying or brushing. Spraying, however, provides better results.

CHECK YOUR UNDERSTANDING

- Designing is the process of finding solutions to human problems, making use of a range of competencies such as creative thinking, written and graphic communication, research investigation, analysis, synthesis and technical skills.

- The main activities of the design process include: designing, making and evaluation.
- The broad stages in designing comprise:

(a) investigating, clarifying and specifying the task or problem;
(b) developing and communicating the design ideas or proposals;
(c) making the final design;
(d) testing, modifying and evaluating the product.

- The three main forms of pictorial drawing are: isometric, oblique and perspective.
- Orthographic drawing is the process of using symbols and diagrams to communicate ideas visually with the aim of conveying the shape, size, material, finish etc. of an engineering or building component.
- A working drawing basically consists of two parts: a detailed working drawing and an assembly working drawing.

REVISION EXERCISES AND QUESTIONS

1. Explain the concept of designing.
2. Make a line diagram to illustrate the design process. Explain briefly the activities of each stage.
3. How would you communicate design ideas to a client in terms of the proposed ideas and the final idea?
4. What are the two main parts of a working drawing?
5. Make a sketch of a bench hook to illustrate the use of both 1st and 3rd angle projections.
6. Briefly describe the steps to be taken in assembling (glue) a large piece of work such as a bookcase.

16 Occupational information

Introduction

This chapter deals with certain basic information aimed at helping students to be aware of a number of ways they could possibly further their study in other post-school training institutions or enter into an entrepreneurial role. The discussion on occupational information will be limited to three areas: post-school training, types of business enterprises and setting up a business.

Post-school training

Generally, entrance into institutions of further or higher education (universities/polytechnics) is restricted and competitive; only those with outstanding results in the final school-leaving examination are considered. This leaves the majority of candidates to look elsewhere to continue their studies. Many countries operate a variety of training schemes to help those who may not be able to enter university and other institutions of higher education to acquire technical/vocational skills and attitudes that would assist them in entering into productive employment.

These schemes range from informal craft/vocational apprenticeship training, to formal craft/vocational training, leading to the award of recognised certificates, such as the City and Guilds Certificate or the Royal Society of Arts Certificate (RSA) or even the General Certificate of Education (GCE). Some of the courses existing in some countries are: Youth Training Scheme (YTS); Technical Vocational Education Initiative (TVEI); Business and Technicial Education Council (BTEC); National Vocational Training Institute (NVTI); Opportunity Industrialisation Centre (OIC).

The sponsorship of these training schemes is undertaken by a number of organisations and institutions from both the public and the private sectors and private individuals. Quite often, governmental ministries such as the Ministry of Education, Ministry of Youth and Social Welfare, the Ministry of Industry and others (according to government policy in the country) may be responsible for training. Government policy on post-school training differs from country to country and your teacher should know about the particular schemes operating in your own country as well as the conditions and requirements for admission into them. Where practicable, personnel from the various organisations and institutions often visit schools to advise the students, especially those in their final year, on their prospects in these areas of training.

Types of enterprises

In every country, citizens need goods and services in order to survive. Individuals and organisations or enterprises are involved in the distribution of the goods and services required.

The establishments or **enterprises** that are involved in buying and selling and distributing raw materials, manufactured goods and providing services belong to one of two sectors: the **public sector** or the **private sector**. Enterprises and organisations that are controlled by government are called public sector enterprises or organisations. These are established for the general benefit of the citizens as a whole. Primarily, profit is not the motivating factor. In some countries, postal services, the health service, the police and the army are public sector institutions. Those enterprises and organisations that are owned and run by private individuals, whether small or large, are regarded as being within the private sector of either industry or commerce. Generally, private enterprises or businesses are run with the ultimate aim of making profits for the owners.

Enterprises within the public and private sectors are broadly classified into three groups: industrial enterprises, trading enterprises and service enterprises. The enterprises within the industrial sector are mainly concerned with the production of a variety of raw materials and processing them into saleable forms or goods for public consumption. Organisations in the construction sector are industrial enterprises.

Trading enterprises are a range of enterprises in the business of buying and selling raw materials, components and various kinds of products made by the industrial enterprises. The enterprises engaged in this operation include: small one-person shops; supermarkets; departmental stores; hypermarkets and wholesalers (middlemen).

Service enterprises are solely concerned with providing some form of service to other enterprises or individuals. The main categories of service enterprises are: warehousing; transport; maintenance and repair; banking and finance; insurance; accountancy; accommodation and catering.

A **small-scale enterprise** (sole proprietor), as the name implies, is an enterprise or business that is usually relatively small and owned and managed by one person. The range of activities may include: trading – both wholesaling and retailing; small-scale manufacturing; small-scale fishing; small-scale farming; metalwork and woodwork crafts; motor mechanics; electricians; radio/tv engineering; painters and decorators; estate agents; insurance agents; restaurants; guest houses and chop bars. The day-to-day running of a small-scale business is normally carried out by the owner. Sometimes, they may engage a manager to run the business on their behalf. It is also a normal practice for family members to get involved. Some owners employ non-family workers as well.

Since few people are engaged in the operation of such a business, and they are mainly relations, it is easier for the manager or the owner to know each worker personally and understand their abilities and shortcomings. Control is therefore easy. In times of illness or in the absence of a worker, the owner/manager may perform the work. Business decisions are taken quickly as the owner does not usually have to consult with anybody else. Personal contact with customers or clients and suppliers is easy. Any profit accruing from the business is singly managed.

However, there may be no one who can step in when the owner is ill. Moreover, the sole-proprietor may not have the time to carry out all the necessary managerial functions. Raising enough funds for starting the business or for expansion later may be a major problem. In times of loss the sole-proprietor bears the whole liability.

A **partnership** is a business in which two or more people have agreed to set up and run an enterprise together with the intent of making profits. Quite often, the partners pool together skills, experience, knowledge, finance or assets or a combination of any two or more of the above factors. A partner who is not involved in the management is called a **sleeping partner**.

It is possible to share the workload and other responsibilities among the partners. There may always be somebody to do the work of a member who is ill or absent for any other reason. Consultation is necessary and decisions must be taken collectively. In partnerships, there is a possiblity of conflict, particularly when one partner is not performing his/her work satisfactorily. Such conflicts can have adverse effects on the operation of the business.

Setting up a business

This is a complex subject, but for purpose of our discussion, we shall consider only three factors which are relevant to the level of our treatment.

Planning

The first thing to do when an individual or a group of persons wish to create an enterprise is to decide the nature of the business and establish its aims. Is the enterprise going to be involved in manufacturing a product? If so, what? Is it to be involved in trading? If so, which goods? Will it provide services? If so, what kind of services? Out of these analyses, the goals or the objectives and aims of the enterprise will emerge more clearly.

The next considerations are: where will the premises be? Is there accommodation ready for use or land to build a store/workshop for the business? Who will form the labour force? Are the necessary skills available among the workers? Are there sufficient funds to start the business?

A written statement incorporating all these objectives and considerations forms the policy of the enterprise.

Forecasting

Another essential factor that should be taken into account, once the policy has been drawn up, is to consider the future, both immediate and long term. In other words, the effects of future events on the operation of the business have to be assessed. For example, what will have to be done if: the general economy of the country should change; new political trends arise; the cost of raw materials or goods rises; new competitors enter the market?

Market research

It is unwise to enter into the production of goods that are not in demand. Likewise, it will be unreasonable to go into trading in commodities that do not have any immediate or future demand. It is therefore necessary to conduct a market survey to find out the general demand for the items the enterprise wishes to produce. The survey should not be restricted to consumer demands; the supply situation must also be assessed.

Final decision

Having prepared the objectives and the general policy, considered the availability of starting capital, made the necessary forecasts and conducted a market survey, you are now in a position to decide whether or not the proposal is viable and worth pursuing.

CHECK YOUR UNDERSTANDING

- Enterprises involved in the distribution of goods and services required by the citizens of every country comprise two sectors: the public sector and the private sector.
- The public sector enterprises are those controlled by the government. Their primary aim is to render services to the citizens as a whole so profit is not the motivation.
- Private sector enterprises are owned and run by private individuals. Generally, their ultimate aim is to make profit.
- Enterprises in both sectors are broadly grouped into industrial, trading and service enterprises.
- An enterprise or business owned by one person is often termed a small-scale enterprise and the owner is called a sole proprietor.
- A partnership enterprise is owned by two or more persons who have agreed to set it up and run it together.
- The three essential processes to be considered when setting up a business are: planning, forecasting and market research.

REVISION EXERCISES AND QUESTIONS

1. State the two types of enterprise and explain the difference between them.
2. Explain the term 'sole proprietor'.
3. With what areas of business is a trading enterprise concerned?
4. State two advantages and disadvantages each of a small-scale enterprise.
5. What is a sleeping partner?

17 Project work

Introduction

This chapter is intended to assist the student to improve his or her competence in the design process as well as his or her skills in the making or manufacturing process. The skills to be acquired in both the design and the making processes can best be achieved if the student is personally involved in these activities. The teacher must therefore ensure that each student is given equal opportunity to participate in all classwork and project work. In order that the desired skills and competencies are acquired, when carrying out the projects the following procedure should be used:

1. Analyse the given brief carefully. If possible carry out further investigation or research to have a clearer understanding of the problem (project).
2. Make pictorial sketches (perspective or isometric) of *at least three* alternative solutions.
3. Select the simplest solution that best satisfies the specifications in the brief or project.
4. Develop it further by considering alternative shapes, forms, materials, etc. of each component part. Add evaluation notes to each proposal. Indicate the chosen ones by ticking them.
5. Make a pictorial assembly drawing, embodying all the chosen parts as the final solution or design.
6. Draw exploded views of the final solution showing how each component part fits to or relate with all the others.
7. Draw orthographic views (3rd angle) of the final solution. Show the overall dimension on each view.
8. Prepare a detail drawing of all the component parts. Add full dimensions to each component part showing both the overall and detail dimensions.
9. Prepare a material or cutting list, including the estimated cost of each item.
10. Write out the procedures that you will adopt to make or manufacture the artefact.

Examples of projects

Three examples are presented here from exercises undertaken by students at the Advanced Technical Teachers' College, Kumasi, Ghana. Suggestions for further projects are given at the end in the form of **project briefs**.

Example 1

This example shows how a towel drier was designed by a student after identifying the problem. The stages covered include:

1. **Situation**: the need to design and make the drier has been identified (Figure 17.1).
2. **Brief**: what is to be designed and made has been clearly stated (Figure 17.1).
3. **Analysis**: relevant questions have been posed to find out more about the problem (Figure 17.2).
4. **Research** and **investigations** have then been carried out.
5. Three **initial ideas** have then been proposed, the last of which has been selected and developed to present a **final solution** (Figures 17.3, 17.4, 17.5).
6. An **exploded view** shows how the components relate to each other in assembly. There is

an **orthographic projection** enabling overall dimensions to be added and a **detailed drawing** to show the parts (Figures 17.6, 17.7, 17.8).

7. **Planning** in the form of a **cutting list** and **costings** have been done (Figures 17.9, 17.10).
8. **Evaluation**: the product meets the design brief, which was quite simple.

Figure 17.1 Example 1: situation and brief.

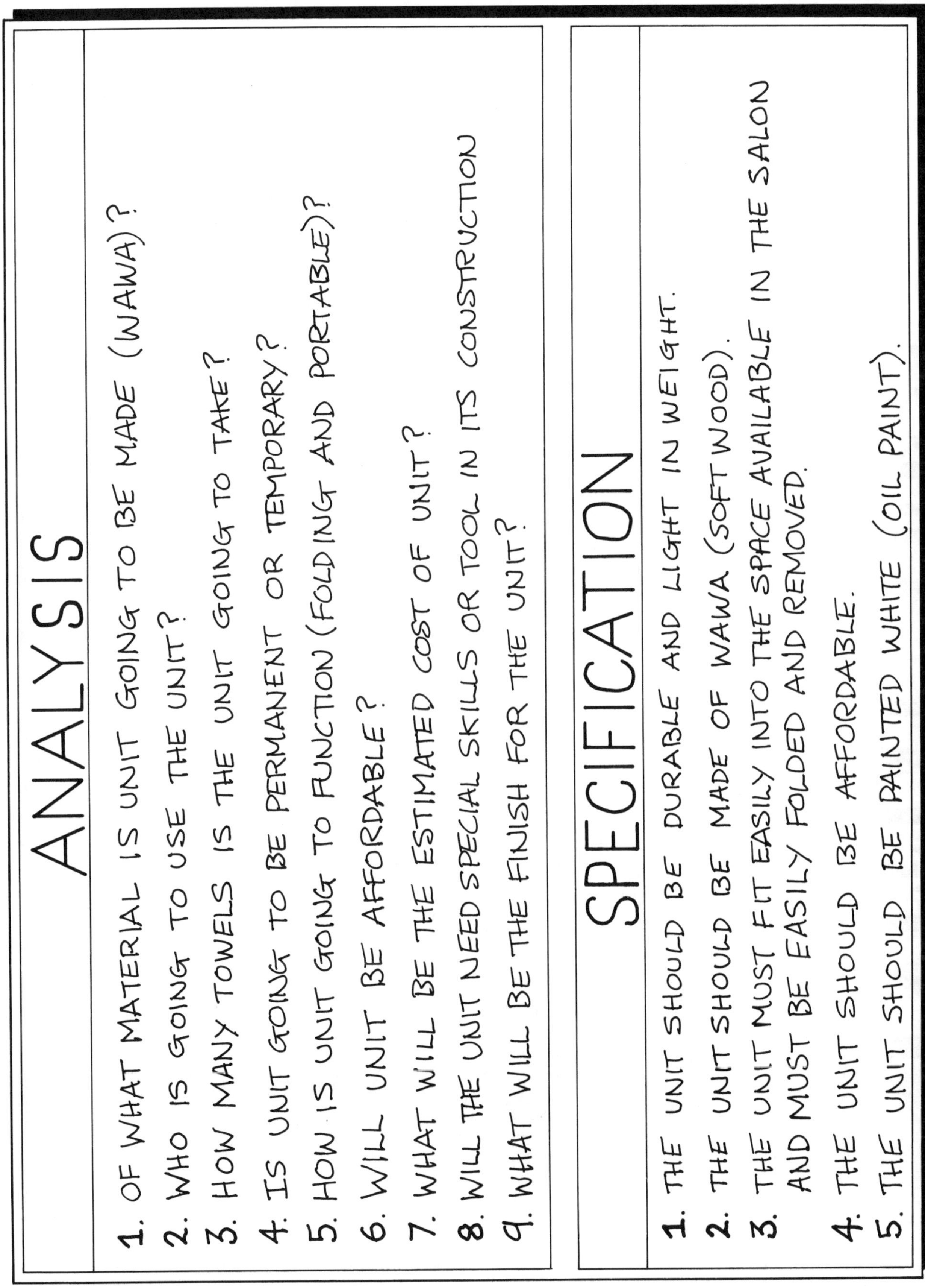
ANALYSIS

1. OF WHAT MATERIAL IS UNIT GOING TO BE MADE (WAWA)?
2. WHO IS GOING TO USE THE UNIT?
3. HOW MANY TOWELS IS THE UNIT GOING TO TAKE?
4. IS UNIT GOING TO BE PERMANENT OR TEMPORARY?
5. HOW IS UNIT GOING TO FUNCTION (FOLDING AND PORTABLE)?
6. WILL UNIT BE AFFORDABLE?
7. WHAT WILL BE THE ESTIMATED COST OF UNIT?
8. WILL THE UNIT NEED SPECIAL SKILLS OR TOOL IN ITS CONSTRUCTION
9. WHAT WILL BE THE FINISH FOR THE UNIT?

SPECIFICATION

1. THE UNIT SHOULD BE DURABLE AND LIGHT IN WEIGHT.
2. THE UNIT SHOULD BE MADE OF WAWA (SOFT WOOD).
3. THE UNIT MUST FIT EASILY INTO THE SPACE AVAILABLE IN THE SALON AND MUST BE EASILY FOLDED AND REMOVED.
4. THE UNIT SHOULD BE AFFORDABLE.
5. THE UNIT SHOULD BE PAINTED WHITE (OIL PAINT).

Figure 17.2 Example 1: analysis.

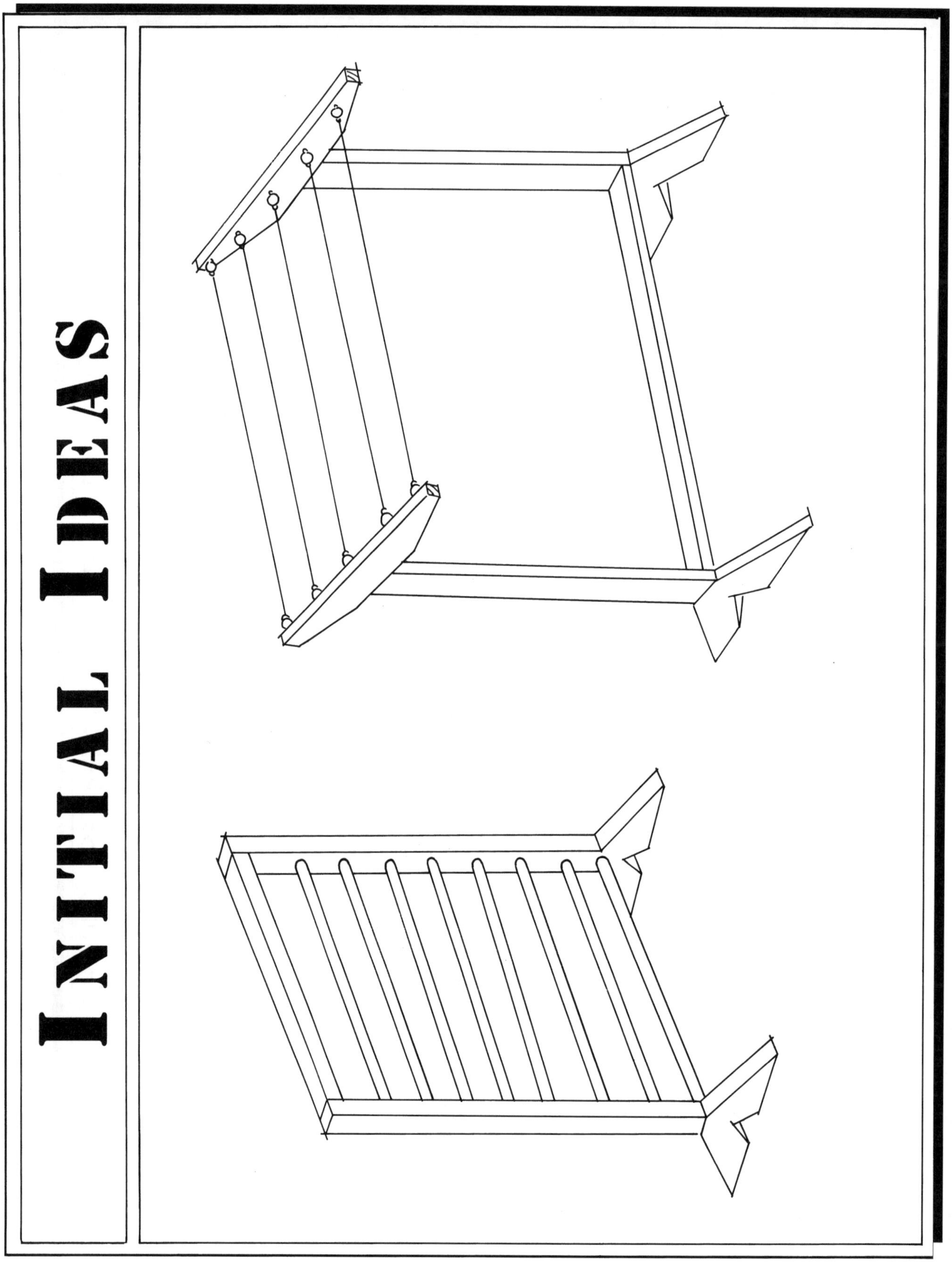

Figure 17.3 Example 1: initial ideas.

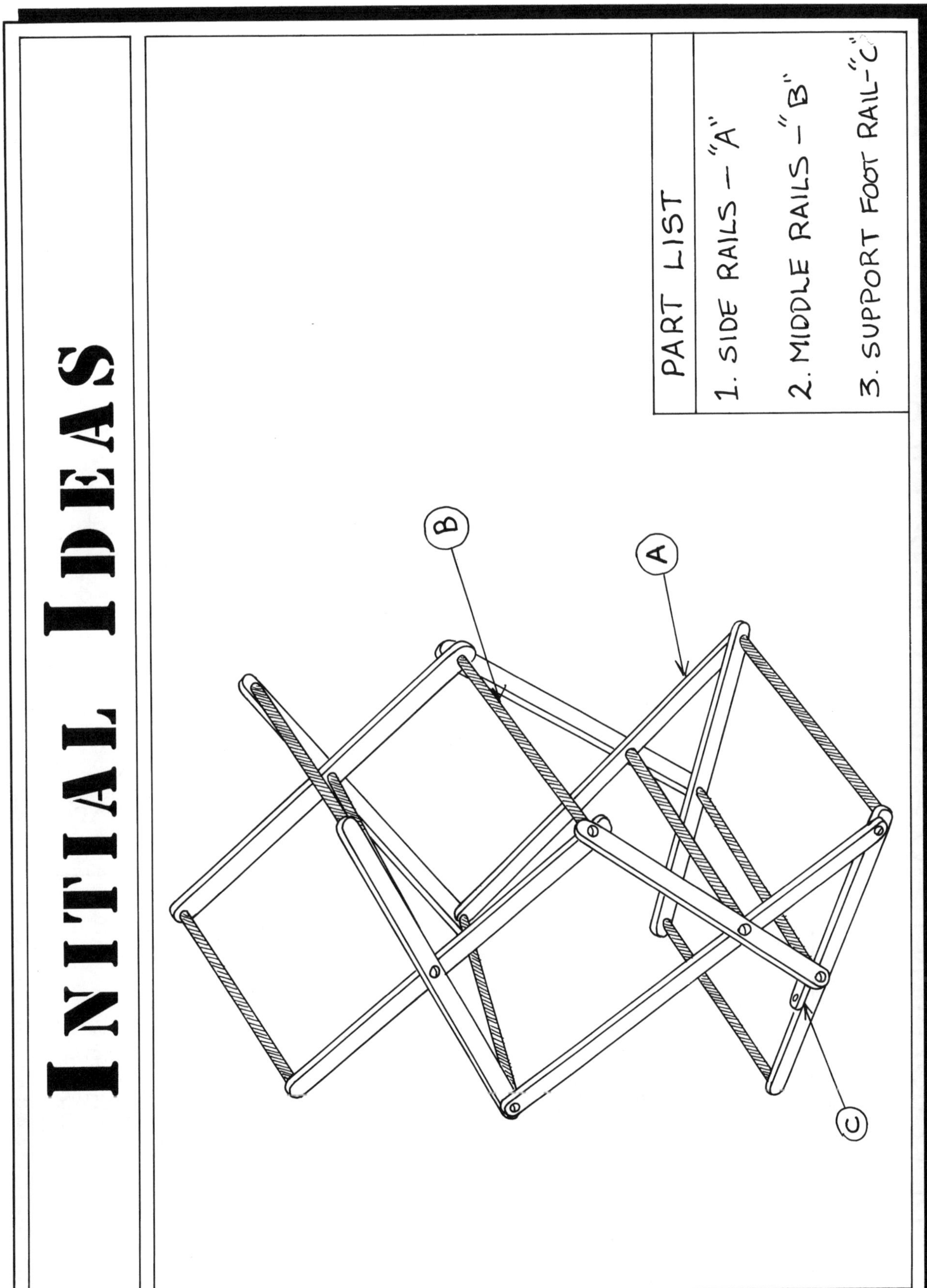

Figure 17.3 (continued)

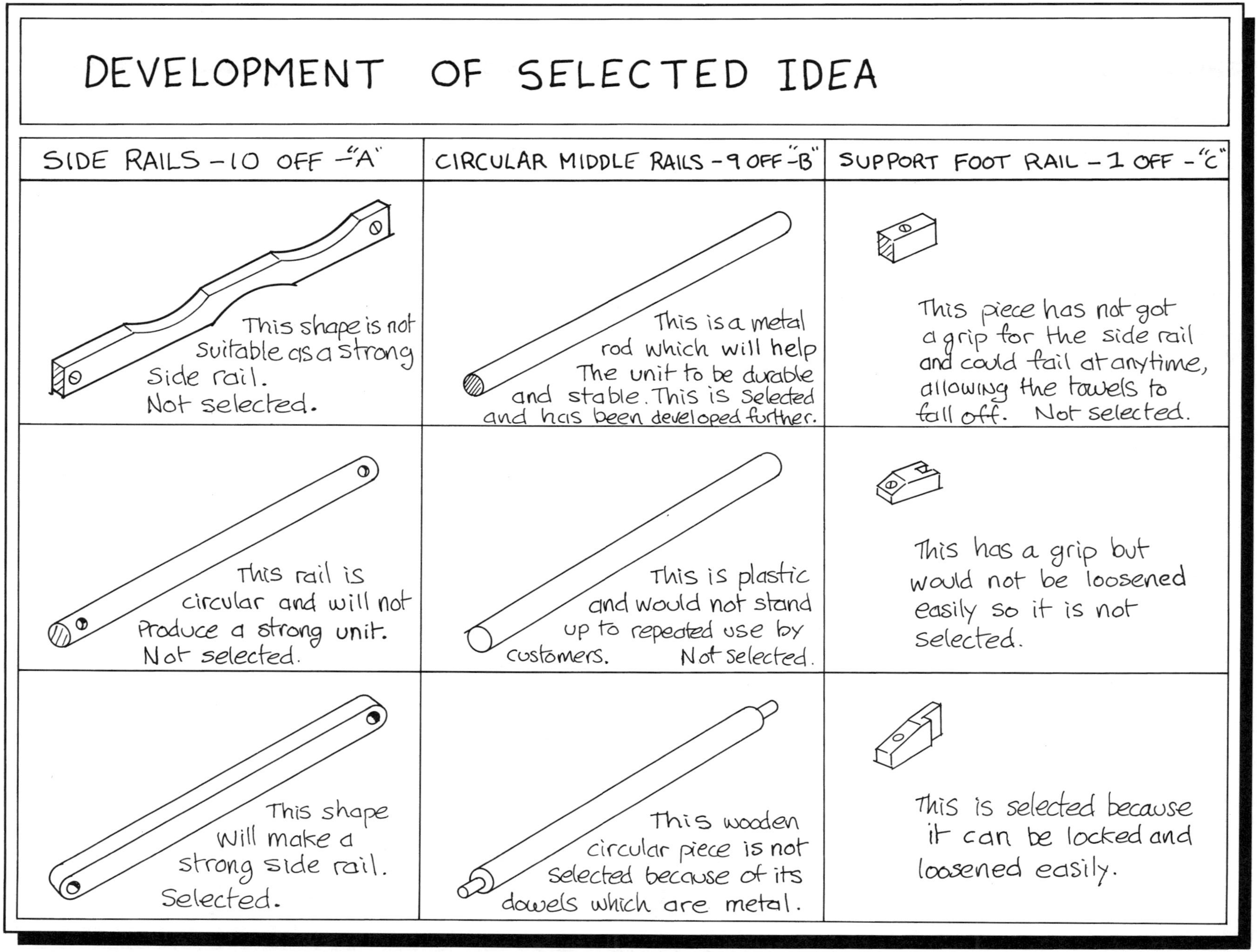

Figure 17.4 Example 1: development of selected idea.

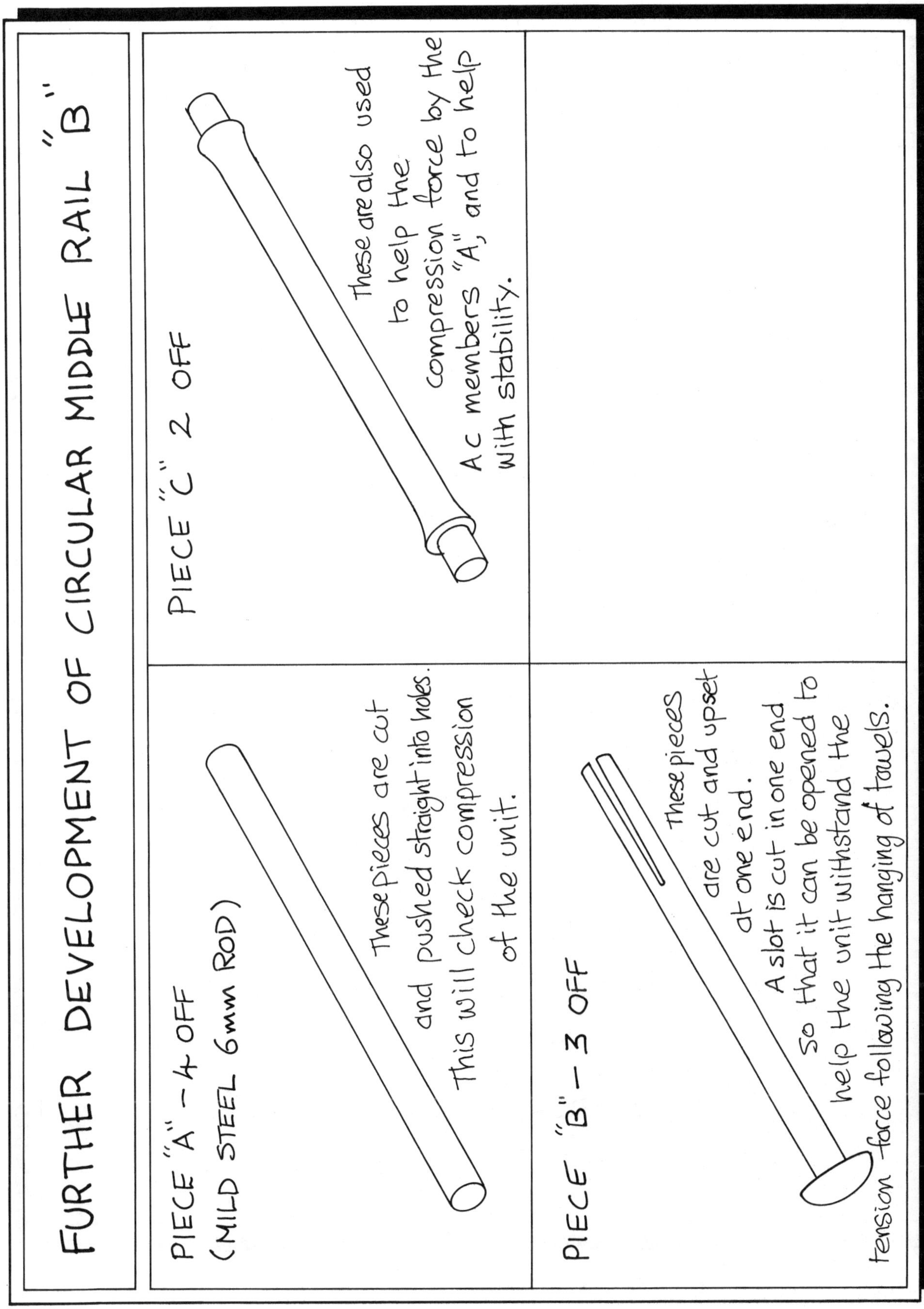

Figure 17.4 (continued)

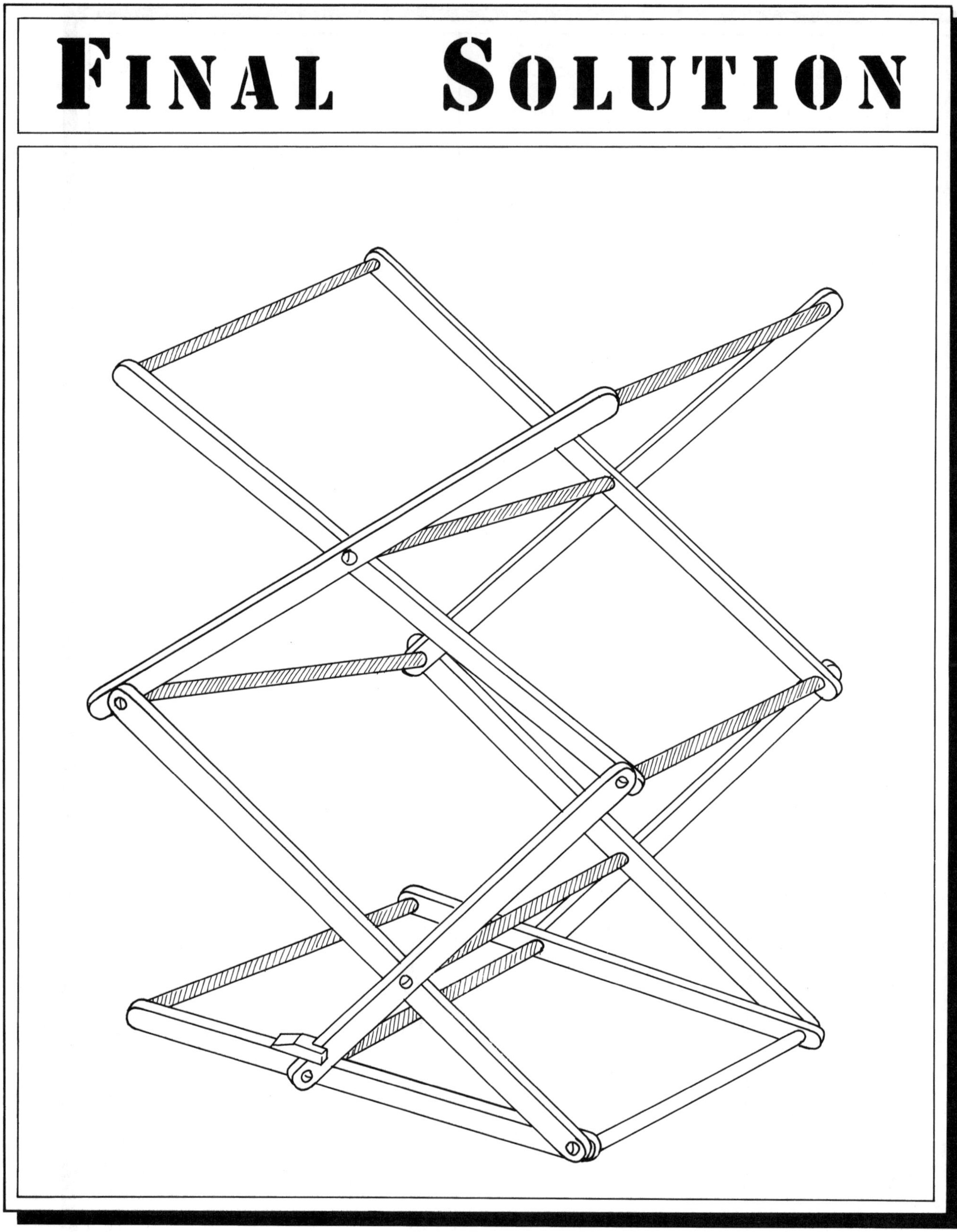

Figure 17.5 Example 1: final solution.

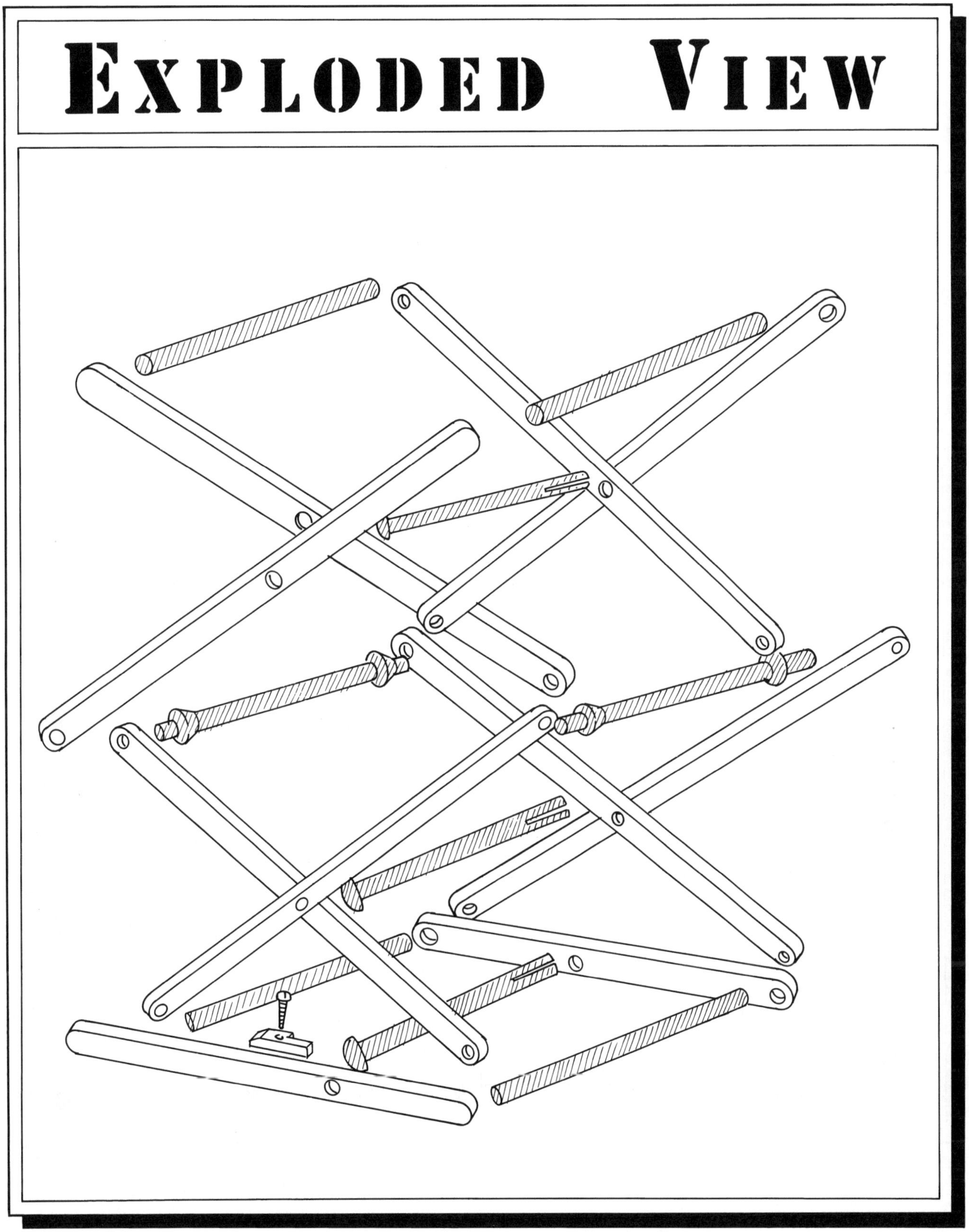

Figure 17.6 Example 1: exploded view.

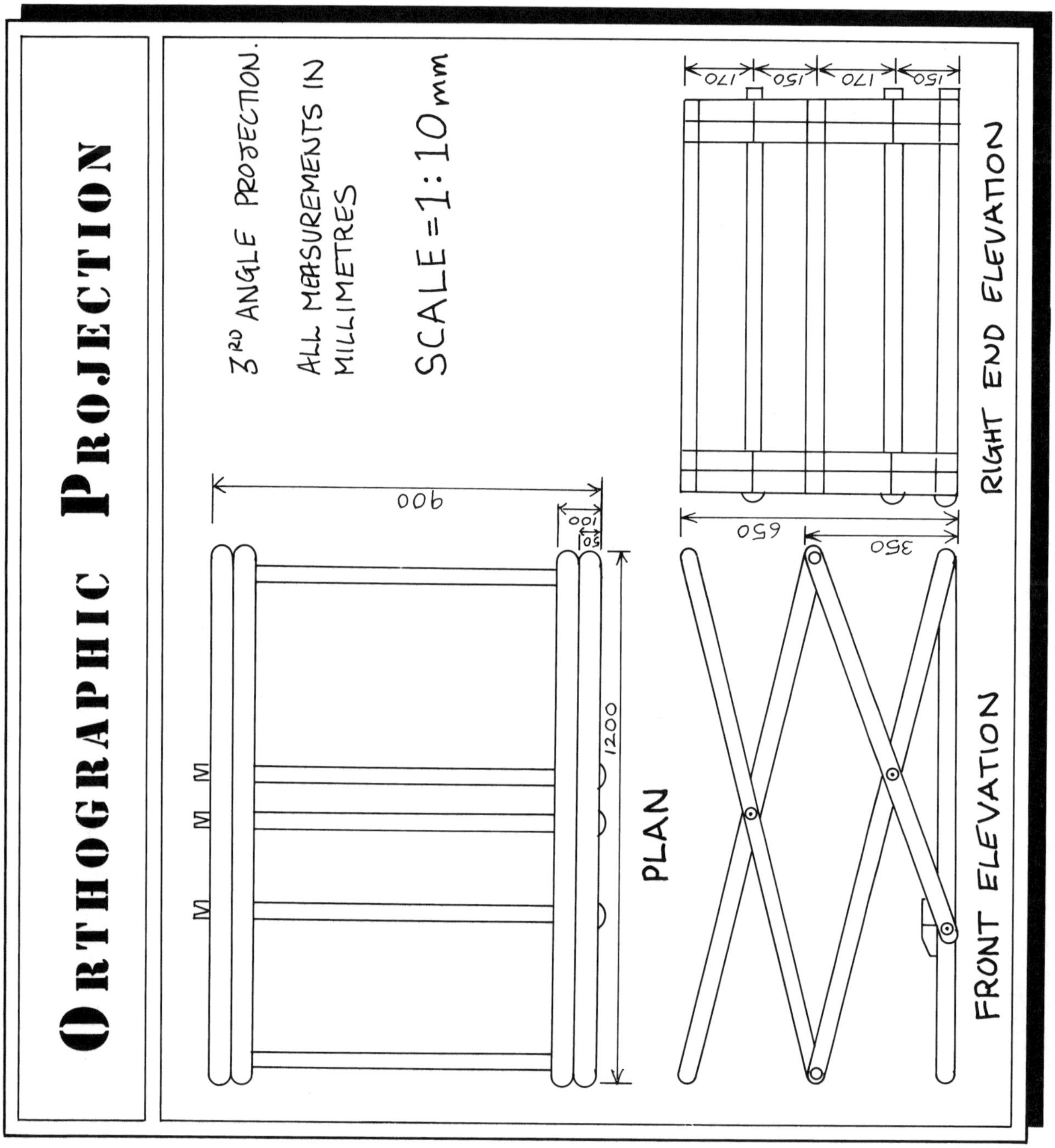

Figure 17.7 Example 1: orthographic projection.

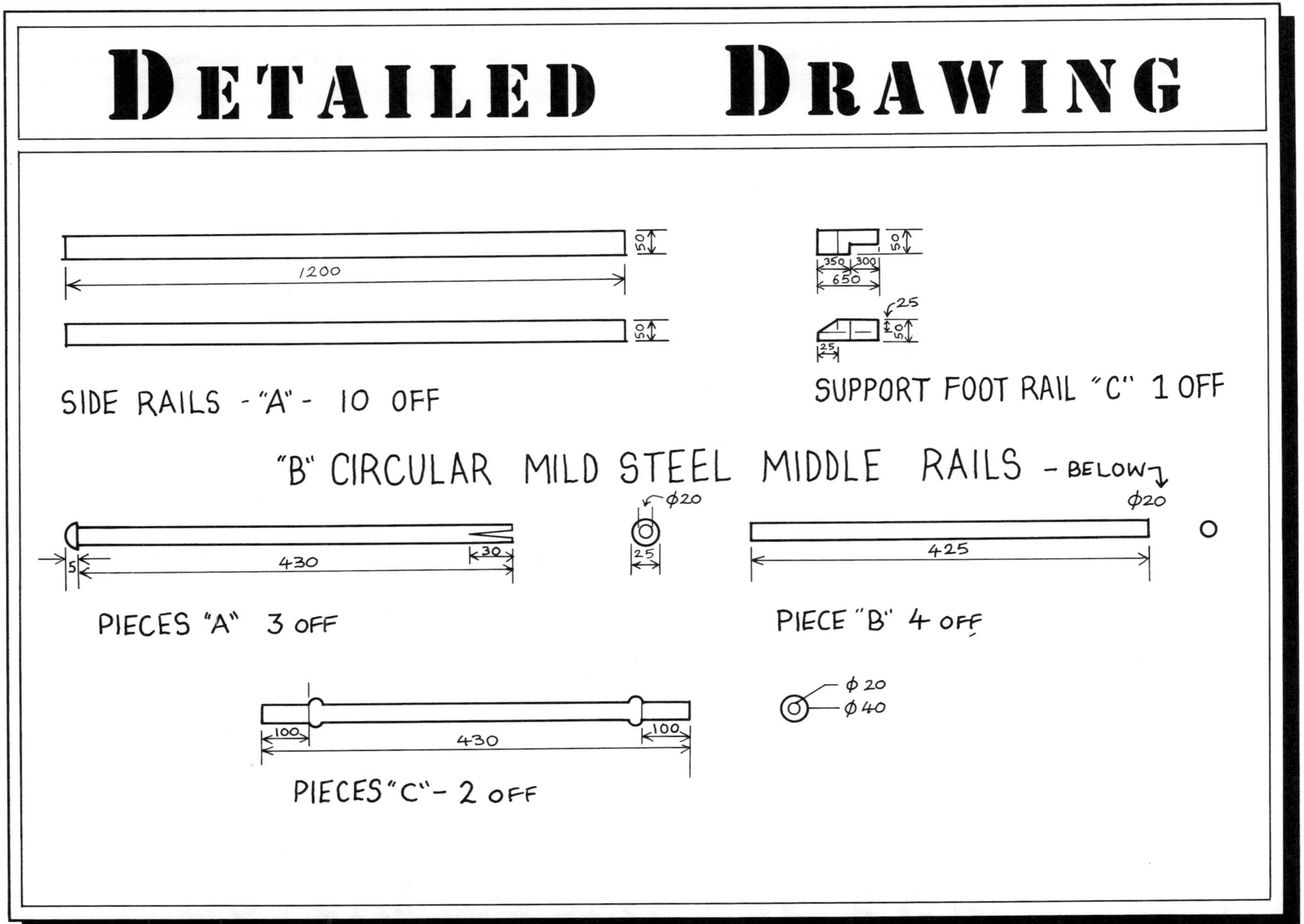

Figure 17.8 Example 1: detailed drawing.

CUTTING LIST

PART No.	DESCRIPTION	No.	DIMENSION (MM)	MATERIAL	FINISH	REMARKS
1.	SIDE RAIL – "A"	10	1205 × 55 × 55	ODUM	WHITE OIL PAINT	
2.	MIDDLE RAIL – "B"	9		MILD STEEL	WHITE OIL PAINT	BREAK DOWN BELOW
	PIECE – "A"	3	435 × ϕ 20			
	PIECE – "B"	4	430 × ϕ 20			
	PIECE – "C"	2	435 × ϕ 20			
3.	SUPPORT FOOT RAIL – "C"	1	655 × 55 × 55	ODUM	WHITE OIL PAINT	

Figure 17.9 Example 1: cutting list.

COSTING

No.	PART No.	DESCRIPTION	No.	UNIT COST (¢)	TOTAL COST (¢)	REMARKS
1	A	SIDE RAIL – 1200×50×50 – ODUM	10	630.00	6,300.00	
2	B	MIDDLE RAIL – MILD STEEL ROD Ø20	9			ANALYSED BELOW
		PIECE "A" – 430 × Ø 20	3	240.00	720.00	
		PIECE "B" – 425 × Ø 20	4	240.00	960.00	
		PIECE "C" – 430 × Ø 20	2	240.00	480.00	
				SUB TOTAL	2,160.00	
3	C	SUPPORT FOOT RAIL – 650×50×50 (ODUM)	1		70.00	
4		WHITE OIL PAINT	1/4 LITRE		450.00	
5		MISCELLANEOUS			1,000.00	
				TOTAL	9,980.00	
6		LABOUR COST			1,996.00	
		TOTAL COST OF UNIT			¢11,976.00	

Figure 17.10 Example 1: costing.

Example 2

This example shows how a seating unit with a built-in notepad support for use in lectures was designed. The stages covered include:

1. The brief and the **design specification** (Figure 17.11).
2. **Initial ideas**: four possible designs are considered and the reasons for rejection or selection given (Figure 17.12).
3. **Development of selected idea**: the best design for each part of the chair is chosen, with reasons (Figure 17.13).
4. Final solution (Figure 17.14).
5. **Working drawings** are presented. The exploded view shows how the parts relate to each other. The orthographic projection shows the overall dimensions. The detailed drawing is clear and thorough (Figures 17.15, 17.16, 17.17).
6. **Evaluation**: the unit is tested and found to be strong and easily operated, thus meeting the design specification. The student gives consideration to the aesthetic appearance of the unit and proposes modifications: in future it should be upholstered and the legs turned on a lathe. A costing of these modifications should be given (Figure 17.18).

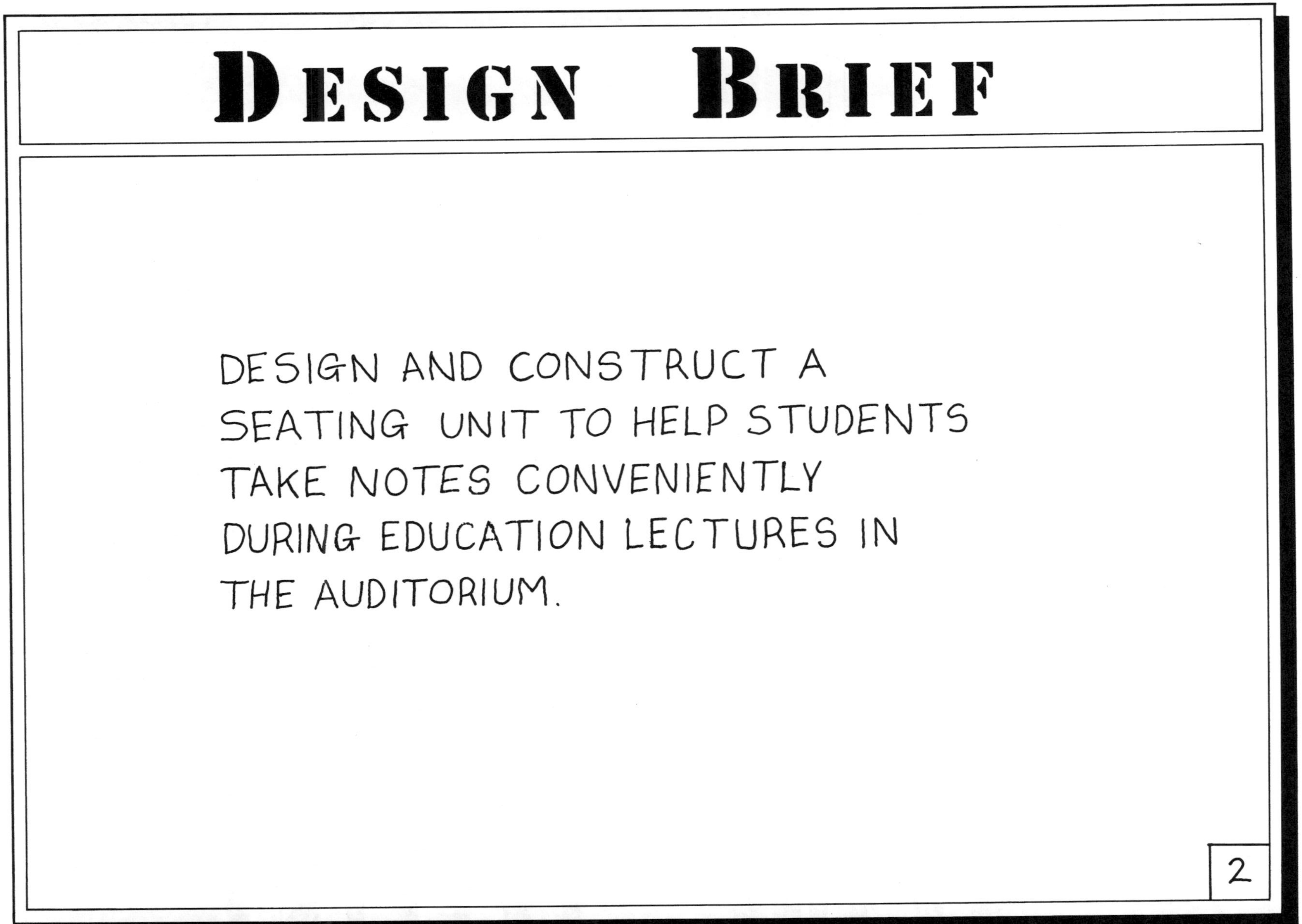

Figure 17.11 Example 2: brief and design specification.

Figure 17.12 Example 2: initial ideas.

Figure 17.12 (continued)

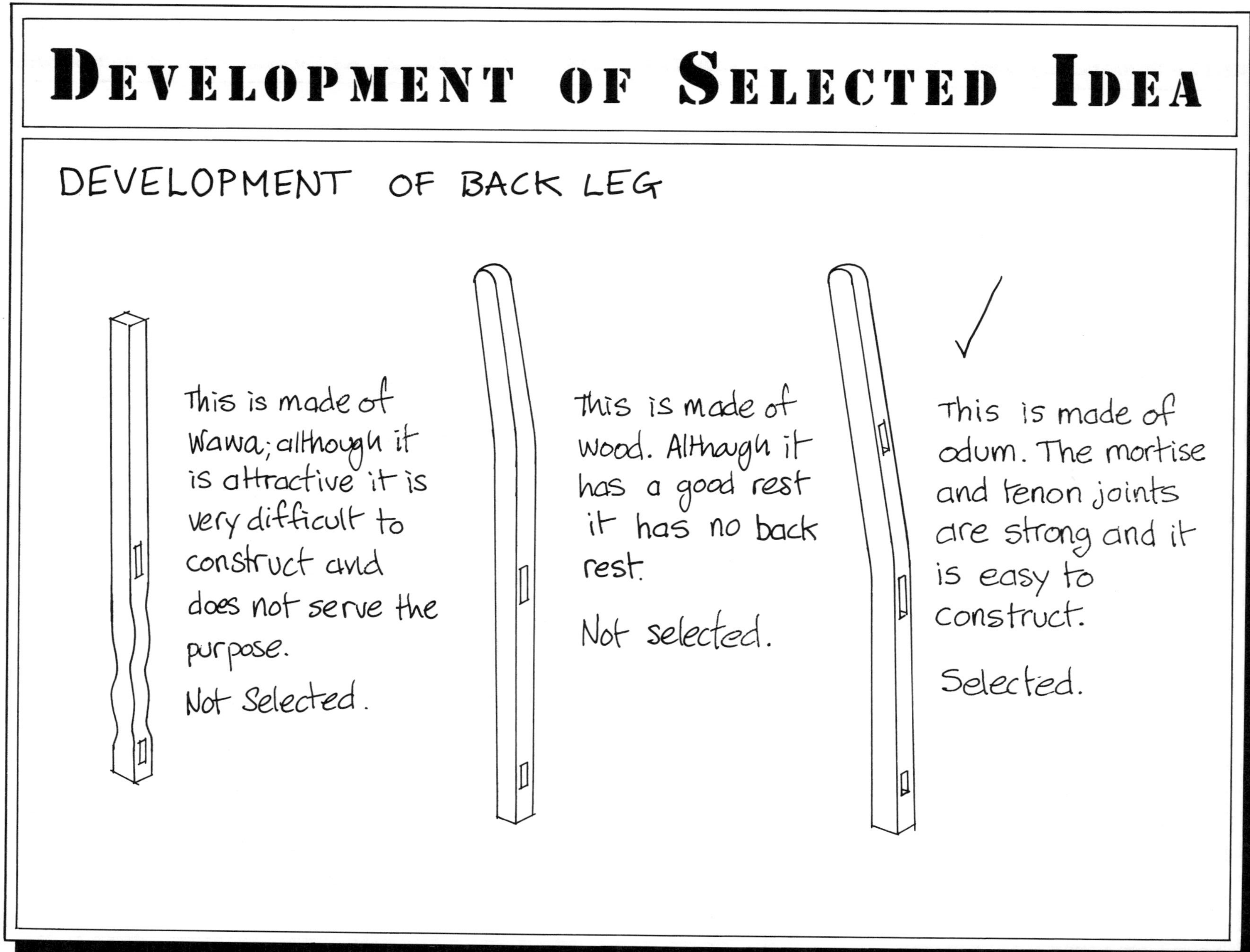

Figure 17.13 Example 2: development of selected idea.

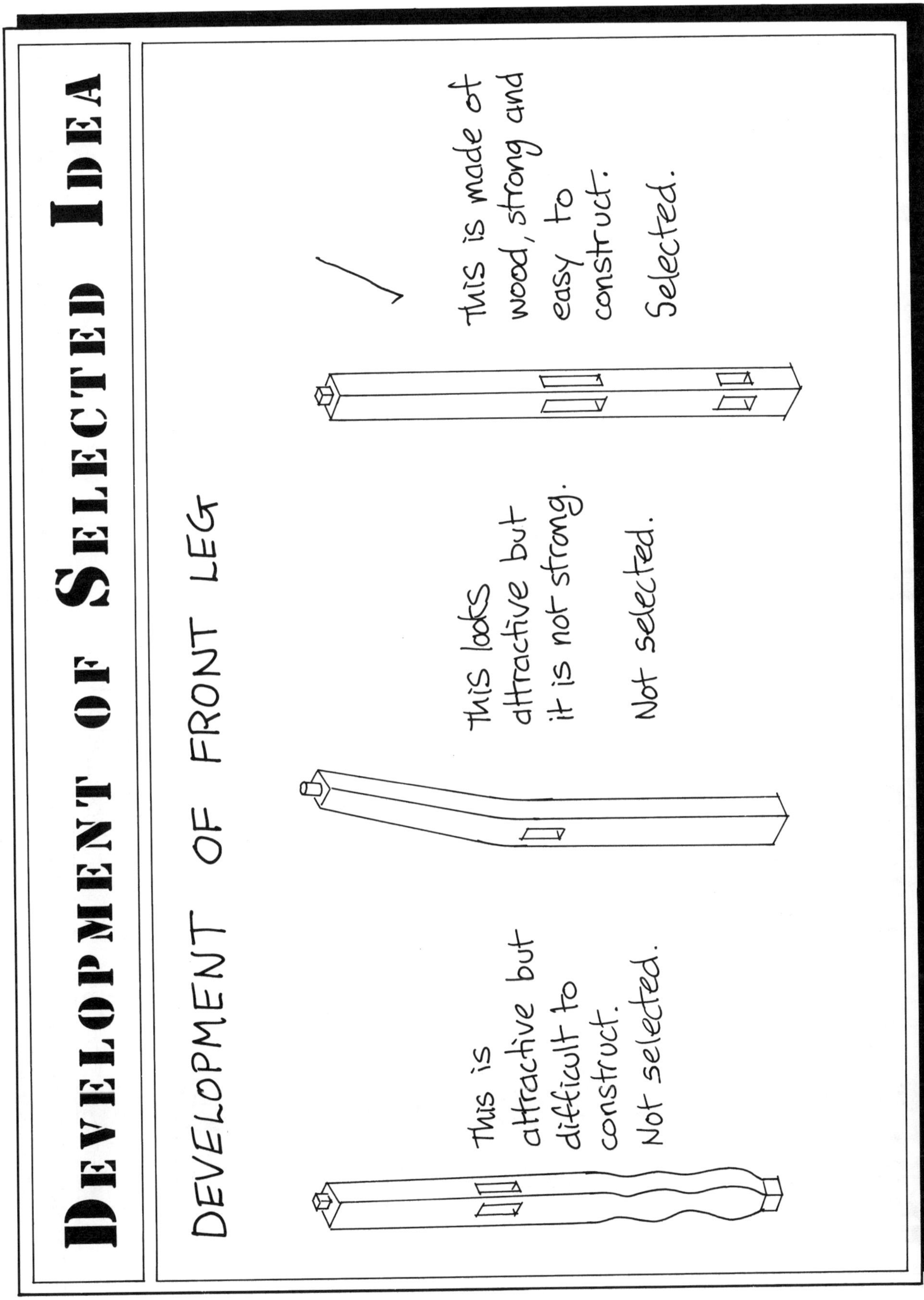

Figure 17.13 (continued)

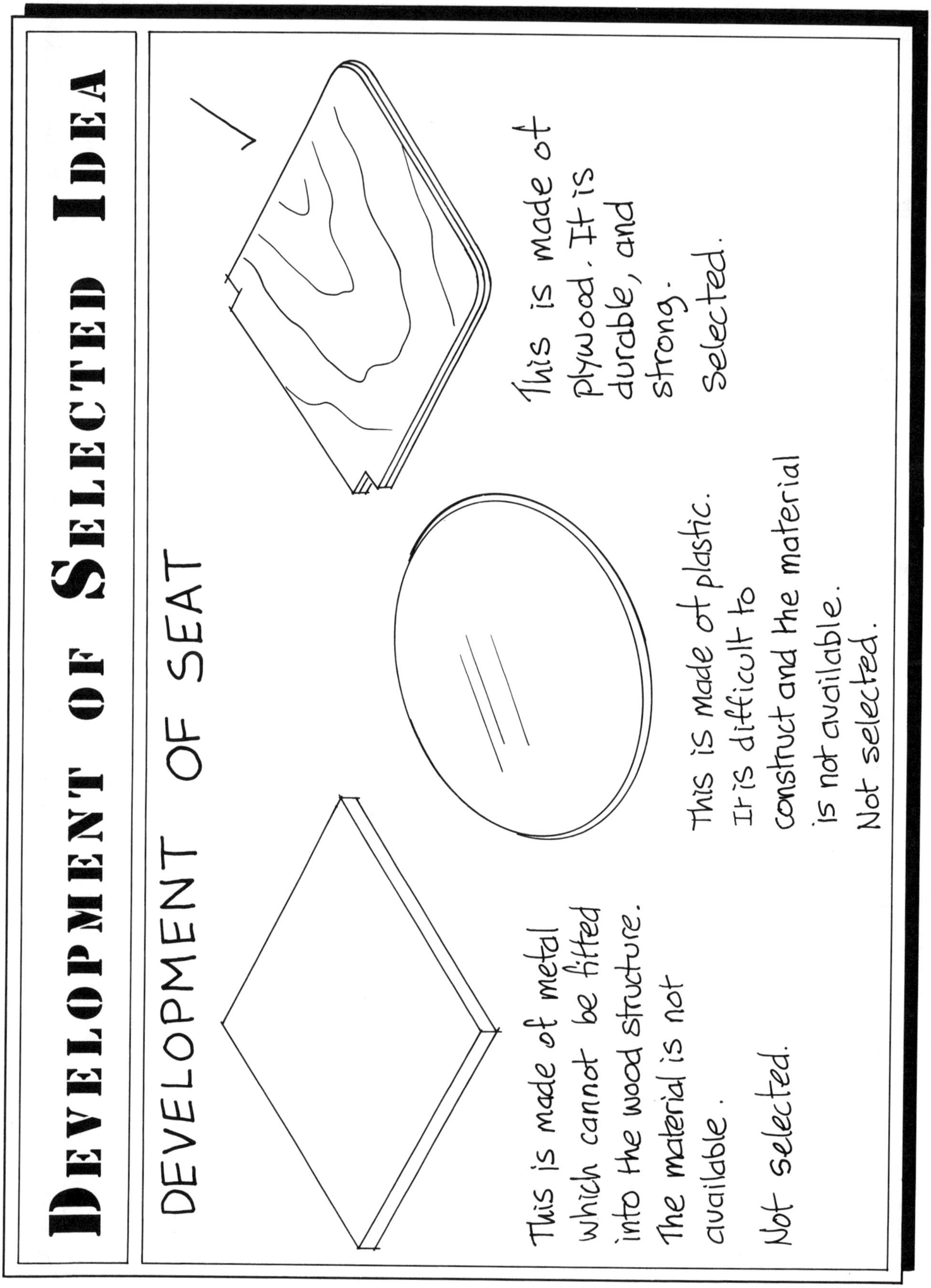

Figure 17.13 (continued)

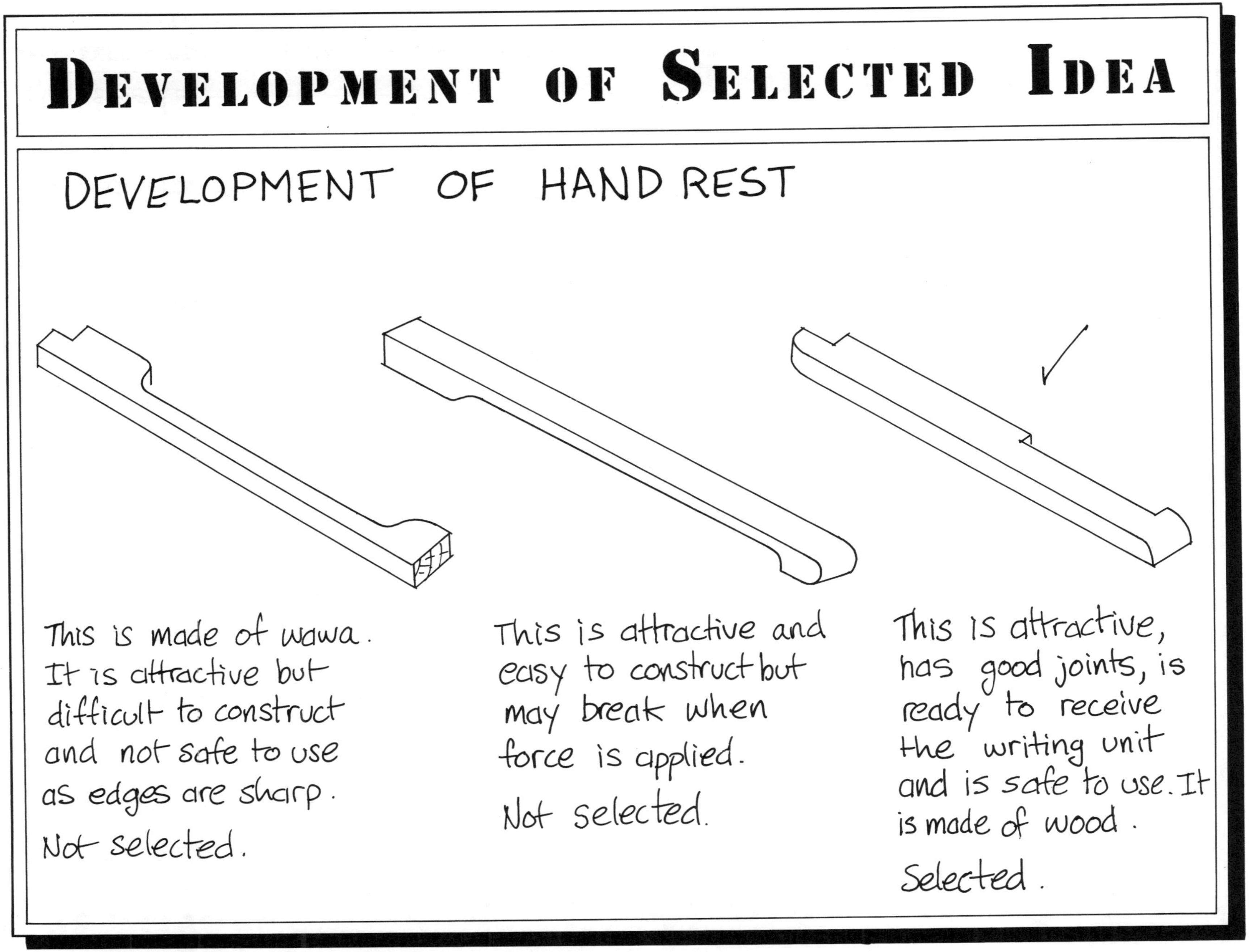

Figure 17.13 (continued)

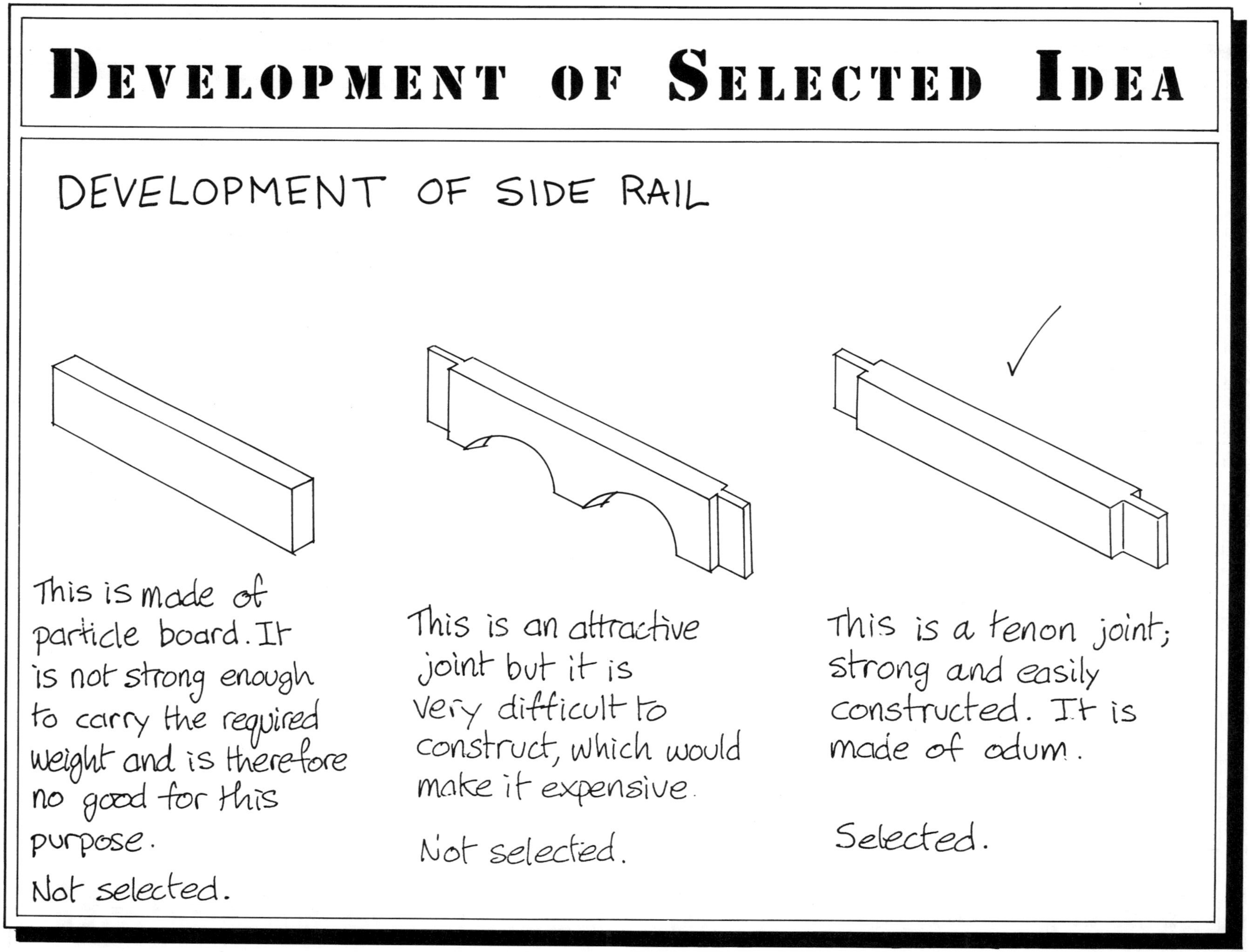

Figure 17.13 (continued)

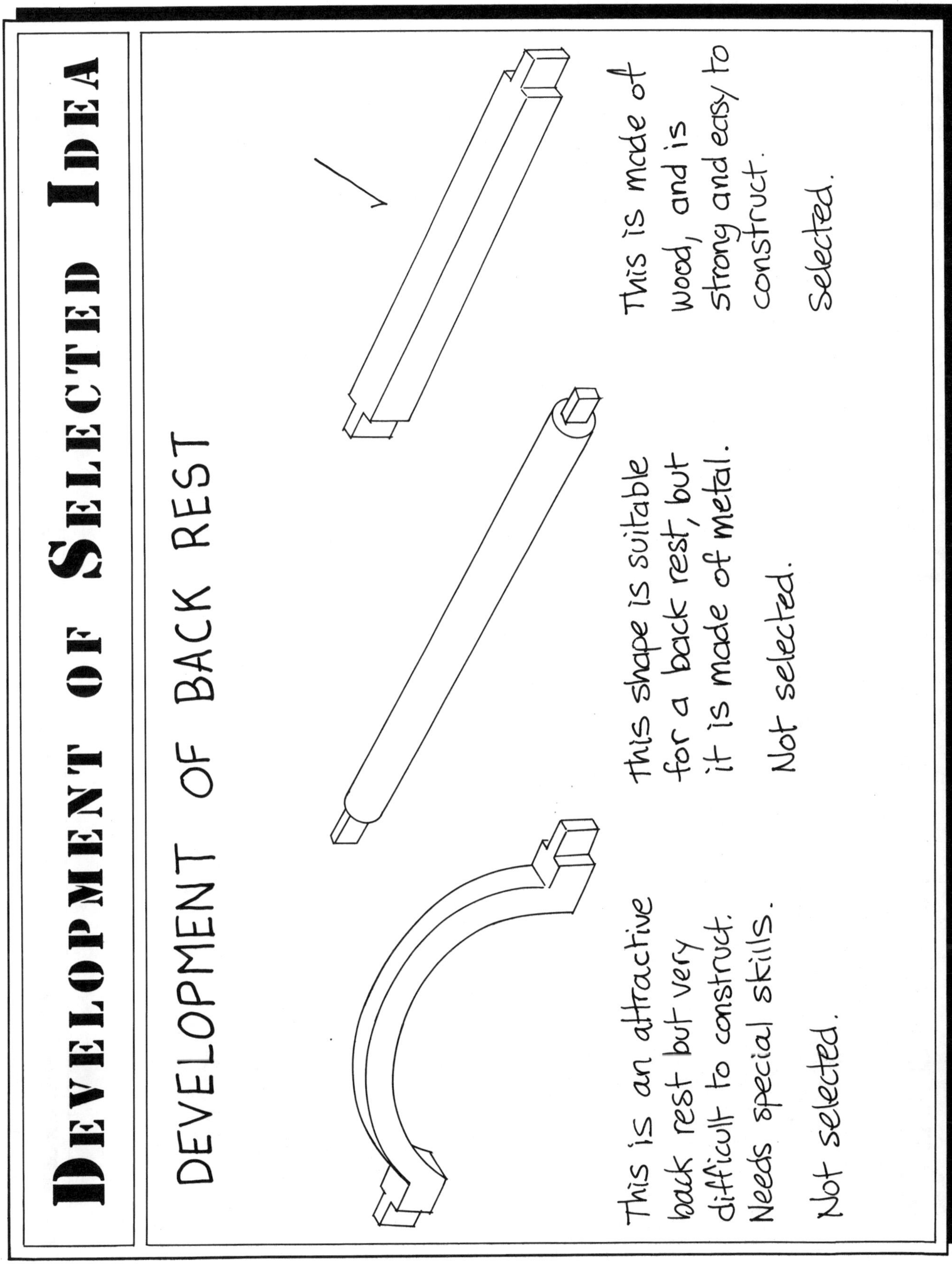

Figure 17.13 (continued)

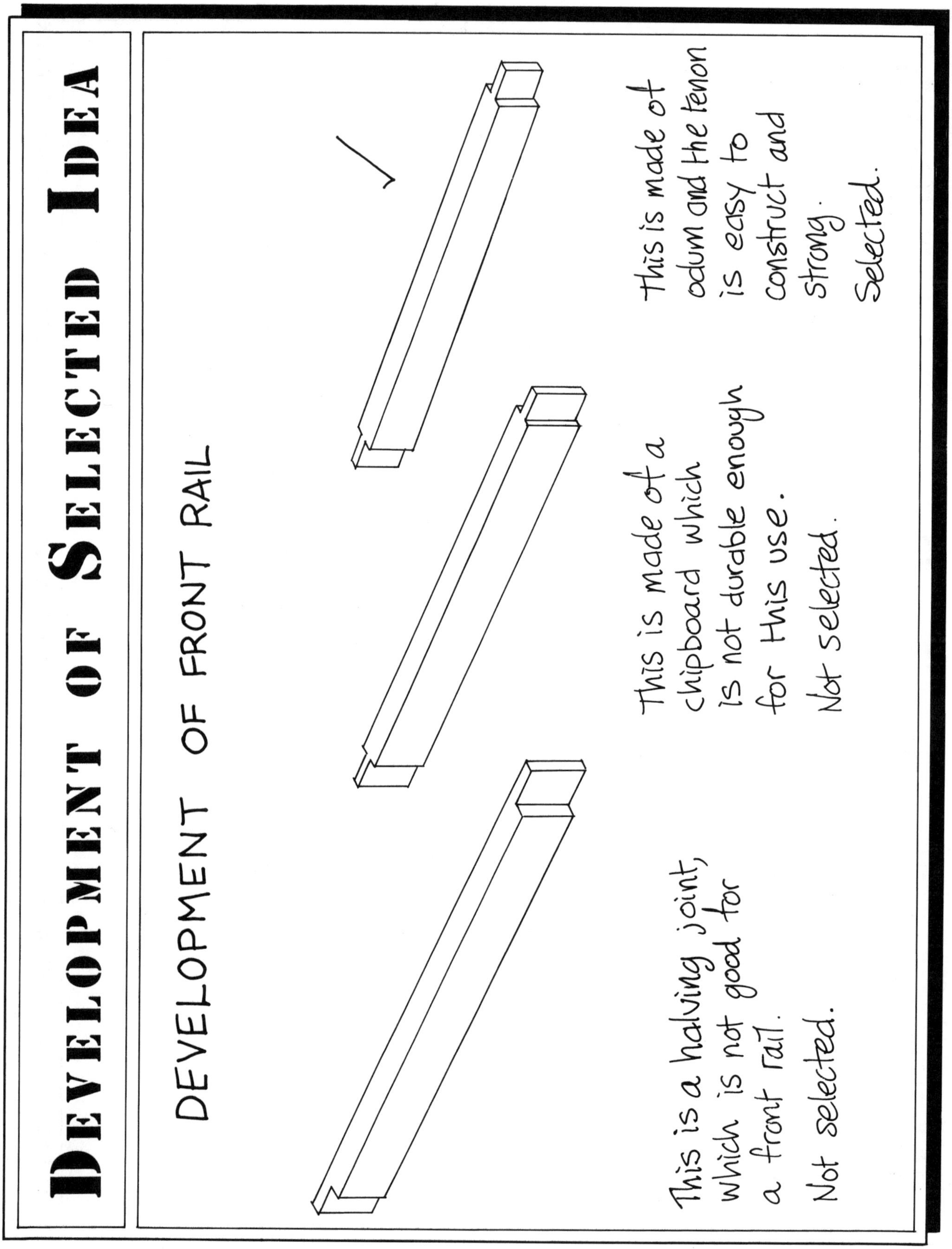

Figure 17.13 (continued)

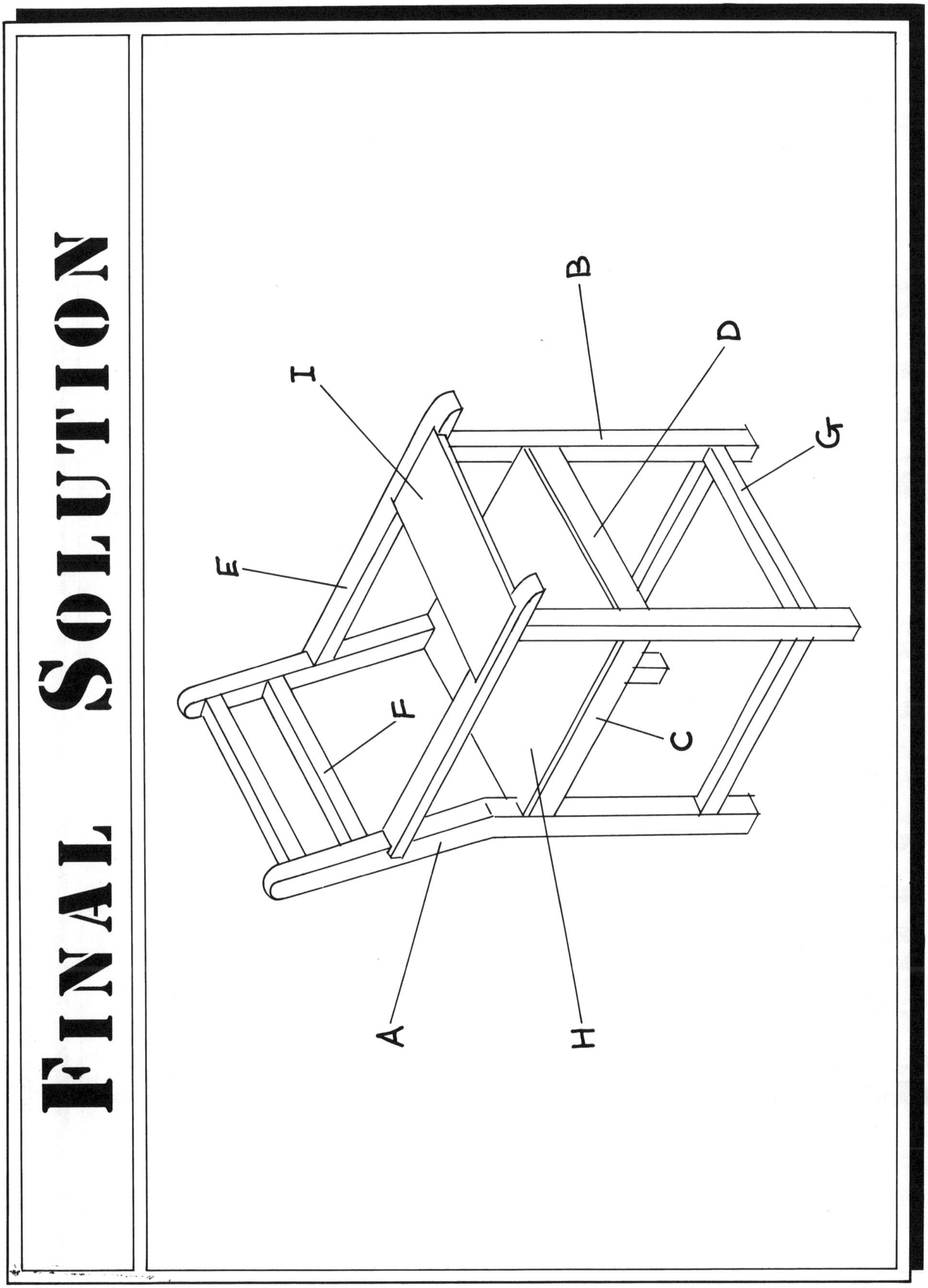

Figure 17.14 Example 2: final solution.

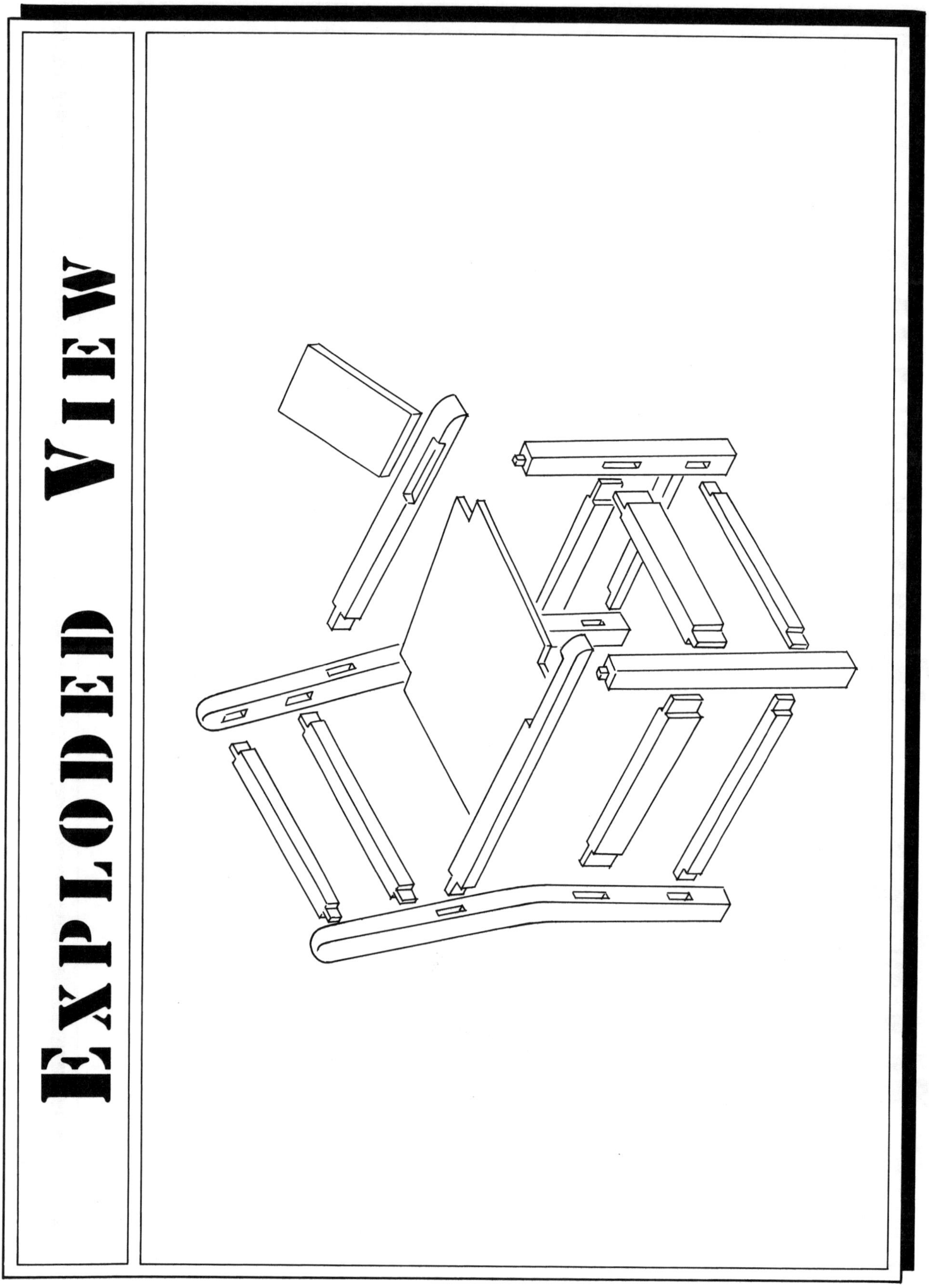

Figure 17.15 Example 2: exploded view of seating unit.

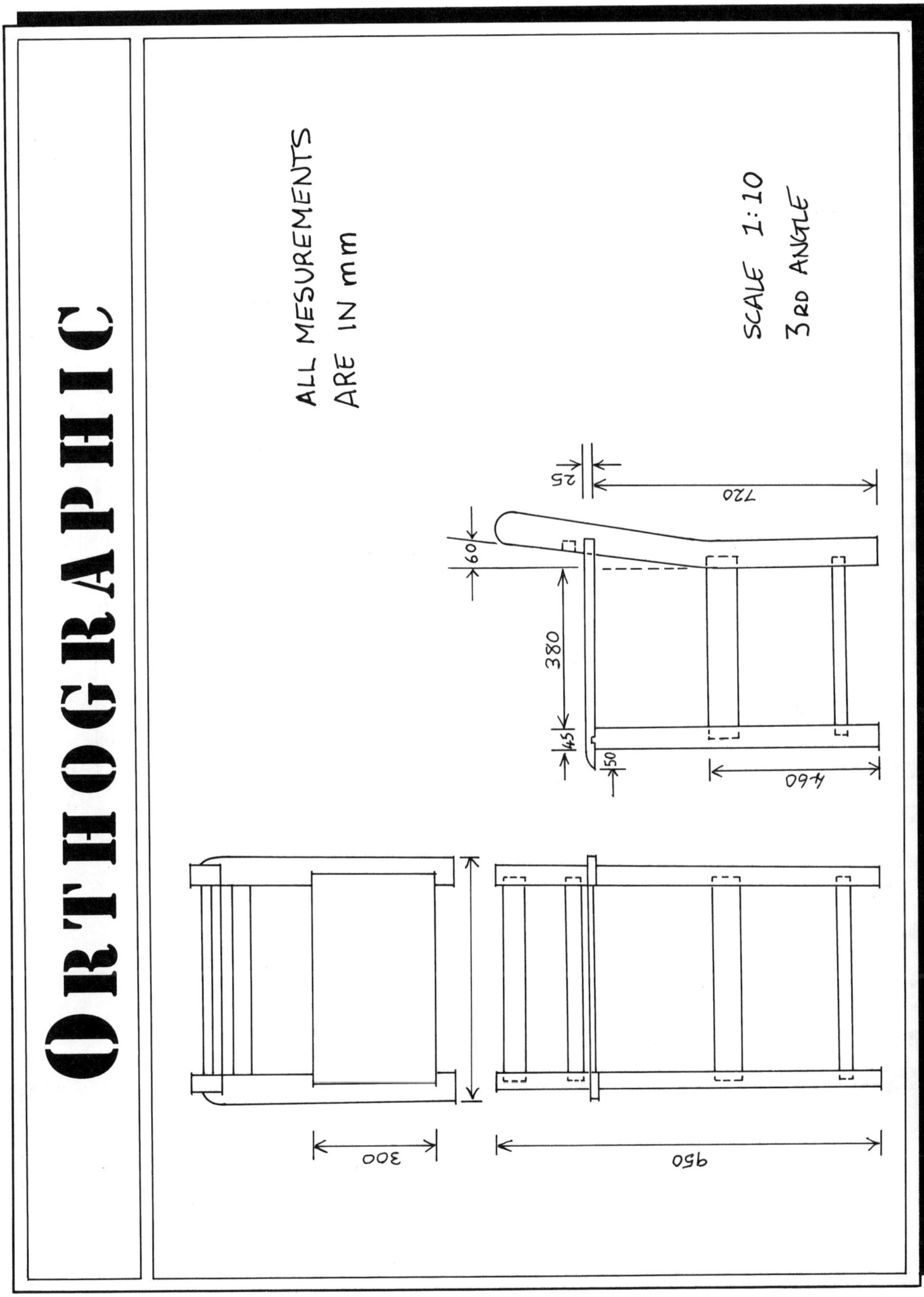

Figure 17.16 Example 2: orthographic projection.

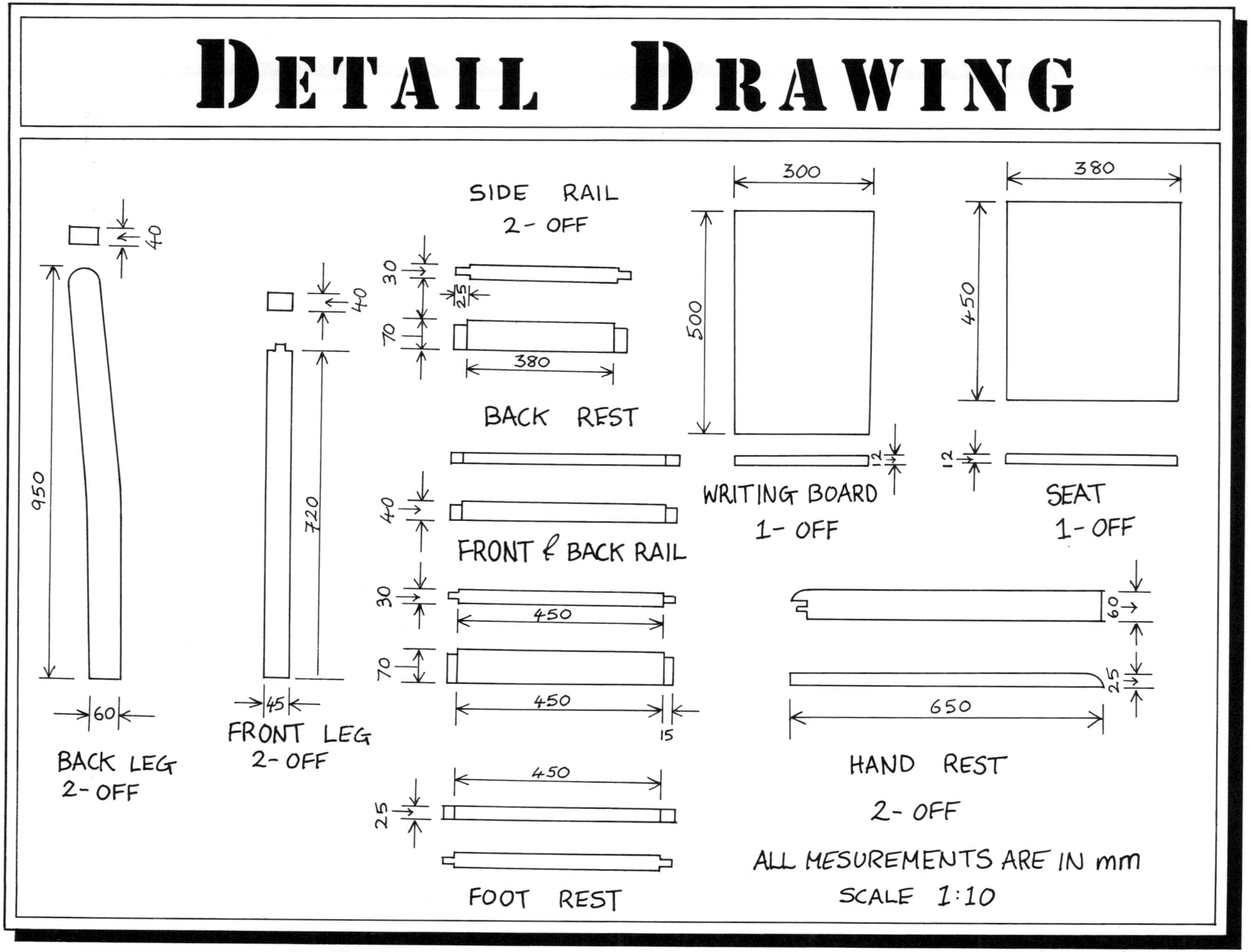

Figure 17.17 Example 2: detailed drawing.

EVALUATION

After designing and constructing the unit in the workshop it was tested and evaluated, the following were the outcomes:

The unit should have been made of odum and plywood. The model is made of a mixture of fully grown wawa, red wood, etc., as they were available at the time of construction. It was first tested for it's durability and strength. A weight was used on the model and it was found to be strong and durable, and able to resist compressive, tensile and sheer force. It was found to be easily operated by the users. The unit was very easy to construct and joints used are strong and serve the purpose for which the unit was made. The unit has a polished finish to protect it against marks and surface damage, and to make it attractive to the user, as it is to be used indoors.

MODIFICATION

The unit was found not to be very attractive so in future the front leg should be turned on a machine to improve it's appearance. The seat of the unit should be upholstered and padded to make it comfortable to the user.

The writing board should be fastened to the hand-rest with a hinge, or a curtain should be made attached to the side to accomodate it when not in use, as the writing board may obstrict users when the unit is being used as a seat only.

Figure 17.18 Example 2: evaluation.

Example 3

This example shows how a container for drawing instruments was designed for use in lectures. The stages covered include:

1. The situation, brief and the design specification (Figures 17.19, and 17.20).
2. Initial ideas: five possible designs are considered and the reasons for rejection or selection are given (Figure 17.21).
3. Development: the choice of fittings is shown in Figure 17.21.
4. Final solution (Figure 17.22): note how the proposed unit is portable but lacks a lid.
5. Cutting list and costings (Figure 17.23).
6. Evaluation: (Figure 17.24): a thorough and critical evaluation is presented. Problems arising from the available materials and the need for special skills in making the dovetail joints are noted. Proposals for modifying the design are suggested, including attaching a lid and replacing the dovetail joints with pin joints.

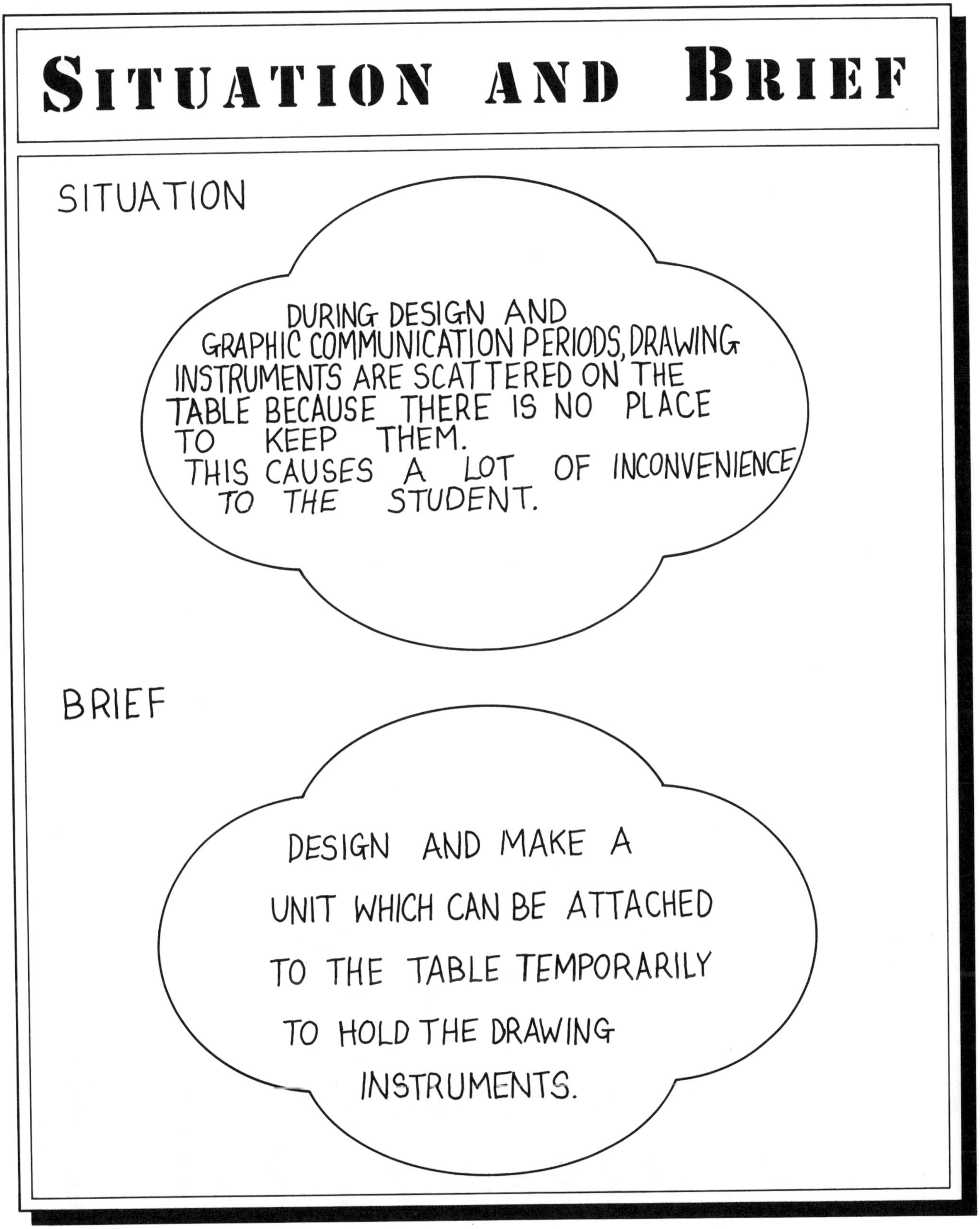

Figure 17.19 Example 2: situation and brief.

SPECIFICATION

1. The unit should be affordable and light in weight.
2. the unit should be within the reach of the user.
3. The unit should be rectangular in shape and safe to use.
4. The unit should be a temporary one which can be locked.
5. The unit should be able to hold drawing instruments.
6. The unit should be 450mm by 300mm in size.
7. The unit should be strong, durable and easily detached.
8. The unit should be made of wood and easily handled.
9. The unit should be polished.
10. The unit should be at 150° to the top of the table when fixed.
11. The materials required to make the unit should be available.
12. Screws and nails should be used.
13. The unit should have dovetail and mortise and tenon joints for the sides and frame.
14. The unit should be easily accessible and visible to the user.

Figure 17.20 Example 3: design specification.

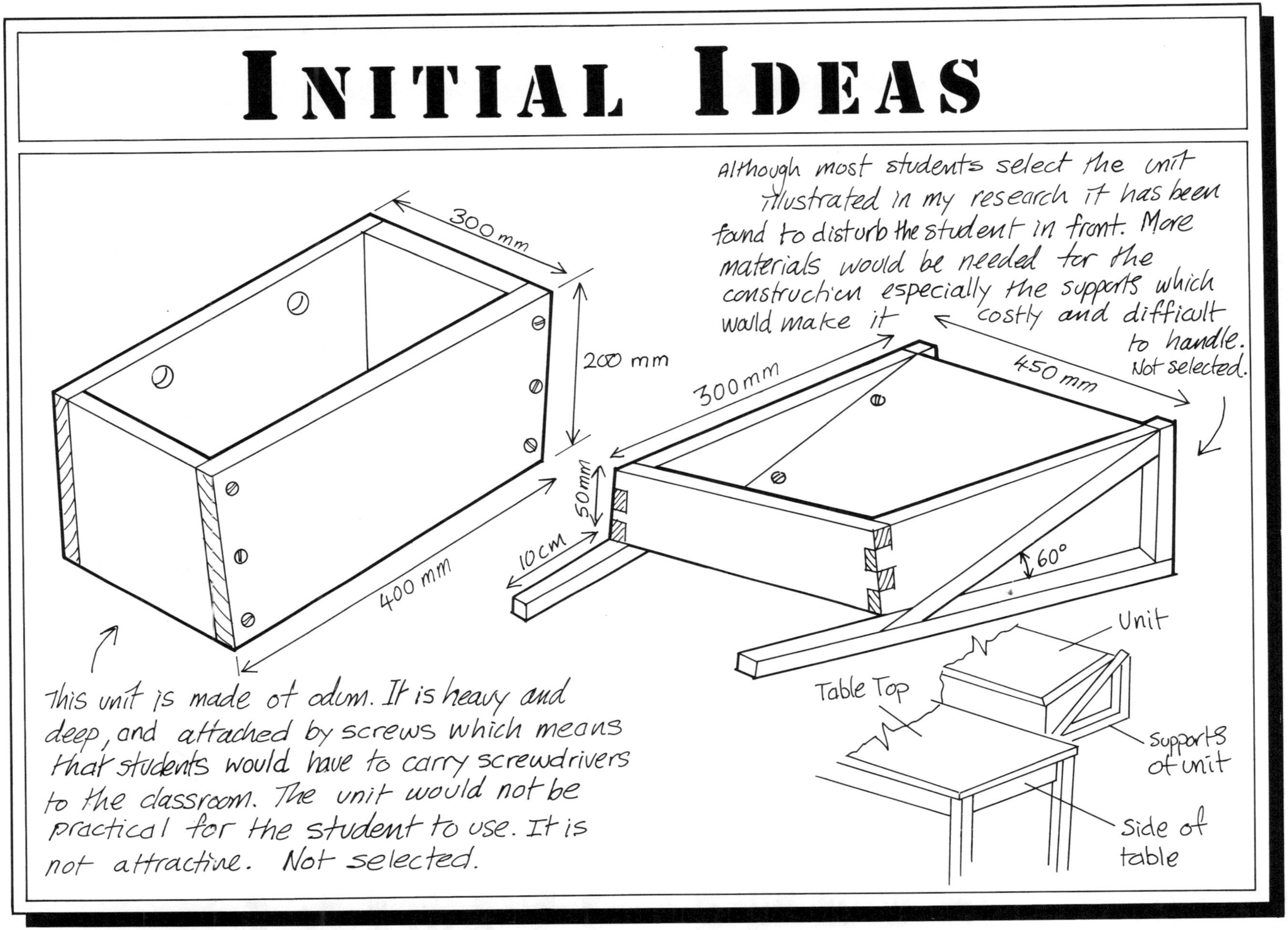

Figure 17.21 Example 3: initial ideas.

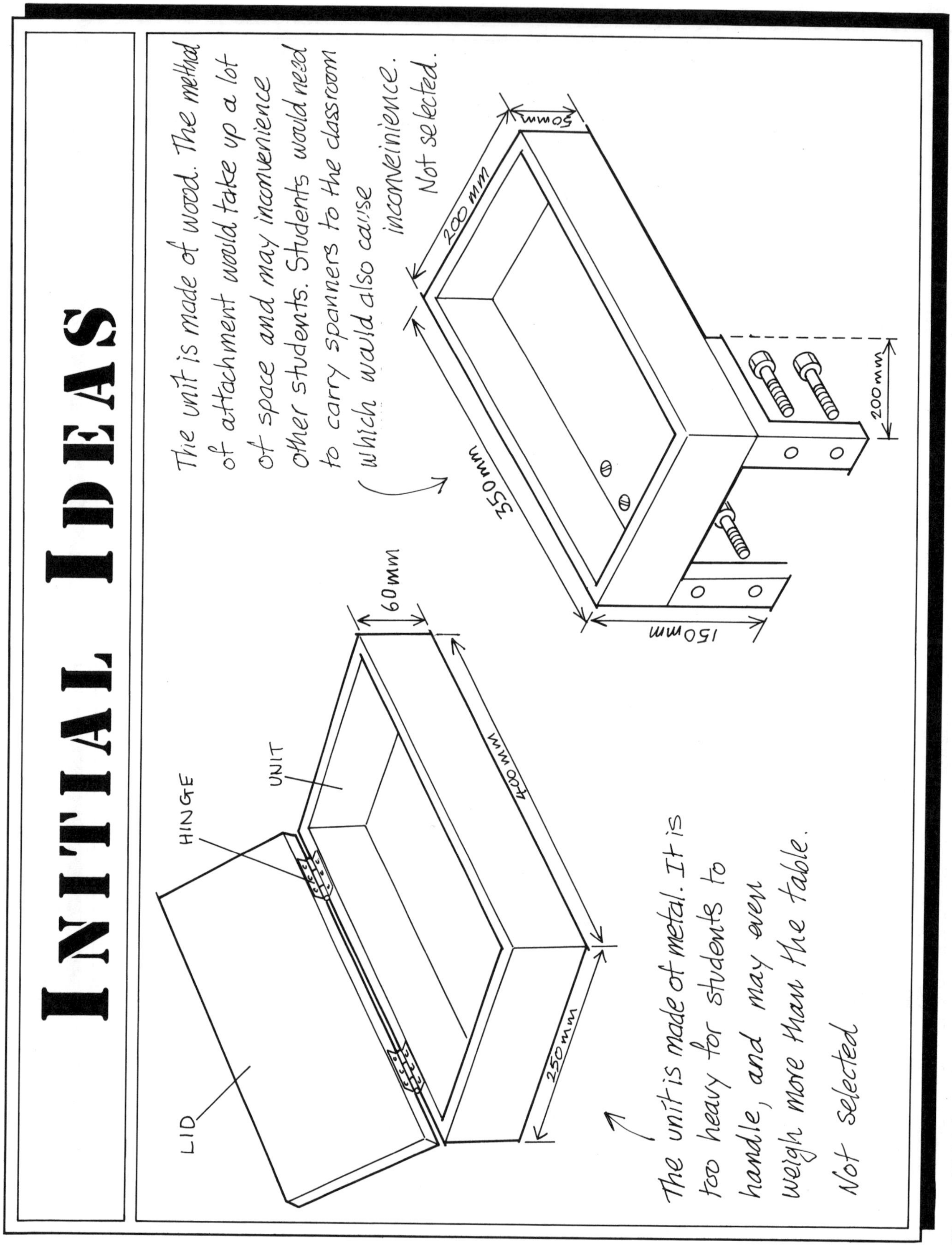

Figure 17.21 (continued)

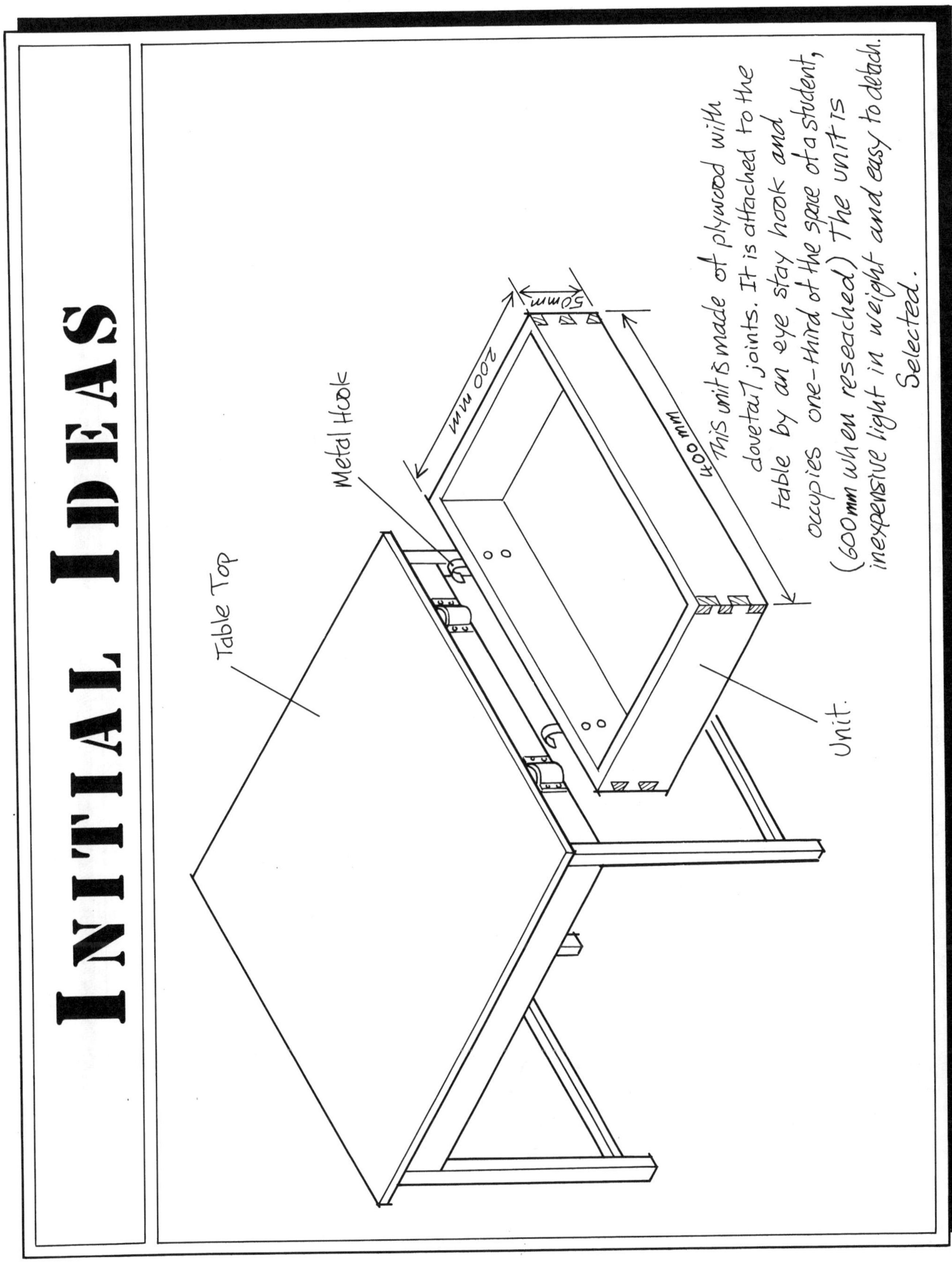

Figure 17.21 (continued)

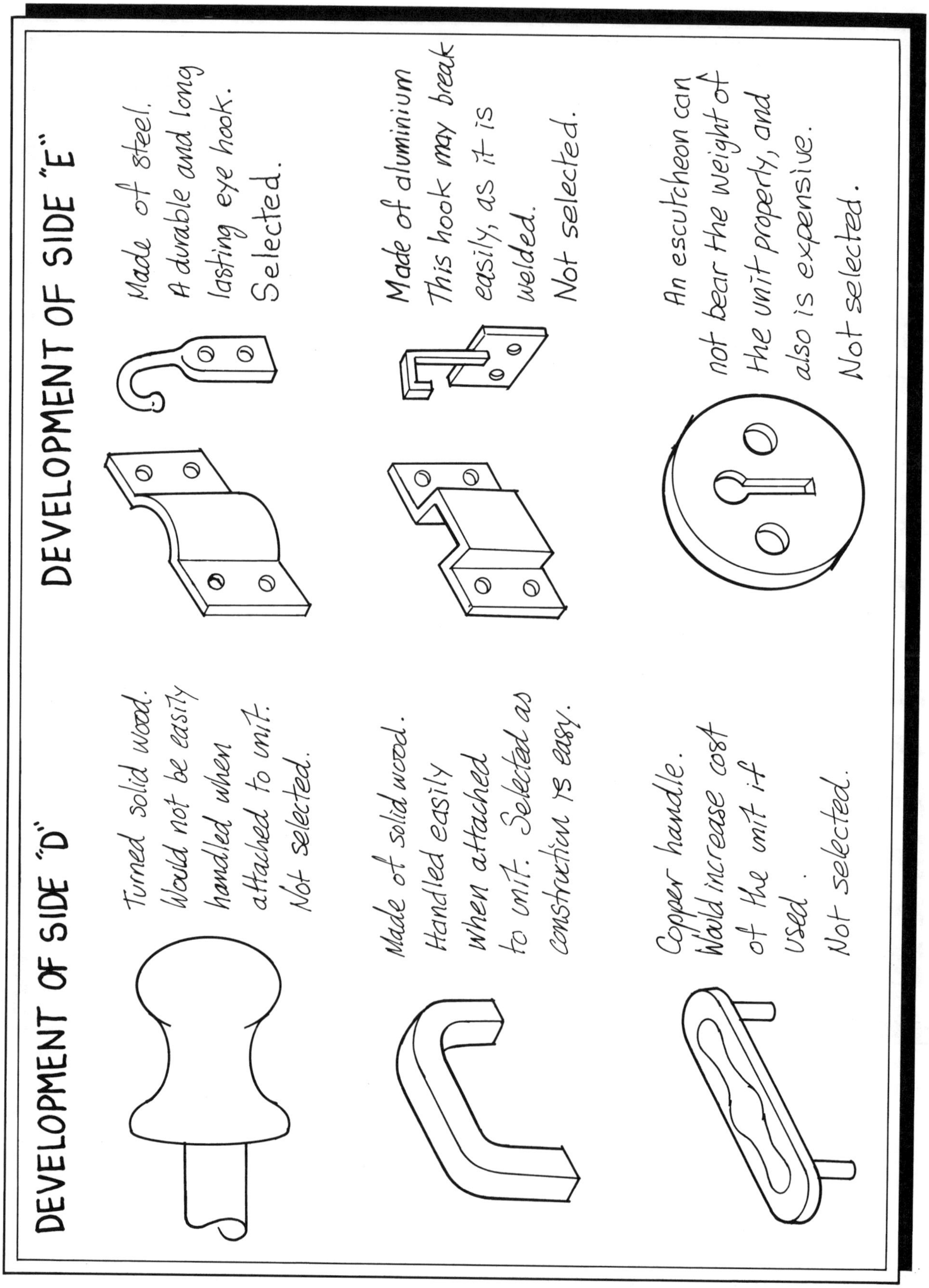

Figure 17.22 Example 3: development: choice of fittings.

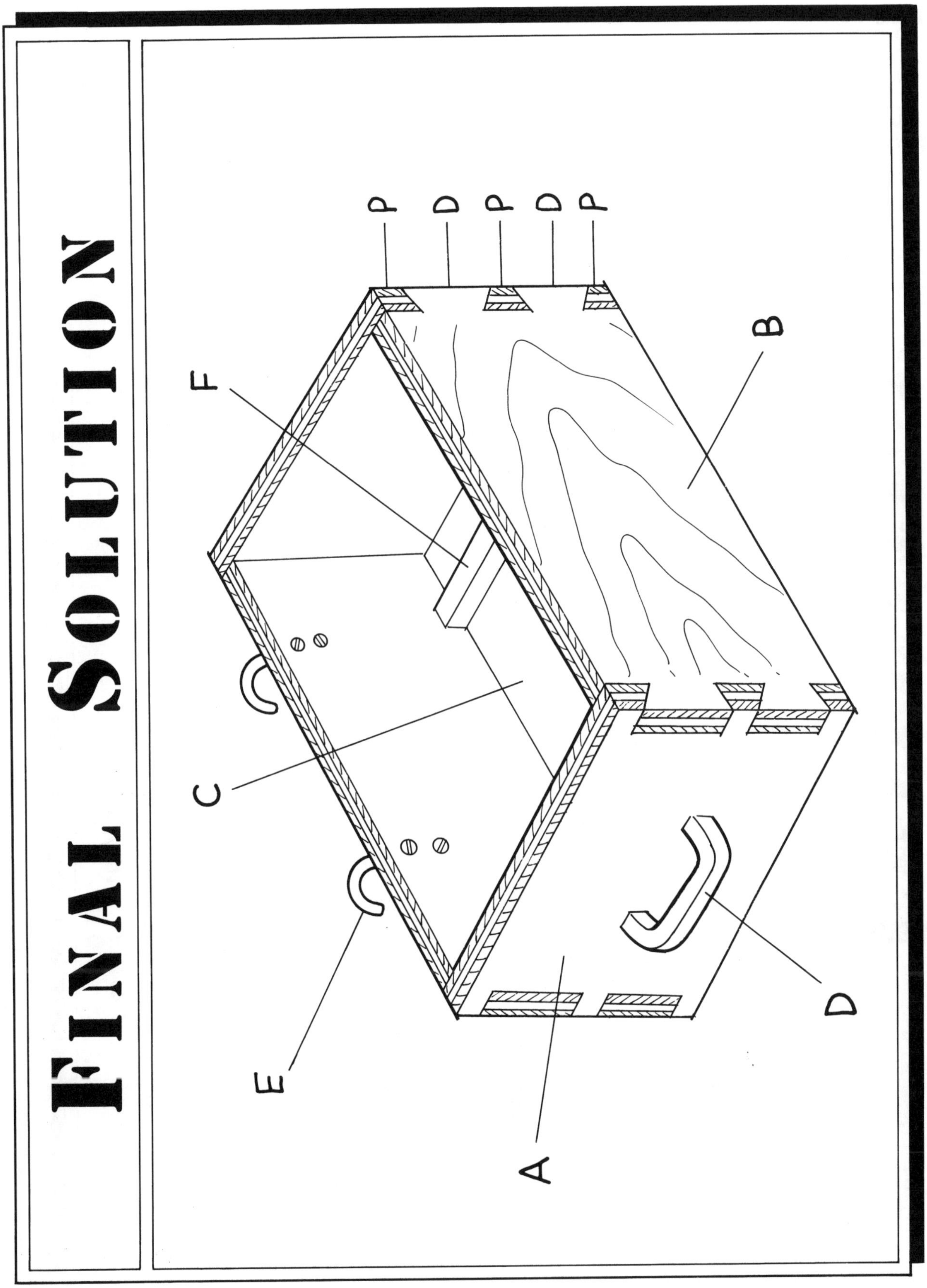

Figure 17. 23 Example 3: final solution.

EVALUATION

When the unit was made up in the workshop it was tesied and evaluated and the findings were as follows:

The drawing instruments in the mathematical set, (set square, pair of compasses, pair of dividers, protractor and pencils) were used. The unit was large enough to take them, and was easy and safe to use. The model is made of plywood; a standard thickness, available during construction. When the unit was attached to the table it was found to be in proportion to the table. The attachment was made of plastic because steel was not available at the workshop. It was discovered that the attachment would function more efficiently when made of steel.

The construction of the dovetail joints had some problems as it needs special skills to construct these joints. The unit was polished to serve as a perservative and to make the unit attractive to the user, as it is to be used indoors.

MODIFICATION

It was found at the unit should be given a lid to enable the users to store instruments in it permonently and to make carrying easier. Pin joints should be used to replace dovetail joints for easier construction and to save time.

Figure 17.24 Example 3: evaluation.

Project 1 – Brief

Design and make a container for a set of cutlery consisting of twelve knives, forks, spoons respectively and two bread knives. It must have a locking device.

Project 2 – Brief

A family of six is unable to have their meals together due to the lack of a dining unit. Design and make a dining unit capable of seating all the members of the family at meal times.

Project 3 – Brief

A new waiting room has been built on to the office of the headmaster of a senior secondary school. Design and make a set of three single lounge chairs for the new waiting room. Each chair must be of the same design.

Project 4 – Brief

Design and make a dressing unit for a young lady. The unit must be able to contain her cosmetics, jewellery and other essential dressing items. It must have a mirror that can be adjusted to different angles.

Project 5 – Brief

Design and make a unit to be used in a dining room to keep items required for meals and drinks in your headmaster's bungalow.

Project 6 – Brief

Parents in the rural areas are often called upon by the school authorities to provide tables and chairs for their children in the nursery and primary schools. Design and make a suitable table and chair that can be used by a child from the nursery school through to primary six, taking into account the growth of the child within the period.

Answers to questions

Chapter 1

1 To avoid unnecessary accidents and dangers.

2 Wear a well-fitting overall, overcoat or a close fitting apron tied at the back and strong upper footwear with a thick sole. This is because loose fitting clothing is likely to get caught in revolving machine parts and a thick sole and strong upper footwear will protect your feet from injury if an object falls on them.

3 Hold the sharp edge in your hand when you are giving the tool to someone. Keep your body, particularly the fingers away from the cutting edge, especially when you are paring with a chisel. The rule is: keep both hands behind the cutting edge and cut away from the body.

4 Always ensure that the necessary guards are in place and are well fitted. Always stand at a safe distance away from the machine and never in line with the revolving parts or work if you are observing an operation.

5 Press a clean piece of cloth or cotton wool on the cut to stop the bleeding. Remove any foreign bodies from the cut if you can do so safely. Wipe the cut gently with cotton wool or a swab soaked in warm, soapy water. Apply a plaster or bandage.
Report the incident to someone in authority.

6 If you are talking to someone operating a machine, he or she is likely to lose concentration and the chance of an accident is increased.

Chapter 2

1 'G' cramp, for holding the work-piece to the work-bench. Bench hook, for supporting work-piece when sawing or chiselling. Bench stop, for supporting a work-piece when planing a piece of work on the bench. Sash cramp, for holding parts being assembled together, especially when gluing.

2 Measuring tools, example: rule used for taking and checking measurements.

Marking out tools, example: marking knife, used for marking a cut line across shoulders of joints.

Setting and testing tools, example: is a try-square that is used for testing squareness of surfaces or edges of a work-piece.

3 Sketch each tool and indicate the angle formed between the stock and blade. The try-square is used for checking an angle of 90° or marking a line at 90° to a chosen edge or surface, while a mitre square is used for marking and checking an angle of 45°.

4 The pin of a marking gauge is like a needle while that of a cutting gauge is ground and sharpened flat like a knife. Hence the marking gauge is for marking a line with the grain and the cutting gauge is for cutting a line across the grain of a piece of wood.

5 (a) Sketch the tool and label the following parts: striking face, for driving home nails; claw end, for removing nails; handle for holding the tool.
b) The advantage of the claw hammer over the Warrington hammer is that it is used for both driving in and removing nails. Very useful when working on a site project.

6 Impelling tools are those used for forcing materials into a piece of wood. Example: a screwdriver used for screwing screws into a work-piece; a nail punch, used in conjunction with a hammer for driving a nail head below the surface of a work; a brace for driving a drill bit into a piece of wood when making a hole.

Chapter 3

1 Bench planes, for planing flat surfaces; curve cutting planes, for producing or planing

curved surfaces and edges; special purpose planes, for carrying out special operations such as rebating, grooving etc.; multi-purpose planes, for carrying out diverse special operations including tonguing and grooving, moulding, matching, beading etc.

2 Escapement – the opening in the stock into which the cutting iron is fixed and through which the shavings escape.
mouth – the cutting edge of the cutting iron is set at the mouth;
wedge – holds the cutting iron in position in the escapement;
cap iron – attached to the blade and breaks the shavings and make them curl.
body – the main part holding all the other parts;
button – for striking loose the cutting iron or for setting.

3 The wooden compass plane has a fixed curved sole while the sole of the metal type is adjustable.

4 The fillister plane has a fence, a depth gauge and a spur, but the rebate plane has none of these.

5 Sketch the tool in an exploded form; label the essential parts: the blade and cutting edge; the tang that fits into the handle; the handle that receives the tang and is used for holding the blade; the ferrule, fitted to the handle to prevent it splitting.

6 The rip saw has chisel-like teeth with the front edge being at 90° to the horizontal line of the teeth. It is used for sawing along the grain of a thick board.

7 Triangular file, for shaping sharp corners and grooves, and for shaping worn out saw teeth. Round file, for smoothing concave curves and enlarging holes. Half round file, for smoothing concave edges and enlarging holes.

8 The particles of abrasive materials imbedded in the piece can dull the cutting edge of the cutting iron.

Chapter 4

1 For efficiency and good quality work.

2 Silicon carbide and aluminium oxide held together by a bonding agent known as a matrix.

3 Set the tool rest close to the wheel.
Switch on the machine. Hold the cutting tool at the correct angle on the tool rest against the revolving wheel. Move the cutting edge across the wheel. Dip the cutting edge into a bowl of coolant from time to time.

4 Ensure that the tool rest is set correctly and all guards are in place. Wear a pair of goggles and a pair of gloves.

5 Store the oilstone in a box to protect it. Keep the surface of the oilstone clean and oiled.

Chapter 5

1 A dressed board is a board that has had all faces and edges planed.

2 Face markings serve as reference or datum from which measurements and gauging are taken to ensure accuracy and squareness.

3 Mark a cut line across the work-piece before you rebate.

4 Common mortise and tenon, used for frame construction where the pieces meet at a distance away from the end; haunched mortise and tenon, used where the pieces meet at the end; double mortise and tenon, for wide stiles or frames; long and short shouldered mortise and tenon, used for rebated panels.

5 Rubbed or butt joint; tongued and groove joint, tongue joint; rebated joint; dowelled joint and self-screw joint. The general term for these joints is widening joints.

6 Top: plain mitre joint, rebated and mitred joint, and secret dovetail joint. Bottom: lapped dovetail joint, housing and shouldered joint. Shelves and partitions: pinning, stopped housing joints, through housing joint.

Chapter 6

1 Roofing nail has dome-shaped head and either round or square twisted shank. It is used for roofing.
Flat head nail, may have either round shank or a square twisted shank. It is used for packing case and box construction.
Diamond head nail, has a round shank and a diamond shaped head. It is used for nailing flooring boards, lining boards, and for packing case and box construction.
Panel pin has a thin shank and an inverted cone-shaped head. It is used for fixing plywood to frames, cabinet backs, moulding etc.

2 State the weight required, length, gauge, type, metal and type of finish.

3 A screw has a stronger holding power than a nail because the thread on the shank cuts and buries itself into the fibres of the wood.

4 The diameter of the pilot hole should be the same as the core diameter so that the thread of the screw cuts into the fibres of the wood.

5 Protein adhesives, example: casein glue used for manufacturing plywood, laminated boards and caul veneering.
Synthetic resin adhesives, example: urea formaldehyde resin glue, used for caul or cold press veneering, plywood and particle board manufacture.
Contact glue, used for bonding together two or more dissimilar (different) materials such as formica, leather, glass, wood, fabric etc.

6 Straight lock, used for doors and drawers where it is screwed flat on the inside face.
Cut lock, fitted into a recess in the inside face of a door or a drawer.
Mortise lock, used for large and thicker doors, where it is fitted into a mortise in the stile.

7 The spring action catch locks and opens by simply pushing and pulling the door or lid.
The non-spring action type can lock by pushing but it requires a button or knob to turn the locking device (latch) before it opens.

8 Bolts may be used for pairs of doors or window for holding one of the pair in position to enable the other pair to be locked.

9 Handles that have long strip with a wide opening to accommodate the hand.
Knobs, having round or circular shapes.
Pulls, having hook or a recess for the fingers.

10 Stay is used for supporting lids or tops when opened. A set of castors is fitted to the legs and bases of furniture for ease of movement.

Chapter 7

1 The saw guard prevents the piece being sawn from kicking back at the operator.

2 Set the fence to the required distance; set the saw to the required depth of cut; switch on the motor; hold the work against the fence and push it forward against the revolving blade to cut off the rebate. Adjust the fence to take off the whole width of the rebate.

3 A narrow blade is able to negotiate sharp on smaller curves where as a larger blade is used cut larger curves.

4 Mortising, shaping or cutting of mouldings, routing and sanding.

5 Always keep your hands away from the cutter; always ensure that the spindle nut is tight; avoid shaping a piece of short length; ensure that all attachments and guards are in place and firmly secured.

6 Face turning, spindle turning, cup chuck turning, boring.

7 Spindle sander, used for sanding concave curves or the edge of the work-piece.
Disk sander for sanding straight edges and convex curves on the edge of the work-piece.
Belt sanders, used for sanding flat surfaces.

Chapter 8

1 The motor housing controls the depth of cut and the angle of the blade.

2 By the shape of the body and the shape of the blade.

3 The regular plane, used for planing longer surfaces; the block plane, used for planing shorter surfaces.

4 Friction clutch, used for driving home screws. Flexible unit, used for drilling in restricted areas and at an angle. Drill stand, for supporting the body of the electric drill.

5 Power hand routers are classified according to the power of the motor and the diameter of the cutter shaft.

6 The capacity to determined by the depth and diameter of the shaping tool to be used.

Chapter 9

1 They are called evergreen because their leaves remain green through out the year. They have tall, slender trunks and narrow leaves.

2 Crown, consisting of the leaves that process the solution of mineral salt and water into food for the tree. Trunk, that supports the crown and from which logs are obtained. Roots support the entire tree and also draw water and mineral salt from the soil.

3 A log may be converted into commercial sizes by through sawing, tangential sawing and quarter sawing.

4 Natural and artificial seasoning to produce a stable workable timber for working.

5 Wood preservation is the treatment of wood with chemical solutions to prevent fungus and insect attack.
Tar-oil, water-based and organic solvent preservatives. Preservatives may be applied by brushing, spraying, hot and cold dipping.

Chapter 10

1 They can be obtained in larger sheets, have uniform strength, are free from shrinkage and warping, can be easily worked and cut to size.

2 This is to prevent warping and ensure uniform strength.
3 Because the manner by which the strips forming the core are laid (heart side of consecutive strips face in opposite directions) counteracts warping or curling.
The strips used for forming the core of blockboard are wider than those of laminboard.
4 Toothing plane, for roughing the surface of ground to provide key for the glue; veneer hammer, for pressing down the veneer in order to force out air bubbles and surplus glue; flat iron, for melting the glue (animal glue) and squeezing out the surplus; cutting gauge, for cutting narrow strips of veneer; taping stick, for pressing a tape into place when joining veneer edge to edge.
5 Particle boards are classified according to the manner of which the structure (core) is composed. The classes are single layer, three-layer and extruded layer.
6 Hardboard and softboard. The difference between them is that the fibres of a hardboard are compressed whilst those of the softboard are not.

Chapter 11

1 Type of timber to be shaped; direction of the grain; strength required; amount of curvature required; and type of finish intended.
2 A piece of wood shaped by sawing may have short grain fibres that tend to break off whereas the one produced by bending has its grains following the shape, thereby providing strength.
3 Prepare the surface to be inlaid. Mark out the portion to be inlaid on the surface. Cut out the groove or recess. Apply glue to the strip, press it into the recess carefully and draw a hammer along it. When set, clean off with a scraper or a finely by set plane.
4 In relief carving the background of the pattern is cut away to raise the pattern above the surface whereas incised carving has the pattern cut into the surface, i.e. lying below the surface.
5 Stringings are plain strips of wood of simple (one) colour whereas bandings are multi-coloured thin strips of various patterns.

Chapter 12

1 Ferrous metals; cast iron and steel. Non-ferrous metals; zinc and copper.
2 Thermoplastics, example cellulose nitrate celluloid and P.V.C. (polyvinyl-chloride). Thermosets, example bakelite used for electrical plugs, switches and radio cabinets; epoxy resins, used as a base for adhesives.
3 Cloth type, ribbon type and tissue type.
4 Glass is made by mixing together soda ash, limestone, sand and cullet heating the mixture in a furnace at a temperature of 1500°C. It is then drained and drawn into sheets.
5 A leather is obtained from animal skins. The surface of a leather can be decorated by using punches or embossed tools.
6 Linen: strong, good conductor, absorbent. Cotton: takes dyes well, resists bleaching and detergents.
7 Wool: warmth and comfort. Silk: elastic, fine appearance, good conductor of heat.

Chapter 13

1 Dress the surface (smooth plane); fill all nail holes and open joints; sand the surface with glasspaper, working from rough to smooth; dust off the surface.
2 Paint is made up of a pigment, a vehicle, a thinner and a drier. Oil paints and latex paints. Oil paints have oil as the vehicle whereas latex paints have a mixture of water and resin as the vehicle.
3 Wood filler may be applied where the timber used for the furniture has pores or open-grain, in order to produce a smooth surface.
4 Polyurethane varnish is very tough, hard, flexible, resistant to chipping, abrasion, solvents and water.
5 When lacquer is applied over stained or painted surface it may peel or lift off.

Chapter 14

1 Latex and plastic foams have good resilience and are durable. Synthetic foams are cheaper and less resilient, slow to regain their shape when compressed.
2 Jute, plastic, burlap, rafia, cane.
3 Padding is the process of covering springs with stuffing materials to form a soft, smooth surface. Examples of padding materials are rubberised hair, rubber foam, and plastic foam.
4 Upholster's tack, webbing tacks, and decorative tacks.

Chapter 15

1 Designing is the process of finding a solution to an identified problem in a given situation through written and graphic communication.

2 SITUATION → BRIEF → INVESTIGATION
EVALUATION ← MAKING ← SOLUTION

3 Through graphic communication: pictorial projection, sketches and detailed drawing.

4 The detailed drawing. The main component parts of an artifact or device are extracted and drawn in orthographic projection showing all the essential details that may be required for the manufacture of the artefact or system. The assembly drawing is where the views drawn show all the component parts are fitted together and give an overall pictures of the artefact. A sectional view may also be drawn.

6 Assemble all parts without glue to try their fitness. Clean up all internal surfaces and edges. Apply glue to parts to be fitted together. Assemble and cramp until the glue sets.

Chapter 16

1 Public sector enterprise and private sector enterprise. A public sector enterprise is owned and controlled by the Government for the purpose of serving the needs of the citizens while a private sector enterprise is owned and managed by a private individual or a number of individuals for sole aim of making profit.

2 Sole-proprietor is an individual who owns and runs a business or enterprise all by himself or herself.

3 Trading enterprises are mostly concerned with the buying and selling of goods of various kinds.

4 Advantages: business decisions are taken quickly; any profit accruing from the business is singly managed. Disadvantages: raising of business capital becomes a burden; in times of loss the sole-owner bears the whole liability.

5 A sleeping partner is a member of a partnership enterprise who is not physically involved in the management of the enterprise.

Key words and definitions

Accelerator A substance that causes a chemical reaction to take place more quickly.
Accident An event that occurs unexpectedly and causes injury or damage.
Angiosperm A botanical name for a deciduous tree (often called a hardwood).
Artefact An article made by man.

Bleeding Emitting of blood from a cut etc.
Burr Ragged edge produced at the edges of cutting tools when sharpening.

Carcase A framework of a cupboard or sideboard without the fittings such as doors, drawers etc.
Catalyst A substance that speeds up a chemical reaction (without itself changing).
Chamfer A level cut at 45° on the edge of a piece of timber.
Chlorophyll The green substance in the leaves of a tree
Clearance hole A drilled hole slightly larger than the shank of a screw or bolt.
Clog Become blocked with something.
Compression A state of an object subjected to two or more parallel forces pressing upon it from opposite ends.
Consistency A degree of thickness of a liquid, especially glue.
Convert To saw a log into commercial sizes.
Cross-cutting Sawing a piece of timber across the grain.

Detergent A substance that removes dirt from the surface of a material.
Dicotyledon A plant that has a seed out of which grow two leaves at the embryo stage.
Diluent A substance that makes a liquid or colour thinner or weaker.
Dressing A process of smoothing off the surface of a piece of work prior to glasspapering.

Enterprise A business venture.
Escapement An opening in the stock of a plane into which the cutting iron is fitted.
Evergreen A coniferous tree which keeps its leaves throughout the whole year

Ferrule A brass ring fitted at the end of tool handles to prevent splitting.

Galvanised Coated with zinc to prevent rusting, e.g. galvanised plate.
Green timber Freshly sawn timber containing moisture.
Gymnosperm A botanical name for a coniferous tree called softwood.

Impel To force an object into another
Injury Physical harm to the body.

Kerf The grove made by a saw cut.
Kicking back A tendency of a piece of timber to be forced back towards the operator when sawing with the circular saw.

Lustrous Having a soft, smooth or shining surface.

Matrix A substance used for bonding together abrasive particles of a grinding wheel.
Matt A surface that is not shiny or glossy but dull.

Paring The process of removing a waste from a piece of wood with a chisel by hand pushing.
Parenchyma A soft tissue of small, thin-walled cells that run vertically in the trunk of a pored tree.
Photosynthesis The process by which green plants convert carbon dioxide and water into food and give of oxygen by the effect of sunlight.
Pilot hole An initial hole drilled in a piece of work to guide a screw or a nail.

Ratchet A mechanism that enables a device to be turned in one direction only E.g. the ratchet brace.
Ripping Sawing a piece of timber along the grain.

Secrete To produce a substance or liquid as a waste material or to use in a body.
Shavings Thin layers of wood fibres that are produced when a piece of timber is planed.
Shrinkage The effect of changes in the weather on a piece of wood causing it to be come smaller.
Simultaneously Occurring at the same time.
Sleeping Partner A member of a business partnership who is not involved in the management of the business.
Species A class of plants or animals within a genus.
Splinter To break off into small, thin, sharp pieces.
Spur A sharp projecting pin of a marking gauge or mortise gauge.
Stagger To arrange objects (e.g. nails, pins, screen) so that they are not in a line.

Starved joint A joint lacking sufficient glue to hold fast the pieces.
Swab A piece of cotton wool used for cleaning wounds.

Tension A state of an object subjected to two or more parallel forces pulling it apart.
Tracheids Narrow, tubular cells that run parallel to the trunk and branches of a non-pored tree.

Trench A recess cut across the grain of a piece of wood.
Trimming Cutting away excess material to make a neat, smooth finish.

Warped Becoming twisted or bent.

Index

Note: page numbers in *italics* denote illustrations where these are separated from their textual reference.